To Amity and Danny,
   Who light up our lives here
in Washington.
             Love,
             Bernard

From the Bottom Up

DIRECTIONS IN DEVELOPMENT
Energy and Mining

From the Bottom Up

*How Small Power Producers and Mini-Grids Can
Deliver Electrification and Renewable Energy in Africa*

Bernard Tenenbaum, Chris Greacen, Tilak Siyambalapitiya, and James Knuckles

THE WORLD BANK
Washington, D.C.

ISBN (paper): 978-1-4648-0093-1
ISBN (electronic): 978-1-4648-0111-2
DOI: 10.1596/978-1-4648-0093-1

Cover image: NASA Earth Observatory image by Robert Simmon, using Suomi NPP VIIRS data provided courtesy of Chris Elvidge (NOAA National Geophysical Data Center). Suomi NPP is the result of a partnership between NASA, NOAA, and the Department of Defense. Image retrieved from NASA's "Visible Earth" website: http://visibleearth.nasa.gov/view.php?id=79765.

Cover design: Debra Naylor, Naylor Design, Inc.

Library of Congress Cataloging-in-Publication Data

Tenenbaum, Bernard William, author.
From the bottom up : how small power producers and mini-grids can deliver electrification and renewable energy in Africa / Bernard Tenenbaum, Chris Greacen, Tilak Siyambalapitiya, and James Knuckles.
  pages cm. – (Directions in development)
ISBN 978-1-4648-0093-1 (alk. paper) – ISBN 978-1-4648-0111-2
  1. Rural electrification—Africa. 2. Rural electrification–Economic aspects—Africa. 3. Small power production facilities—Africa. 4. Distributed generation of electric power—Africa. 5. Remote area power supply systems—Africa. I. Greacen, Chris. II. Siyambalapitiya, Tilak. III. Knuckles, James. IV. World Bank. V. Title. VI. Series: Directions in development (Washington, D.C.)
  HD9688.A352T46 2014
  333.7932096—dc23
                                                                                                    2013043831

Contents

Boxes

Figures

Tables

Foreword

This guide was written for the Africa Electrification Initiative (AEI). The objective of the AEI program is to create and sustain a living body of practical knowledge and to establish an active network of Sub-Saharan African practitioners who work on the design and implementation of rural, peri-urban, and urban on-grid and off-grid electrification programs. These practitioners include individuals who work for electrification agencies and funds, government ministries, regulators, and state, community, or privately owned utilities.

More than 170 of these practitioners came together in June 2009 in Maputo, Mozambique, for a three-day workshop on electrification sponsored by AEI. Of the 42 countries represented, 32 were in Sub-Saharan Africa. The workshop was carefully designed to emphasize practical implementation issues rather than general policy discussions. At its conclusion, one participant observed, "Most conferences fly at 35,000 feet, but here we were down at ground level."

Neither the 2009 Maputo workshop nor a 2011 follow-up workshop in Dakar, Senegal, followed the typical conference format. In particular, the sessions were not limited to the standard 20–25 minute PowerPoint presentations followed by a few minutes for audience questions. Instead, they used a mix of formats to maximize informal interaction. For example, the Maputo workshop included 50 presentations by experts and practitioners delivered in 12 plenary sessions, 17 breakout discussion sessions designed to enable participants to pursue follow-up questions with one another, and 3 structured half-day clinics. The two workshops covered a wide range of topics—among them different institutional models for grid and off-grid electrification, the role of small power producers and mini-grid operators, the design and implementation of capital and consumption subsidies, pricing for grid and off-grid suppliers, operation of rural electrification agencies and funds, low-cost electrification techniques, design and implementation of connection charges, microfinance, carbon finance, prepaid meters, alternative service and maintenance models, and economic and technical regulation. (The presentations and handouts from the two workshops can be found at http://go.worldbank.org/WCEDP90SZ0.)

Two recommendations that came out of the workshops were relevant to this guide. The first was that AEI's postworkshop activities should continue to emphasize real-world implementation issues. The second recommendation was that, because the creation of a clear and credible regulatory system is important for attracting private sector investment, AEI should undertake follow-up work focused on ground-level regulatory implementation for small power producers that could promote both renewable energy and electrification.

This guide was written in direct response to these recommendations. Its four authors—Bernard Tenenbaum, Chris Greacen, Tilak Siyambalapitiya, and James Knuckles—have worked on scaling up access to electricity through on-grid and off-grid small power producers in more than 15 countries. Because the guide represents a synthesis of this multicountry experience, we believe that it constitutes an important contribution to the United Nations' recently announced 2014–24 Decade of Sustainable Energy for All.

Meike van Ginneken and Lucio Monari
Sector Managers, Africa Energy
The World Bank

Acknowledgments

This guide could not have been written without the assistance of many colleagues inside and outside the World Bank who gave freely of their time and shared with us their invaluable knowledge and insights on small power producers and mini-grids in both developed and developing countries. On more than one occasion, they graciously repeated explanations when we did not quite grasp them the first time. So it is with heartfelt appreciation that we acknowledge the advice and assistance of Rajesh Advani, Javaid Afzal, Christopher Aidun, Pär Almqvist, Pedro Antmann, Beatriz Arizu, Tonci Bakovic, Sudeshna Bannerjee, Pepukaye Bardoille, Douglas Barnes, Mikul Bhatia, Ky Chanthan, Nazmul Chowdhury, Joy Clancy, Vyjanti Desai, Neeltje de Visser, Koffi Ekouevi, Sunith Fernando, Pradit Fuangfoo, Hari Gadde, Isabella Gawirth, Defne Gencer, Ben Gerritsen, Vanessa Lopes Janik, Balawant Joshi, Daniel Kammen, Ralph Karhammar, Bozhil Kondev, Prayad Kruangpradit, Jeremy Levin, Guy Marboef, Frank Mejooli, Rob Mills, Mohua Mukherjee, Stephen Mwakifwamba, Sreekumar N, Monali Ranade, Kilian Reiche, Miguel Acevedo Revolo, Sebastian Rodriguez, Robert Schloterrer, Jakob Schmidt-Reindahl, Arsh Sharma, Binod Shrestha, Ruchi Soni, Payomsarit Sripattananon, Rauf Tan, Fabby Tumiwa, Hung Tien Van, Jim Van Couvering, Richenda Van Leeuwen, and Harsha Wickramasinghe.

We also appreciate the efforts of those who took the time to review our first full draft: Gabriela Elizondo Azuela, Anton Eberhard, Katharina Gassner, Bikash Pandey, Tjaarda Storm van Leeuwen, and Gunnar Wegner. Just when we thought that we had successfully covered the subject, they pointed out what we had missed or misunderstood. We thank them for their patience and gentleness with us.

We first discussed the possibility of writing this guide over breakfast in Tanzania nearly three years ago. We owe a special debt of gratitude to our colleagues there. As one reads the guide, it will become clear that many of the ideas and recommendations it contains come from our experience in Tanzania, where we were fortunate to have the privilege of working with many talented and diligent colleagues. At the Energy and Water Utilities Regulatory Authority (EWURA), Norbert Kahyoza, N'ganzi Jumaa Kiboko, Edwin Kiddifu, Haruna Masebu, Anastas Mbawala, and Matthew Mbwambo offered many insightful comments on the ground-level decisions needed to establish a viable small power producer (SPP) regulatory framework. We also benefited from the insights and experience of several knowledgeable colleagues at the Rural Energy Agency: Justina Aisso, Bengiel Msfofe, Lutengano U. A. Mwakahesya, George M. J. Nchwali, and Boniface Gissima Nyamo-Hanga. And within Tanzania's growing SPP community, we were fortunate that those on the "receiving end" of regulation, both as buyers and sellers, were willing to explain the real-world implications of proposed rules and processes. These include Giuseppe Bascaglia, Mayank Bhargava, Nizam Codabux, Fabio De Pascale, Mark Foley, Mike Gratwicke, Erik Haule, Maneno Katyega, Sister Yoela Luambano, Roselyne Mariki, Menus Mbunda, Godfrey Ndeuwo, Evodia Ngailo, Pascal Petiot, Gyan Ranjan, Jem Rigall, Charles Shayo, Luca Todeschini, and Patrice Tsakhara.

The need for this guide was raised at workshops of the Africa Electrification Initiative (AEI, http://go.worldbank.org/WCEDP90SZ0). Several members of the AEI network went above and beyond the call of duty in responding to many questions in what must have seemed like an endless stream of e-mails and Skype calls. These include: Alassane Agalassou, Mansour Assani-Dahouenon, Zachary Ayieko, Alphadio Barry, Jörg Michael Baur, Ray Holland, Ousmane Fall Sarr, and Nava Touré.

At the World Bank, we received strong initial encouragement from Vijay Iyer, the former manager of the Africa Energy Group. Meike van Ginneken and Lucio Monari, the current managers of the Africa Energy team, continued to support us in innumerable ways. We also benefited from the many insights of our colleagues who worked side by side with us over the last three years in providing ground-level assistance to Tanzania's SPP initiative. These include Anil Cabraal, Jon Exel, Stephen Ferrey, Richard Hosier, Stephanie Nsom, and Krishnan Raghunathan. Finally, we would be remiss if we did not highlight the critical support that we received from our two World Bank task managers. Dana Rysankova took a leap of faith in deciding that we had something useful to say and that it was worth sharing with a broader audience. Raluca Golumbeanu continued that support by giving us more time to complete the guide and the benefit of her detailed knowledge of World Bank operations throughout Africa. Finally, our two capable editors, Steven Kennedy and Fayre Makeig, continually reminded us that the goal of the guide was not just to put words on paper but to communicate ideas and information in a way that would be of genuine benefit to our readers.

A project of this size has both a professional and personal dimension. Our wives and partners—Ellen Tenenbaum, Chom Greacen, Namalie Siyambalapitiya, and Silvia Camporesi—went to extraordinary lengths above and beyond the call of duty in supporting us in both dimensions. For this, we will always be grateful.

Finally, we should point out the obvious. None of these individuals, who did their best to educate us, should be held responsible for any errors of fact or interpretation that remain. Any remaining errors are solely the responsibility of the authors.

Funding for this guide was provided under the first phase of the Africa Renewable Energy and Access Program (AFREA), the principal assistance program for Africa of the Energy Sector Management Assistance Program (ESMAP), a global knowledge and technical assistance program that the World Bank administers. AFREA was established in 2009 to meet the special needs of Sub-Saharan Africa, namely the urgent need to develop scalable, innovative solutions to close Africa's energy access gap. AFREA's overall objectives are to help Sub-Saharan Africa meet its energy needs and to widen access to energy services in an environmentally responsible way.

In addition to the AFREA funding, separate direct financial assistance came from ESMAP. ESMAP provides analytical and advisory services to low- and middle-income countries to increase their know-how and achieve environmentally sustainable energy solutions that can reduce poverty and stimulate economic growth. ESMAP is governed and funded by a Consultative Group of official bilateral donors and multilateral institutions representing Australia, Austria, Denmark, Finland, Germany, Iceland, Lithuania, the Netherlands, Sweden, the United Kingdom, and the World Bank Group.

Finally, the guide also received financial support from the Global Partnership on Output-Based Aid (GPOBA). GPOBA is a global partnership program administered by the World Bank. GPOBA was established in 2003 to develop output-based aid (OBA) approaches across a variety of sectors—among them infrastructure, health, and education. To date, GPOBA has signed 35 grant agreements for OBA subsidy funding for a total of $157 million. GPOBA projects have disbursed $67.6 million based on independently verified outputs, directly benefitting 3.3 million people. The program's current donors are the United Kingdom's Department for International Development (DFID), the International Finance Corporation (IFC), the Directorate-General for International Cooperation of the Dutch Ministry of Foreign Affairs (DGIS), the Australian Agency for International Development (AusAID), and the Swedish International Development Cooperation Agency (Sida).

About the Authors

Bernard Tenenbaum (btenenbaum2002@gmail.com) is an independent energy and regulatory consultant. Before retiring from the World Bank in 2007, he served as a lead adviser on power sector reform and regulation projects in Brazil, China, India, Mozambique, Nigeria, and Tanzania. Prior to joining the World Bank, he served as the associate director of the Office of Economic Policy at the U.S. Federal Energy Regulatory Commission. He is the author or coauthor of *Regulation by Contract: A New Way to Privatize Electricity Distribution?; Governance and Regulation of Power Pools and System Operators: An International Comparison; Electrification and Regulation: Principles and a Model Law; Regulatory Review of Power Purchase Agreements: A Proposed Benchmarking Methodology;* and *A Handbook for Evaluating Infrastructure Regulatory Systems.* He is a member of the editorial board of the *International Journal of Regulation and Governance.* He has a PhD in economics from the University of California at Berkeley.

Chris Greacen (chrisgreacen@gmail.com) works on policy and hands-on implementation of renewable energy projects from the village to the national level. As codirector of the nonprofit organization Palang Thai he helped draft Thailand's policies on very small power producers and conducted studies in support of the country's feed-in tariff program. He has worked on renewable energy mini-grid projects and programs in Cambodia, China, Democratic People's Republic of Korea, the Lao People's Democratic Republic, the Federated States of Micronesia, Republic of the Union of Myanmar, Nepal, Thailand, and Vanuatu, and on Native American reservations in the United States. As a World Bank consultant he has worked since 2008 with the Tanzanian Energy and Water Utilities Regulatory Authority to help develop Tanzania's small power producer program. He has a PhD in energy and resources from the University of California at Berkeley.

Tilak Siyambalapitiya (tilak-rma@sltnet.lk) is an electrical engineer who has worked on power generation planning and economics for 30 years. He initially worked in power utility planning in Sri Lanka and Saudi Arabia, followed by 12 years of consulting work for many international agencies and national agencies in Sri Lanka. His recent work includes renewable energy development, pricing, and technical guidelines in Sri Lanka, Tanzania, and Vietnam. Additionally, he works on energy efficiency and power sector regulatory support, as well as in project formulation and evaluation in power generation, transmission, and distribution. A graduate of the University of Moratuwa in Sri Lanka, he earned his PhD from the University of Cambridge in electrical power engineering.

James Knuckles (jamesknuckles@gmail.com) works at the intersection of business, energy, and international development. Most recently, he has developed new business models, financial models, and business plans for energy-related startups in Monterey, Port-au-Prince, Sydney, and Tegucigalpa. Previously, with the Council on Competitiveness in Washington, DC, he helped launch a series of cross-border clean energy projects between Brazil and the United States. Knuckles holds an MBA and a master's in international environmental policy from the Monterey Institute of International Studies. In fall 2013, he began a PhD program at Cass Business School in London to study the business models, operations, and financing mechanisms of small energy companies doing business in developing countries.

Abbreviations

AAC	automatic adjustment clause
AC	alternating current
AEI	Africa Electrification Initiative
AFREA	Africa Renewable Energy and Access Program (World Bank)
AICD	Africa Infrastructure Country Diagnostic
AMADER	Agence Malienne pour le Développement de l'Energie Domestique et de l'Electrification Rurale (Mali's rural energy agency)
AMR	automatic meter reading
ASER	Agence Sénégalaise d'Électrification Rurale
BERD	Bureau d'Électrification Rurale Décentralisée (Bureau for Decentralized Rural Electrification, Guinea)
CBO	community-based organization
CDM	Clean Development Mechanism
CEB	Ceylon Electricity Board
CER	certified emission reduction
CFL	compact fluorescent light/lamp
CO_2	carbon dioxide
CPI	consumer price index
DC	direct current
DDG	decentralized distributed generation
DG	distributed generation/generator
DNO	distribution network operator
DSCR	debt service coverage ratio
EA	engineering assessment
EAC	Electricity Authority of Cambodia
EEPCO	Ethiopian Electric Power Corporation
EGAT	Electricity Generation Authority of Thailand
EPIRA	Electric Power Industry Reform Act (the Philippines)
EPPO	Energy Policy and Planning Office (Thailand)

ERC	Energy Regulatory Commission (the Philippines)
ESCO	energy service company
ESMAP	Energy Sector Management Assistance Program (World Bank)
EWURA	Energy and Water Utilities Regulatory Authority (Tanzania)
FITs	feed-in tariffs
GB	gigabyte
GET FiT	Global Energy Transfer Feed-In Tariff
GPOBA	Global Partnership on Output-Based Aid
GW	gigawatt
GWh	gigawatt-hour
Hz	hertz
IBF	input-based franchise
IBT	increasing block tariff
IFC	International Finance Corporation (World Bank)
IG	induction generator
INENSUS	Integrated Energy Supply Systems
IPP	independent power producer
IRR	internal rate of return
KPLC	Kenya Power and Lighting Company
kV	kilovolt
kVA	kilovolt-ampere
kW	kilowatt
kWh	kilowatt-hour
kWp	kilowatt peak
LCOE	levelized cost of electricity
LCPD	least-cost power development
LDU	local distribution utility
LED	light-emitting diode
LOI	letter of intent
LRMC	long-run marginal cost
MCB	miniature circuit breaker
MEA	Metropolitan Electricity Authority (Thailand)
MV	medium voltage
MW	megawatt
MWh	megawatt-hour
NCRE	nonconventional renewable energy
NDA	nondisclosure agreement
NEA	Nepal Electricity Authority

NEPC	National Energy Policy Council (Thailand)
NERSA	National Energy Regulator of South Africa
NGO	nongovernmental organization
NOx	nitrogen oxide
NPV	net present value
O&M	operation and maintenance
OBA	output-based aid
OMC	Omnigrid Micropower Company
OSINERGMIN	Organismo Supervisor de la Inversión en Energía y Minería (Peruvian electricity regulator)
PAC	project approval committee (Sri Lanka)
PBS	Palli Bidyuit Samity (rural electric cooperatives, Bangladesh)
PCASERS	Projets de Candidatures Spontanées d'Electrification Rurale (Mali)
PCC	point of common coupling
PEA	Provincial Electricity Authority (Thailand)
POA	program of activity
POI	point of interconnection
POS	point of supply
PPA	power-purchase agreement
PRG	partial risk guarantee
PUCSL	Public Utilities Commission of Sri Lanka
PV	photovoltaic
QTP	qualified third parties
REA	rural electrification agency; Rural Energy Agency (Tanzania)
REE	rural electrification enterprise
REF	rural electrification fund
RESCO	renewable energy services company
RF	revenue franchise
RGGVY	Rajiv Gandhi Grameen Vidyutikaran Yojana (electrification program, India)
ROI	return on investment
RPS	renewable portfolio standards
SAIDI	System Average Interruption Duration Index
SAIFI	System Average Interruption Frequency Index
SCADA	supervisory control and data acquisition
SEA	Sustainable Energy Authority (Sri Lanka)
SEB	state electricity board (India)
SG	synchronous generator

Sida	Swedish International Development Cooperation Agency
SLEF	Sri Lanka Energy Fund
SOE	state-owned enterprise
SOx	sulphur oxide
SPD	small power distributor
SPP	small power producer
SPPA	standardized power-purchase agreement
SREP	Scaling-Up Renewable Energy Program
SRMC	short-run marginal cost
TANESCO	Tanzania Electric Supply Company
tCO_2e	tonnes of carbon dioxide equivalent
THD	total harmonic distortion
TOD	time of day
TOU	time of use
UL	Underwriters Laboratories
UNFCCC	United Nations Framework Convention on Climate Change
V	volt
VBT	volume-based tariff
VECSs	Village Electricity Consumer Societies
VHPs	village hydro projects
VSPP	very small power producer
WACC	weighted average cost of capital

Money amounts are in U.S. dollars unless otherwise noted.

Overview

Rural Africa's low level of electrification is a topic of much discussion. One widely cited estimate is that only 14 percent of rural households in Sub-Saharan Africa (excluding South Africa) have access to electricity (IEA 2012). As a first step to improving access, most governments in the region have developed national electrification strategies. Virtually every one of those strategies recommends a two-track approach to providing greater access to grid-based electrification.

The Centralized and Decentralized Tracks to Electrification

On the centralized track, electrification is undertaken by national governmental entities such as the state-owned national utility, a rural electrification agency (REA), or the ministry of energy, acting alone or together. Electrification occurs primarily through extension of the national grid. By contrast, on the decentralized track electrification is generally carried out through nongovernmental entities such as cooperatives, community user groups, or private entrepreneurs. These entities will usually construct and operate isolated mini-grids—small-scale distribution networks typically operating below 11 kilovolts (kV) that provide power to one or more local communities and produce electricity from small generators using fossil fuels, renewable fuels, or a combination of the two.

Although there is widespread agreement on the need for a two-track approach, most national electrification strategies contain few, if any, details on how the two tracks should be implemented. In this guide, our emphasis is on how to implement the decentralized track, with particular emphasis on how to create commercially viable small power producers (SPPs) and mini-grids in rural areas. If the decentralized track is going to be workable, SPPs will need to invest in and operate equipment to produce and distribute electricity to customers such as households, businesses, public institutions, and, in some instances, the national utility. And they are unlikely to invest unless regulations and policies are clear and credible.

This guide focuses on the regulatory and policy decisions that African electricity regulators and policy makers must make to create a sustainable decentralized track and how the decentralized track can complement the traditional central-ized track. For many decisions, the guide provides specific recommendations. If no specific recommendation is given, we present several options and factors to consider in choosing from among the alternatives. While our principal focus is on SPPs that use renewable energy or cogeneration technologies, most of the regula-tory decisions will also apply to SPPs that use fossil fuel or a combination of fossil fuel and renewable energy—that is, hybrid SPPs. Hybrid generating technologies are generally a lower-cost option than diesel generation for serving customers on isolated rural mini-grids.

The guide assumes that the national government has made a policy decision to promote SPPs and the decentralized track and that the national electricity regulator (among others) must now implement that decision. To make the guide as useful as possible for African regulators and energy policy makers, we drill down to actual questions that will need to be answered to achieve commercially sustainable outcomes. The focus is on ground-level economic and technical regulatory questions that routinely confront electricity regulators in Sub-Saharan Africa and elsewhere. Many of the required decisions are inherently controversial because they directly affect the economic interests of investors and consumers. The guide highlights rather than hides these real-world controversies by present-ing candid comments from key stakeholders—national utility managers, mini-grid operators, government officials, and consumers.

What Are Small Power Producers and Distributors?

SPPs are independently operated electricity providers that sell electricity to retail customers on a mini-grid or to the national utility on the main grid or on an isolated mini-grid, or to both. SPPs are usually defined by their size (for example, less than 10 megawatts [MW]), the fuel they use (for example, diesel and biomass), or their technology (for example, solar photovoltaic). In some countries SPPs are referred to as distributed generators, mini-grids, or community-level mini-utilities.

Small power distributors (SPDs) are a related but different type of entity. In contrast to SPPs, SPDs do not generate electricity. Instead, their primary business is distribution. They buy power at wholesale, typically from a national utility, and resell it at retail to households and businesses. The guide examines how an entity that initially operates as an isolated SPP could convert to an SPD once the main grid is extended to the SPP's service area.

It is important to understand the various kinds of SPPs. We classify them into four principal cases, as shown in table O.1, according to who their custom-ers are and whether or not they are connected to the main grid. SPPs can also operate as combinations of these cases. For example, an SPP may sell at whole-sale to the national utility on the main grid (Case 4) but at the same time also sell at retail to households and businesses on new mini-grids that are electri-cally connected to the main grid but operate as separate distribution businesses (Case 3).

Table O.1 Four Basic Types of Small Power Producers (SPPs)

		Location of generation	
		Connected to isolated mini-grid	*Connected to main grid*
Nature of customers	Selling retail (directly to final customers)	Case 1: Isolated SPP selling directly to retail customers	Case 3: SPP connected to main grid selling directly to retail customers
	Selling wholesale (to utility)	Case 2: Isolated SPP selling wholesale to utility	Case 4: SPP connected to main grid selling wholesale to utility

Regulating SPPs

If government policy makers decide that SPPs must be regulated in some way—and they usually do—they will need to decide who should regulate them. Options include:

- A department within an existing government ministry of power or energy
- A separate national electricity regulator
- Rural energy or electrification agencies
- Communities and community organizations
- Local governments

When we think of electricity regulation, we usually think of the first two options: a governmental department or separate national electricity regulator. However, it is also possible—and sometimes more efficient—for other entities such as REAs or community organizations to support or replace, at least initially, the work of a national electricity regulatory agency.

In Sub-Saharan Africa, more than 15 REAs have been created to promote rural electrification. The reality is that the typical "business plan" review that REAs conduct before making connection grants is very similar to a traditional "cost of service" review that a regulator would undertake in setting tariffs. The purpose of the REA review is to ensure that the SPP's revenues are high enough to ensure financial viability but not so high as to allow the SPP to earn monopoly profits at the expense of its customers. In addition, most REAs have a legal mandate to maximize the number of new households that will receive electricity. Therefore, most REAs are already acting as "quasi-regulators" since they are required to balance commercial viability against the affordability of the service that will be supplied by the SPPs applying for REA grants. This suggests that some REAs could take over some regulatory functions, especially for isolated mini-grids that have received grants from the REA.

Types of Regulatory Decisions Affecting SPPs

SPPs are affected by three basic types of regulatory decisions: technical, commercial and economic, and process. Examples of each are shown in box O.1.

Of the three types of decisions, technical and economic decisions usually get the most attention because they tend to be more visible and have an obvious impact. For example, it is clear that few, if any, main-grid-connected SPPs will

---

Box O.1 Examples of the Three Types of Regulatory Decisions That Affect SPPs

Examples of Technical (Engineering) Decisions
- Voltage, frequency, and power quality standards for grid-connected small power producers (SPPs)
- Regulations to provide safe and robust electrical connections between the national utility and a grid-connected SPP
- Distribution-system safety standards for both grid-connected and isolated SPPs

Examples of Commercial and Economic Decisions
- Price that the SPP is allowed to charge its retail customers
- Determination of who pays the cost of the interconnection between an SPP and the national grid operator so that the SPP can sell to the national grid or to an existing mini-grid
- Price that a grid-connected SPP receives for the power that it sells to the national or regional utility (the so-called feed-in tariff)
- Whether power-purchase agreements (PPAs) should be standardized for main-grid-connected SPPs and the provisions that should be included in the PPAs
- Price charged to the mini-grid for backup power because of planned or unplanned maintenance on its system

Examples of Process Decisions
- Whether the regulator consults with some or all stakeholders before making a technical or economic decision
- What information and approvals must be provided to obtain a license or permit
- Whether the consultation is conducted publicly or privately
- The time the utility has to respond to a request for interconnection by an SPP

---

be created if the price that they will be paid for electricity sold to the national utility is set below the SPPs' costs of supply. But even if the regulator sets a price that ensures economic viability for SPPs, the regulatory system may still fail if the decision-making process involves too many steps, if government entities ignore one another's responsibilities, or if the regulator fails to enforce its decisions in a timely manner. As one SPP developer observed, "[b]y the time the regulator gets around to enforcing his decision, I will be bankrupt." So an effective regulatory system for SPPs requires both fair and efficient technical and economic decisions as well as timely processes for making and enforcing those decisions.

Complying with regulatory rules costs time and money. With SPPs, regulators need to be especially conscious of the costs of regulation because many SPPs operate on the edge of commercial viability. This is especially true of new SPPs that intend to serve isolated rural communities. SPPs that propose to create rural mini-grids are not likely to develop unless the regulator makes a conscious effort to create a light-handed regulatory system.

In practice, light-handed regulation should:

- Minimize the amount of information a regulator requires
- Minimize the number of separate regulatory processes and decisions
- Use standardized documents or similar documents created by other agencies, with all documents available on the Internet
- Where possible, rely on related decisions made by other government or community bodies

For example, in Tanzania, SPPs generating less than 1 MW are exempt from applying for a license. Instead, they register with the regulator so that the regulator knows of their existence. Registration, unlike licensing, is solely for information purposes. It does not require the approval of the regulator. For very small power producers (VSPPs) with an installed capacity of 100 kilowatts (kW) or less, the Tanzanian regulator requires no prior regulatory review or approval of proposed retail tariffs. However, the regulator does reserve the right to review the VSPP's tariffs if 15 percent of its customers complain.

Light-handed regulation should not be blindly adopted in every situation. For example, Nepal and Sri Lanka became overwhelmed by applications from SPPs to sell wholesale electricity to the utility's main grid (Case 4). This occurred because application fees were too low, deadlines were too easy, prefeasibility studies were not required or could be copied, and project milestones between deadlines were not monitored.

Deregulation?

Some have argued that small, private SPPs selling electricity to previously unserved rural communities should be deregulated. The proponents argue that small, private, rural operators cannot be regulated in the same way as a large national or regional utility. And even if the regulator consciously starts with light-handed regulation, it is likely to get heavier over time.

Total price deregulation could produce a strong political backlash even if abused by just one or two SPPs. However, we see considerable merit in allowing an initial grace period of five years or so, during which private operators of small mini- and micro-grids in rural areas could experiment with different delivery models without obtaining the national regulator's approval for their retail tariffs or a full license to operate. (But they should still be subject to safety regulation.) The proposed grace period for pricing should be combined with prespecified backstop measures to protect village consumers. Those measures would include the following:

- Annual reporting
- Tracking of customer complaints
- Registration rather than licensing
- Review after five years

* * *

The second half of the guide offers implementation advice for three of the most important scenarios that regulators and policy makers are likely to face when working with SPPs: how to regulate main-grid-connected SPPs, how to regulate SPPs that sell electricity to retail customers, and how to prepare for the arrival of the main grid in an area currently served by an isolated SPP.

Regulating Main-Grid-Connected SPPs

If SPPs are going to be able to connect to and sell electricity to the national utility (Tanzania) or other operator of the main grid (Uganda), regulators and policy makers must make clear decisions on purchase agreements between the utility and the SPP, the feed-in tariffs (FITs) designed to promote the use of clean or renewable energy, and the technical and economic requirements for grid interconnection.

PPAs and Buying SPP Power

SPPs are business entities that must earn revenue to survive. For SPPs that wish to connect with and sell electricity to a utility-owned grid, the utility is usually the SPP's only customer and main source of revenue. The contract that enables this relationship between the SPP and the utility is called a power-purchase agreement (PPA).

The guide recommends that PPAs between SPPs and the utility that operates the main grid should have the following three features:

- **Standardization across all SPPs.** There are three reasons to prefer standardized PPAs: First, to reduce lopsided negotiations. As one SPP developer observed: "without standardized PPAs, we would live in a world of never-ending negotiations." Second, the regulator can conduct a single major review of one PPA rather than separate reviews of many different PPAs. Third, having a single PPA document for all projects facilitates due diligence for local banks that lend to SPPs.

- **Sufficient duration to repay project debt and no shorter than the period of availability of the FIT.** SPPs cannot get loans from banks if the duration of the PPA is shorter than the loan term. The PPA should also be at least as long as the availability of the FIT (see below). If the PPA were shorter, it would create an anomaly: the SPP would be offered a specified price, but the national utility would have no legal obligation to purchase at that price once the PPA expires.

- **Obligation on the part of the utility to purchase all of the SPP's power output.** PPAs for main-grid-connected SPPs should have a "must-take" clause that obligates the buying utility to purchase all of the SPP's electrical power output. Renewable energy (and to a lesser extent fossil-fueled cogeneration) are typically intermittent and thus not dispatchable, yet their contribution of electricity, when available, offsets the need for dispatching electricity from some other generator to meet demand at that moment.

Apart from these three core features, two other tariff provisions—so-called deemed energy clauses and the price of backup power—are frequently in dispute in the PPAs sought by SPPs in Sub-Saharan Africa.

Deemed Energy Clauses

Deemed energy clauses are designed for situations in which a main-grid-connected SPP seller is able to produce electricity, but the buyer is unable to receive it. The clause obligates the buyer to provide compensation for electricity that the SPP was capable of producing but the buyer was unable to receive. SPPs argue that such clauses are necessary to compensate them for lost revenue. They contend that if there is no deemed energy clause in the PPA, then "the buying utility's obligation to take the energy that I produce is nothing more than a joke."

It is not unreasonable for SPPs to be concerned about lost revenues. However, we recommend that the PPAs for main-grid-connected SPPs should not include a deemed energy clause because they are difficult to administer and can greatly increase regulatory transaction costs. In many cases, it could take considerable time and effort to determine whether the buying utility or the selling SPP was responsible for an interruption in sales. But it is reasonable to try to reduce risk for the SPP. Therefore, we recommend that the purchasing utility should be required to provide historical data on the frequency and duration of interruptions at the substation to which the SPP wishes to connect. This will prevent unnecessary cost and time burdens for all parties involved and give the SPP additional information on which to make its initial investment decision.

Backup Power

A backup or standby tariff compensates the national utility for providing electricity to an SPP when the SPP is not generating electricity, or not generating enough electricity to meet its loads. The SPP may need to buy backup power for one of several reasons:

- The SPP's generator is too small to meet its own or its retail customers' demand.
- The SPP's generator may need an external source of power to start or restart after it was shut down because of a planned or unplanned outage.
- The SPP's retail customers and/or the SPP facility's own load consume power while the SPP is not generating for whatever reason.

Disputes often arise over payments for backup power caused by unplanned outages on the buying utility's system. For example, in Tanzania, power quality problems on the buying utility's network frequently trip the SPP's protection relays, taking the SPP offline. To get back online, the SPP may have to purchase several hundred kW of electricity from the utility for several minutes. From the perspective of a grid-connected SPP, this is unfair. The SPP is forced to shut down because of a problem with the utility's system and is then forced to pay the utility high charges for electricity needed to restart the SPP's generator. In this situation, SPPs may be forced to pay high backup charges under traditional backup tariffs that include an energy and a demand charge that is "ratcheted" (that is, continues to be paid monthly for several months after the triggering event). When determining a backup tariff for SPPs, African utility regulators should consider granting a special, lower backup tariff to backup customers whose import load factor is less than 15 percent. Countries in which SPPs trip offline because of either instability on the national grid or insufficient overall generating capacity should consider implementing a backup power tariff with no demand charge, but with a charge for energy (kWh) that is higher than for regular customers.

Setting Feed-In Tariffs

The most common way for regulators and policy makers to set the wholesale price for electricity produced by grid-connected renewable energy SPPs—whose electricity is often more expensive than that of fossil fuel generators—is with an FIT. An FIT is a tariff-support mechanism for renewable energy generators or cogenerators in which the generator is guaranteed a payment, usually over a long period, for every kWh it feeds into the grid. The guide describes the two methods that regulators typically use to set FITs and offers a recommendation for a two-phase approach that can be used by African regulators and policy makers in setting FITs for SPPs.

The first method that regulators use to set FITs is often referred to as the *avoided-cost approach*. This approach is based on the costs that the utility and/or society avoids by not having to produce the amount of electricity that the SPP proposes to produce. Calculations may include costs that the utility avoids (for example, building a new generator), costs that the economy avoids (for example, building transmission lines with taxpayer money), and costs that society avoids (for example, local environmental damage). In practice, calculations rarely include social or even economic costs because they are difficult and expensive to measure. Because this approach is technology neutral (that is, all generators get the same FIT), if only the utility's avoided costs are included, many renewable energy SPPs will not be commercially viable, because their FIT will not cover their costs of generation.

The second and more common method that regulators use to set FITs is a *standardized, cost-reflective, technology-specific calculation*. This approach is based on calculating the levelized generating costs of each different type of renewable energy generator. Standardized calculations are made for each technology, such that a hypothetical well-run electricity generator can earn a reasonable profit after paying for its costs. What constitutes a reasonable profit is a decision for the regulator or policy-making entity. In theory, this approach allows more renewable energy SPPs to be commercially viable. Assumptions must be made concerning several parameters: capital structure, capacity factor, cost of capital equipment, interest rates for loans, and so on. The assumptions are often contested because they affect the level of the FITs that will be paid.

At first glance, the second approach may appear better suited for maximizing the production of electricity from renewable energy. In fact, it is already the most common approach on a worldwide basis. However, it may not be feasible for many developing countries because it requires the buying utility (usually a cash-strapped government-owned utility) to buy electricity from renewable energy SPPs at premium prices (that is, prices higher than the country's own financial-avoided costs).

In cases where lack of funds precludes the second approach right away, we recommend a two-phased approach:

- In phase 1, FITs are set approximately equal to (or below) the buying utility's avoided costs, recognizing that some types of renewable energy generation will not be commercially viable with these FITs.
- In phase 2, some of the FITs are allowed to exceed the buying utility's avoided costs when funds for the incremental costs of these higher tariffs become available.

This two-phased approach allows a country to "walk up the renewable energy supply curve" as more money becomes available. But where will the phase 2 funds come from? They could come from the country itself—for example from taxpayers or cross-subsidies—but this is unlikely to occur in low-income countries. A more viable option is to seek funding from external donors to cover the gap between the utility's avoided cost and the FIT. This guide discusses the issues involved in creating such a top-up fund, drawing from recent experiences in Uganda (appendix G).

Addressing Interconnection Issues

The term interconnection refers to the physical equipment needed to connect a new generator to an existing grid.

Regulators should set standardized rules and standards for SPPs that wish to interconnect to a regional or national grid. Such rules are needed because a utility that is opposed to buying SPP power may be tempted to create an unclear or lengthy application process or may attempt to set unduly stringent technical parameters.

The necessary rules should govern both the application process and technical and engineering standards in interconnection. The rules governing the application process should include specifications of responsibility for analyzing and approving the application, for the payment of fees, and for overseeing construction, as well as guidelines for sharing information.

Regulators should set and enforce technical and engineering standards for interconnection pertaining to the capacity and quality of equipment, measures to protect equipment and the grid from over- and undervoltage, overcurrent, unintentional "islanding" (or isolation of the SPP's generation), and safety hazards (including lightning).

If the main grid is unstable, the regulator should give SPPs the option of intentionally isolating themselves from the main grid—known as intentional islanding.

In addition to the physical issues inherent in connecting an SPP to the utility's grid, regulators must also make decisions on economic issues that arise from

Table O.2 Typical Cost Allocation of Interconnection Assets

Physical assets	Paid by	Cost sharing?	Built by
Transformer(s), switchgear, and line up to where all other equipment and lines are the responsibility of the utility	SPP	None	SPP
Equipment to protect grid from adverse effects of SPP and vice versa	SPP	None	SPP
Energy meter and metering equipment at the point where the SPP sells power to the utility's grid	SPP	None	Utility
Other lines necessary to SPP operations beyond the point where the SPP connects with the grid, up to a certain designated point	SPP	Possible, with another SPP	Utility or SPP
Other equipment necessary to SPP operations beyond the point where the SPP connects with the grid, up to a certain designated point	SPP or utility's customers	Possible, with another SPP, customers, or the utility	Utility or SPP

Note: SPP = small power producer.

interconnected operations. Table O.2 summarizes who builds and pays for interconnection assets, as generally observed in developing countries, when a previously isolated or new SPP connects to a buying utility's grid.

To ensure the smooth interconnection of an SPP to the buying utility's grid, regulators can take two key actions related to the cost and ownership of assets. These are:

- Regulators should allow SPPs to retain ownership of their assets up to the point where the SPP's network interconnects with the grid, while insisting that ownership of SPP-paid assets upstream of this point be transferred to the utility at zero cost. The utility can claim depreciation on these assets but should not earn a profit on them.
- Regulators should require or encourage new SPPs or users that wish to connect to interconnection facilities paid for by another SPP or user to reimburse the initial SPP(s) or customer(s) for the facilities they seek to use. In these cases, the utility must maintain accurate records, charge the new customers/SPPs a *pro rata* share of the initial capital cost, and provide reimbursement to the first customer/SPP.

Regulating SPPs and Mini-Grids That Sell to Retail Customers

SPPs also sell electricity to retail customers if they operate a mini-grid. Here, the two key regulatory concerns are setting maximum tariffs and establishing minimum quality-of-service standards.

Ideally, tariffs should be cost-reflective, which simply means that the SPP operator can reasonably expect that the total revenues received from the tariffs paid by its customers will recover total operating and capital costs for both generation and distribution. If cost-reflective tariffs are not allowed because the SPP operator's tariffs are capped at a lower level (either because of informal political pressures or a formal legal requirement for a uniform national tariff), the SPP will not be commercially sustainable and will soon disappear.

Ensuring Commercial Viability of Isolated SPPs

Rural electrification is expensive, and many SPPs serving isolated communities experience a gap between their costs and revenues. In some cases, the cost-revenue gap arises because a law or regulation prohibits SPPs from charging tariffs that are high enough to cover their costs. In other cases, SPPs cannot charge cost-recovering tariffs because the national utility operating on the centralized track has created a nationwide *de facto* price ceiling by charging its customers below-cost retail tariffs, thereby making it seem to potential customers that the electricity provided by the SPP is too expensive.

Even if SPPs are allowed to charge tariffs that are high enough to cover their costs, they still may not be able to sign up many customers if the initial customer connection charges are high. And even if an SPP solves that problem, it may still operate at a loss if the average consumption of its customers is too low to produce enough revenue to cover the SPP's operating costs.

So, how can regulators help SPPs close their cost-revenue gap and achieve commercial viability? The first step is to measure the size of the gap. Regulators, REAs, and SPPs should use a common financial analysis tool to measure the gap between the costs of supplying electricity to rural communities and the revenues that can be collected. The next step is to close the gap.

In closing the gap, regulators can take measures relating to tariffs to help SPPs become commercially viable. These measures include:

- **Allowing SPPs to charge tariffs above the uniform national tariff** if required to recover efficient operating and capital costs.

- **Allowing SPPs to cross-subsidize different customer groups.** Cross-subsidies exist when an electricity provider charges one group of customers a higher tariff in order to subsidize lower tariffs for other customers. Many African national utilities subsidize the tariffs applied to residential customers who consume small amounts of electricity by charging higher tariffs to their commercial customers. The same opportunity should be available to SPPs.

- **Requiring SPPs to charge tariffs that include depreciation on equipment financed through grants.** A piece of equipment will eventually have to be replaced by the SPP even if it was originally paid for by a donor or government grant. SPP tariffs should build in depreciation costs to generate funds that can be used eventually to replace that equipment. (By contrast, the SPP should not be allowed to earn an equity return on any equipment that was financed by an external grant.)

- **Allowing SPPs to enter into power sales contracts with business customers without obtaining prior or after-the-fact regulatory approval of the terms of the contract.** Most village businesses can self-generate. While self-generation is not a perfect substitute for SPP-supplied power, it places a limit on an SPP's ability to charge monopoly prices. These businesses can serve as "anchor customers" that will make it easier for an SPP to obtain bank loans. In the Indian state of Uttar Pradesh, for example, mobile-phone towers are the initial anchor customers that make it possible to serve rural villages that would otherwise not be appealing to private operators.

- **Allowing SPPs to recover the costs of making loans to actual and potential customers to allow them to connect to the SPP system and to buy appliances and machinery for productive uses.** Sometimes an SPP is not viable because consumption and sales of electricity are too low. Rural households and businesses may want to become customers or increase their consumption once they are connected, but they lack access to financing to purchase appliances and machinery. If SPPs are given explicit regulatory approval to provide financing that might not otherwise be available, this will help SPPs to increase their sales revenues and become commercially viable sooner than they otherwise would. Donors could provide seed money to finance revolving funds from which SPPs could finance customers' purchase of appliances and machinery. The SPP's loans to its customers can be repaid through extended payment plans implemented through on-bill financing.

- **Granting SPPs flexibility in deciding on the tariff structures that work best for their technology and business models.** Our general recommendation is that regulators should give SPPs the freedom to devise tariff structures that are most suitable for their own project. Options include standard "kWh-based" tariffs, as well as less standard approaches such as subscription-based tariffs in which users pay a flat fee per month based on the capacity of a load-limiter that restricts power consumption from exceeding a certain threshold.

A second set of measures that regulators can take to help SPPs reach and maintain commercial viability relates to subsidies and revenue earned from carbon credits.

Governments usually mandate or authorize subsidies to meet a social objective such as promoting electrification or encouraging renewable energy—that is, policy issues. Just as it is the government's job to make policy, the regulator's job is to implement government policy.

We recommend that if a subsidy is authorized or mandated by the government, the regulator should not take actions that would nullify or reduce the effect of the subsidy. Instead, the regulator should take actions that help to ensure that the subsidy is delivered to its intended target (either the SPP or its customers) as efficiently as possible. The regulator also should periodically inform the government of the costs and benefits of the subsidy so the government can decide whether it is achieving its intended purpose.

The case of grants to lower connection charges illustrates how a regulator's decisions can determine whether a subsidy achieves its stated purpose. African national utilities have some of the highest connection charges in the world. Governments and donors often make grants to enable poor rural households to obtain a connection to an electrical grid. A regulator can support the government's objective by making sure that the grant money is actually reflected in lower connection charges paid by new customers and by allowing the national utility or mini-grid operator to take depreciation on the equipment that was paid for by the grant.

Revenue earned from certified emission reduction (CER) credits is not a subsidy. Instead, CER credits are payments for the provision of a service by the SPP: the reduction of global carbon emissions against a calculated business-as-usual benchmark. Regulators are not typically involved with this process or the allocation of these revenues. This guide recommends that regulators should not modify a previously agreed allocation of revenues from CER credits. If asked to pass judgment on an allocation formula, they should adopt the principle that CER revenues should go to the SPP developer that supplied the equity and assumed the risk of developing the project rather than counting the CER revenue as an imputed credit that reduces the overall revenue that the SPP is allowed to recover through tariffs.

Setting Quality-of-Service Standards

Even if a regulator decides to deregulate the retail tariffs of isolated SPPs on a trial or permanent basis, it should still set minimum quality-of-service standards to ensure safety, quality, and reliability of SPP operations. These standards generally fall into three categories:

- **Quality of product:** How useable is the electricity? Are there wide variations in voltage or frequency that damage customer appliances?
- **Quality of supply:** How available is the electricity? Is it available only at inconvenient times, and how frequent are unplanned blackouts?
- **Quality of commercial service:** How good is the SPP's customer service? How long does it take the SPP to resolve a complaint?

The standards should not be prohibitively costly for SPPs or their customers and should be relatively easy to monitor and enforce. Initially, it is easier for regulators to establish standards for inputs (equipment, materials, and so on) rather than outputs (reliability of electricity), but over time it is preferable to focus on output-based quality-of-service standards and let the SPP operator have more discretion over the inputs needed to meet those standards. In countries where the REA has specified quality standards in its grant agreements with mini-grid operators, the agency and the regulator should agree on a single set of standards.

Preparing for the Arrival of the Main Grid

Regulators need to prepare for the moment when the top-down and bottom-up approaches to electrification meet—that is, when the main grid arrives in the service area of an existing SPP. In the absence of regulatory certainty as to what happens "when the big grid connects to the little grid" developers and

investors are unlikely to invest in SPP projects. And when private and cooperative investors are reluctant to invest in isolated mini-grids, rural villages suffer
because they are denied the chance to receive electricity and instead must wait
years or decades for the national grid to arrive, if indeed it ever does. Chapter
10 of the guide presents five post-connection options and the regulatory issues
associated with each. It also describes the fate of the SPP's physical assets under
each option.

Option 1: The SPP Stops Generating and Becomes a Pure Distributor

Under this option, once the SPP connects to the main grid it ceases generating
electricity in favor of purchasing it wholesale from the utility and reselling it at
retail to its customers. This business model is common in Asia but not in Africa.
For example, in Cambodia more than 80 SPPs have now converted to SPDs. In
rural areas, most Asian SPDs are able to achieve higher operating and commercial
efficiencies than the national utility.

The crucial component of ensuring commercial viability under the SPD
option is the distribution margin (the difference between the SPD's average
retail price and its average bulk purchase price) that the SPD retains on its retail
sales of electricity. International experience suggests that a distribution margin of
at least 4–5 U.S. cents per kWh may be needed to ensure commercial viability
for medium-size SPDs in Africa.

If the government formally or informally requires uniformity in retail tariffs,
most SPDs will need operating subsidies to survive. Those subsidies can come
from the government's general budget, from REAs, or through mandated discounts on the price paid by SPDs for wholesale power purchases. Without such
subsidies, the SPD will face a price squeeze because its distribution margin will
be too small to allow it to survive.

In addition, if regulators allow SPP mini-grids to become SPDs, care must be
taken to ensure that the distribution system is built to a standard that is sufficiently high to accommodate interconnection with the national grid. If the SPP
developer cuts corners to save money in the cost of installing the initial distribution system, then that system will need to be upgraded before the isolated mini-
grid connects to the main grid.

Option 2: The SPP Stops Distributing and Sells the Power It Generates to the Main Grid

A second option is for the SPP to stop selling its electricity to its retail customers,
and instead limit itself to selling electricity at wholesale to the main grid.

Whether an SPP can remain financially viable under this option depends
crucially on three factors:

- The cost of electricity production by the SPP
- The FIT that the SPP (now connected to the main grid) will receive for sales
  to the national utility

- The capacity factor (actual output compared to maximum possible output) at which the SPP will be able to operate

Under this option, regulators and utilities must also address technical issues related to the transition from off-grid to grid-connected power generation, especially with regard to control of the generator's frequency.

Option 3: The SPP Operates as a Combined SPP-SPD

Under this business model, the SPP moves from operating an isolated mini-grid to functioning as an SPD that buys electricity at wholesale from a national or regional utility and resells it at retail to local customers. It also maintains an existing or new small generator as a backup and also possibly as a source of power to sell to the main grid and retail customers.

This business model should be encouraged in countries that face shortages of generation capacity on their main grids *and* the challenge of extending rural electrification services to more of the population. It should also be favored in areas where the local distribution grid is weak and brownouts or blackouts are common.

Option 4: The Utility Buys the SPP

A fourth option is for the utility company to purchase and operate the SPP's mini-grid distribution network—and possibly the generator as well if it is of sufficient quality and capacity.

This option may make sense if the mini-grid is built to engineering standards comparable to those used in the utility's own distribution assets and if the utility has the staffing capacity to operate the new acquisition—including bill collection, new hookups, maintenance, and dispute resolution.

The details of which assets will be sold and at what price must be worked out on a case-by-case basis. In principle, the sale price would reflect the depreciated value of the assets that remain serviceable. A further consideration in determining a price is whether, and to what extent, the mini-grid and generator were originally subsidized or paid for entirely by the government or a donor. A private operator should not be compensated for the portion of the investment that was paid for by outside grants.

Option 5: Abandonment

In some cases where the SPP's generator or distribution are of poor quality or were not built in such a way as to be compatible with the main grid (as would be the case, for example, for an SPP operating a photovoltaic-powered direct-current distribution system), it may not be cost-effective to repair or replace the necessary components. In this case, the utility will have to scrap the SPP's assets and build a new distribution system. This is not necessarily a bad outcome, assuming the SPP was able to supply some of its customers' electricity needs and earn a profit on its investment before the main grid arrived.

From Broad Strategy to Ground-Level Implementation

In both rural electrification and renewable energy, the African landscape is littered with the remains of many "strategies" and "programs." All too often, these initiatives are proposed with grand pronouncements from both donors and government officials. But sadly, it is not uncommon for them to achieve far less than originally envisioned because of lack of attention to the practical aspects of implementation.

In this guide, we have consciously tried to stay away from high-level strategies and instead focused on the practical, day-to-day implementation issues that confront African regulators and policy makers. Early evidence shows that private and community-owned and -operated SPPs and mini-grids can succeed with a decentralized approach. In the process, they can play an important role in promoting both renewable energy and electrification in rural Africa and elsewhere. But if this is to happen, one key element (though not the only element) is a rational and supportive system of economic regulation that helps to achieve commercial sustainability while protecting rural consumers. This guide describes in detail what such a system should look like.

The Arrival of Electricity in Rural Texas, 1939

… and as they neared their farmhouse, something was different. "Oh my God," her mother said. "The house is on fire!"

But as they got closer, they saw the light wasn't fire. "No, Mama," Evelyn said. "The lights are on."

—FROM ROBERT A. CARO, *THE PATH TO POWER*, VINTAGE BOOKS, NEW YORK, P. 528

CHAPTER 1

Introduction

… And I have to say, those who are involved in this process, they continually tell us the problem is not going to be private-sector financing. The problem is going to be getting the rules right, creating the framework whereby we can build to scale rapidly.

—U.S. President Barack Obama, speaking on electricity
access in Africa, Dar es Salaam, Tanzania, 2013

We need megawatts not megawords.

—Zambian government official, 2010

… most failures happen at delivery.

—Jim Yong Kim, President, World Bank Group, October 2012

Abstract

Rural Africa's low level of electrification has been discussed at many conferences and workshops. One widely cited estimate is that only 14 percent of rural households in Sub-Saharan Africa (excluding South Africa) have access to electricity (IEA 2012). As a first step to improving access to electricity, most governments in Sub-Saharan Africa have developed national electrification strategies. And virtually every one of them contains a similar recommendation for scaling up rural access to grid-produced electricity—namely, that greater access to grid-based electrification is best accomplished using a two-track approach: a centralized track and a decentralized track.

Africa's Two-Track Approach to Rural Electrification

The centralized track is a top-down approach because electrification is undertaken by one or two national government entities such as the state-owned national utility, a rural electrification or energy agency (REA), or the ministry of energy, acting alone or together. The centralized track is also sometimes referred to as "grid electrification" because electrification occurs primarily through extension of the existing high- and medium-voltage grid.[1] By contrast, the decentralized track is a bottom-up approach because electrification is generally carried out through nongovernmental entities such as cooperatives, community user groups, or private entrepreneurs. The various names given to the decentralized track reflect the fact that these nongovernmental entities will usually construct and operate isolated mini-grids.[2] These names include "off-grid electrification," "decentralized distributed generation" (India), "stand-alone systems" (India), and the "small power producer approach" (Tanzania and Kenya).

In this guide, we explore the tracks in terms of the technology and institutions involved in each. With respect to technology, the main distinction is whether electrification is accomplished through extension of the main grid or through the creation of electrically isolated mini-grids. With respect to institutions, we find it is best to describe the institutional arrangements in terms of assigned functions or responsibilities. In other words, *institutional arrangements can best be understood not by the names given to the institutions, but by knowing who does what.*

- Who decides which communities will be electrified?
- Who funds the construction of the facilities?
- Who builds the facilities?
- Who owns and operates the facilities once they are built?
- Who regulates (that is, *controls*) the actions of the operator by deciding on maximum and minimum prices and minimum quality-of-service standards?

Our Purpose and Approach

Although there is widespread agreement on the need for a two-track approach, most national electrification strategies contain few, if any, details on how the two tracks should be implemented. In this guide, our focus is on implementing the decentralized track, with particular emphasis on how to create commercially viable small power producers (SPPs) in rural areas that will invest in and operate renewable generators, hybrid generators, or cogenerators to produce electricity distributed over mini-grids to various customers such as households, businesses, and public institutions, or sold onto the main grid to the national utility.[3]

We examine regulatory and policy decisions that African electricity regulators and policy makers must make to create a sustainable decentralized track (one in which operations are commercially viable over the long term) and how it can complement the traditional centralized track. We see little point in setting up a regulatory system that creates SPPs that operate for a few years and then collapse because they are not financially viable. If this happens, the outcome is "anti-poor" rather than "pro-poor."

SPPs and mini-grids represent only one type of decentralized generation. Where even mini-grids are too large to be sustainable, other decentralized options may be viable. These might include battery- and solar-charged flashlights and lanterns, portable solar kits, mounted solar systems for individual houses and institutions such as schools and clinics, and small individual diesel generators. Regulatory and policy decisions pertaining to these options are not covered in this guide. While these other technologies can provide critical transitional sources of electricity, rural households and businesses almost always express a strong preference for the "real thing": reliable and affordable grid-supplied electricity. Recognizing this preference, our focus is on how to bring grid-supplied electricity to rural areas both quickly and sustainably using SPPs connected to the main grid or to isolated mini-grids.[4]

For many of the regulatory and policy decisions, the guide contains recommendations on how governments can encourage both. If no specific recommendation is given, we present several options and factors to consider in choosing from among the alternatives. While our principal focus is on SPPs that use renewable energy or cogeneration technologies, most of the regulatory decisions will also apply to SPPs that use fossil fuel or a combination of fossil fuel and renewable energy (that is, hybrid SPPs). In fact, we recommend promoting hybrid generating technologies as the best available option for promoting renewable energy for isolated rural mini-grids.

Our assumption is that the national government has made a policy decision to promote SPPs and the decentralized track and that the national electricity regulator[5] (among others) must now implement that decision. To make this guide as useful as possible for African regulators and energy policy makers, we drill down to actual questions that will need to be answered to achieve a successful outcome. The focus is on ground-level economic and technical regulatory questions that have confronted practicing electricity regulators in Sub-Saharan Africa and elsewhere.[6] These are sometimes referred to as "second-level" regulatory and policy implementation questions. For each category of implementation questions, we provide, first, a description of the regulatory or policy question that needs to be addressed, and, second, our comments and recommendations on specific regulatory actions or decisions that can or should be taken to answer this question.

We do not consider our discussion of the implementation questions that need to be addressed and the actions that need to be taken as the last or the best words on these subjects. Instead, we view this guide as simply one point in an ongoing conversation among regulators, policy makers, developers, and investors (and those who advise them). This conversation will continue long after this guide is published. Our hope is that it will help to make the conversation more focused and more productive.

What Are the Typical Starting Conditions in Africa?

Regulatory systems do not exist in isolation. A system that works in one country may fail in another. In designing SPP regulations for the 26 electricity regulators in Sub-Saharan Africa, one must always be conscious of the economic and political conditions that constrain regulatory options.[7] Much has been written recently about the current state of electricity sectors in Sub-Saharan Africa.[8] We will not repeat that discussion here. But it is important to highlight several key ground-level realities that limit the options of a typical African electricity regulator:

- *Higher political priority for electrification* than for renewable energy or cogeneration. As one African regulator observed, "Look, we have so many rural communities that don't have any electricity at all. We don't have the luxury of saying that electrification should only be done with green electricity. Our villages are desperate for electricity—they don't care whether the electrons are green, purple, or black."[9]
- *Limited coverage of the national grid* and patterns of rural electrification that include isolated mini-grids as well as large expanses of unelectrified and sparsely populated areas.
- *National utilities that may not have financial incentives to connect rural households as customers.* They usually will lose money on most of the power sold because of high generation costs and artificially low social tariffs—even if the initial connection costs are heavily subsidized by outside grants.
- *National utilities that are insolvent*, creating risks for financers and developers of SPPs that payments for electricity supplied may come late or not at all.[10]
- *Politically mandated uniform national retail tariffs* that make it difficult for SPPs selling to rural customers to break even unless they receive significant capital and operating-cost subsidies.
- *New rural energy agencies that provide grants and technical assistance to new rural electricity providers* and whose actions and policies may overlap with the responsibilities of the national electricity regulator.

Whose Regulatory Decisions?

We pay particular attention to decisions made by regulators in Sub-Saharan Africa. We will highlight recent decisions made by EWURA (the Tanzanian electricity regulator) in creating a regulatory system for grid and off-grid SPPs. EWURA was chosen because, at the time of this writing, Tanzania has probably made more progress than any other African country in developing a comprehensive SPP regulatory system. (For ease of exposition, we use "Africa" and "Sub-Saharan Africa" interchangeably.) All the guidelines and rules of the Tanzanian regulatory system are available to the general public on the EWURA website (www.ewura.com).

These guidelines and rules are more than just words on paper. Tanzania's regulatory system has produced results. As of March 2013, eleven standardized power-purchase agreements (PPAs) had been signed between Tanzania Electric Supply Company (TANESCO), Tanzania's government-owned electric utility, and five SPPs. More important, three of these SPPs, with a total installed capacity of 14.5 megawatts (MW), are now selling electricity to TANESCO under the agreements. A fourth is expected to become operational in 2014. It is too early to know whether these initial SPPs will be commercially sustainable given the financial insolvency of TANESCO (the buyer) and whether there will be significant additional investments in other Tanzanian SPP projects.

Tanzania is not the only workable regulatory model for Africa. In fact, a somewhat different SPP regulatory system is being created next door in Kenya. And governments in developing countries outside of Africa have developed other regulatory systems for promoting SPPs. In Asia, both Sri Lanka and Thailand had considerable success in promoting investments in grid-connected SPPs even before their national regulators came into existence. *In other words, both Sri Lanka and Thailand created successful regulatory systems in the absence of a national electricity regulator.*

To illustrate these different possible regulatory approaches, we compare and contrast EWURA's decisions in Tanzania with the comparable regulatory decisions made in Thailand and Sri Lanka. We also draw upon the recent SPP experiences of Nepal, Cambodia, and India. These cross-country comparisons reflect our view that it is dangerous to espouse a single best practice for all countries at all times. The best practice for any country, whether in Africa or elsewhere, can be determined only after taking a close look at the country's starting conditions, examining the regulatory approaches that have been adopted elsewhere, and then making informed decisions as to which elements of these other approaches can be successfully transferred.

Acknowledging Controversies and Understanding Different Vocabularies

A regulator's SPP decisions are often controversial because they significantly affect the economic interests of national utilities, existing or proposed SPPs, and the customers of both. They also affect the political fortunes of numerous politicians. So it should not be a surprise that regulators are always subject to political pressures, whether open or hidden. In most studies of rural electrification, regulatory policies are often described in neutral, analytical terms that obscure or ignore the underlying controversies. This approach is not very helpful. *If progress is going to be made, real-world controversies need to be highlighted rather than hidden.* And it is not enough to characterize an issue as controversial. We have taken a different approach. With respect to several controversial regulatory issues involving SPPs, we have included statements that reflect the different and often conflicting positions of national utilities, SPPs, government officials, and consumers. While we present these statements in the form of quotes, we do not attribute them to specific individuals for two reasons. In some instances, they were made "off the record," and we must respect the confidentiality of the discussions. In other instances, they are not direct quotes but instead represent our paraphrasing of statements made to us by one or more individuals with similar views in different conversations.

In addition to these fundamental disagreements, people often fail to communicate because they use different vocabularies. For example, specialists at the World Bank and other development organizations tend to talk in abstract, high-level concepts like "institutional and regulatory ecosystems," "enabling environments," "non-cost-reflective tariffs," or "flawed governance mechanisms."

In contrast, developers usually talk in a more direct and less abstract way. They will say "regulatory rules," "problems in getting loans," "tariffs that are too low," and "government officials who want bribes." This guide is written for both audiences. Therefore, we will, whenever possible, use words that are somewhere in between these two ends of the vocabulary spectrum. And when the words are not self-evident, we will define them.

Regulation: The Problem or One Part of the Solution?

SPP developers often have different views on regulation. An SPP that has been trying to sell electricity to a reluctant national utility is quite happy when the regulator orders the national utility to buy the SPP's electricity at a price that covers their costs and allows them to earn a profit. In this instance, the SPP considers the regulator to be a friend and an ally. In contrast, an SPP, especially one that is privately owned and wants to sell at retail, will sometimes complain that its dealings with the regulator are exasperating or even dangerous. The most common complaint is that regulators impose significant costs and are very slow in making needed decisions. Complaints about the speed of regulation are not limited to developing countries. As one U.S. utility executive observed, "there is real-world time and there is regulatory time, and the second always takes much longer than the first." In fact, the greater the transparency of the regulatory processes, the slower the regulation. And in many countries there is also the privately expressed concern that the more regulatory approvals that are required, the more likely it is that the developer will be asked for a bribe in return for the needed piece of paper.

Hence, from the perspective of many SPP developers, regulation is at best a necessary evil. The SPP needs a license or some similar document that gives it the legal right to operate. This gives the SPP "official" status that, in turn, is needed to obtain a bank loan or some other source of financing that makes the project possible. The ideal for most SPP developers would be to get this initial regulatory approval without any further substantive regulation. Regulators have a different perspective. In their view, if they are granting a legal monopoly by giving an exclusive right to serve one or more communities, they are legally obligated to protect those customers, especially captive customers who do not have any other viable options.

Not surprisingly, our view is that there must be balance. We recognize and accept that consumers must be protected. But we also know that regulation is never free. Compliance costs time and money for those who make the regulations and for those who must comply with them. Regulations that create excessive information burdens or that impose contradictory and confusing requirements can keep viable and worthwhile projects from moving forward. Therefore, throughout this guide we have tried to identify areas where regulatory requirements are unnecessary and counterproductive and to propose workable, light-handed alternatives. In some instances, we recommend full or partial deregulation as the best solution. We strongly believe that regulators should be continually reminded that regulation is not an end in itself but simply a means to an end, with the end being reliable, grid-based electricity supplied to unserved rural villages at the lowest possible cost as soon as possible.

Two Other Important Success Factors

Even though regulation is the focus of this guide, it is important to recognize that good regulation alone will not produce investments in SPPs. *At best, good regulation can only create fertile ground.* It does this by providing certainty to investors, whether private or community based, that their investments will be protected (that is, that property rights will be created and honored) and to consumers that they will get value for their money. But if SPPs are going to take root in this fertile ground, seeds must be planted and fertilizer must be spread. The seeds are financial capital and the fertilizer is human capital.

Financial Capital

Nothing will grow unless someone is able to provide initial equity to get an SPP project started. Historically, equity capital for many SPP projects in Sub-Saharan Africa has come in the form of grants from governments, donors, or nongovernmental organizations. But such "charitable capital," while given with good intentions, is not a reliable or sustainable source of funding. Donors come and go because their funding sources are unstable and their priorities often shift abruptly. Moreover, donor grants, whether given directly or channeled through an REA, are rarely sufficient to cover the total capital costs of more than just a few SPP installations. Any equity capital, whether a gift from donors or an equity investment from private sources, must almost always be supplemented by loans.

Human Capital

Even if equity and debt financing are available, another scarce commodity is human capital. Building and operating an SPP is not a familiar task for African villagers. It is always possible for donors to provide the outside know-how to build an SPP in the occasional "pilot" village. When the project is inaugurated, it provides a good photo opportunity for an ambassador from a developed country or for a country's president before an election. But in the words of one observer, this is nothing more than "boutique electrification." It does not lead to sustained and significant electrification. If SPP-based electrification is to make a real difference, it requires both private capital and business know-how that is replicable and can be easily scaled up. If available, this human capital is the fertilizer that allows SPPs to take hold and flourish on more than a pilot basis. Without financial and human capital, the permits, licenses, and rules of a regulatory system, no matter how carefully written, are just "pretty words on pieces of paper."

From the Bottom Up • http://dx.doi.org/10.1596/978-1-4648-0093-1

Organization of the Guide

This guide is organized as follows:

- Chapter 2 explains key terms—SPP, small power distributor (SPD), and electrification—that are used throughout the guide. It describes four basic types
  of grid and off-grid SPPs as well as combinations of the four types that are
  emerging in Africa and elsewhere. It analyzes similarities and differences in
  SPPs that operate on isolated mini-grids in three countries (Cambodia, Mali,
  and Sri Lanka). It also evaluates the level of difficulty of achieving financial
  sustainability for different types of SPPs. It discusses the meaning of electrification and why grid-supplied electrification, whether obtained from an SPP
  or by extension of the main grid, is best defined in terms of the quantity and
  quality of electricity supply rather than the traditional approach of counting
  physical connections.

- Chapter 3 describes the three principal kinds of regulatory decisions—technical
  decisions, commercial or economic decisions, and process decisions—with
  examples of how each can affect SPPs. It examines the concept of "light-
  handed regulation" and explores how it can be implemented for SPPs. But the
  chapter also analyzes how light-handed regulation can sometimes produce
  unintended consequences. It explains why a form of tariff deregulation may
  be the preferred option for some rural electricity business models. Finally,
  the chapter addresses the question of who should regulate SPPs: the national
  electricity regulator, an REA, or some other entity.

- Chapter 4 discusses the regulatory approvals that an SPP needs to obtain in
  order to operate. The focus is on the design of regulatory processes—who
  approves what, when, and how—for SPPs that wish to connect to a national or
  regional grid and sell to the national utility or to another buyer that is legally
  obligated to purchase from the SPP. After describing regulatory processes in
  Sri Lanka, the chapter compares and contrasts the Sri Lankan system with the
  one that exists in Tanzania and one proposed in Kenya. While recognizing that
  particular features will inevitably vary from one country to another, this chapter
  recommends six required features of a good regulatory system for SPPs.

- Chapter 5 examines the regulatory treatment of different subsidies that are
  often given to SPPs and SPDs. Particular emphasis is placed on capital-cost
  subsidies received from rural electrification funds and donors to lower the cost
  of connecting new rural households. It also describes the different channels by
  which SPPs may earn carbon credits and suggests how regulators should treat
  revenues from these credits.

- Chapter 6 examines relevant regulatory decisions for grid-connected SPPs.
  In particular, it analyzes the terms and conditions of standardized PPAs that

can be provided to SPPs and how these differ from the PPAs used for larger independent power producers (IPPs). It addresses the issue of "deemed energy" clauses that are a particular concern in Sub-Saharan countries that have weak transmission systems, insufficient overall generating capacity, or both. It also considers the design and implementation of back-up tariffs that must exist when an SPP connected to the main grid buys power from a larger national or regional utility.

- Chapter 7 presents the two principal methods for setting "feed-in tariffs" (FITs) for grid-connected SPPs and how the ground-level implementation of these tariff-setting methods has worked in Tanzania and Sri Lanka. It provides a detailed analysis of key FIT implementation issues: capacity payments, periodic adjustment mechanisms, and which entity should calculate FIT values. It then provides an analysis of the problems that can arise in Africa if the FIT is calculated using a technology-specific, cost-reflective approach. Finally, it recommends a two-phase approach for implementing FITs in Africa based on a strategy of "walking up the supply curve."

- Chapter 8 presents the basic technical and commercial issues that arise in connecting SPPs to the national grid or an existing isolated mini-grid and operating them. It provides a primer for nonengineers on basic engineering terms and concepts that are relevant for interconnected SPPs. The chapter describes issues that arise in obtaining an interconnection and in allocating interconnection costs.

- Chapter 9 discusses the regulatory issues for SPPs that serve retail customers. It includes an analysis of different approaches in setting the level and structure of retail tariffs for isolated and connected mini-grids. It presents a simple financial spreadsheet tool for analyzing the effect of external grants and different tariff structures on the financial viability of a typical isolated mini-grid in Tanzania. The chapter concludes with a discussion of quality-of-service and safety standards for mini-grids.

- Chapter 10 discusses business model and regulatory options that exist when "the big grid connects to a little grid." With the arrival of the national grid to an area previously served by an isolated mini-grid, options include purchasing power from the main grid for retail sale to end-use customers (with the SPP becoming an SPD), the SPP generator selling electricity at wholesale to the national grid (becoming a grid-connected SPP), or a combination of the two. The financial and technical implications of these options are discussed.

- Chapter 11 presents some final thoughts on key factors other than regulation that are critical for successful SPP programs. It examines the issue of whether SPPs that propose to serve isolated mini- and micro-grids should have their retail tariffs regulated in their initial years of operation. It closes with recommendations on what countries that do not have SPP programs can do to launch them, what countries with SPP programs in place can do to improve them, and what the international development community can do to support the development of SPPs.

Chapter Highlights

Key Definitions

- The *centralized electrification track* is a top-down approach in which electrification typically occurs through expansion of medium- and high-voltage power grids built and operated through the separate or joint actions of a national or regional power company, a ministry, or an REA. The *decentralized track* is a bottom-up approach in which grid electrification occurs through the creation of isolated or connected mini-grids operated by private, cooperative, or community-based organizations (CBOs).

Key Observations

- This guide will focus on the decentralized track to expanding rural access to grid-produced electricity, emphasizing how SPPs can fit within a national rural electrification strategy. It will emphasize the ground-level questions that must be decided by national regulatory agencies and other entities responsible for promoting SPPs. It will examine how the decentralized and centralized tracks can be designed and implemented to complement each other.
- Regulation can help or hinder SPPs. It is necessary (but not sufficient) for the success of a decentralized track. Two other key factors for success are access to financing and access to technical and business know-how.
- Regulatory decisions are not made in a vacuum. In Sub-Saharan Africa, a regulator's decisions will be constrained by certain economic and political realities that do not exist in developed regions. The regulator's job is to make decisions in full recognition of those constraints.

Key Recommendations

- To improve rural access to grid-produced electricity, governments in Sub-Saharan Africa should follow both the centralized and decentralized tracks.
- Since regulation creates costs for the regulator and the entity being regulated, regulators should avoid imposing overly burdensome regulation while recognizing the need to protect consumers from unnecessarily high prices and unsafe practices. Whenever feasible, regulators should pursue light-handed regulation and some forms of deregulation.

Notes

1. The best descriptions and analyses of the centralized approach can be found in Barnes (2007). Barnes and his colleagues examined major centralized electrification initiatives in Thailand, Tunisia, Mexico, and Ireland.

2. An isolated mini-grid is a small-scale distribution network, typically operating below 11 kV, that provides power to a local community and produces electricity from small generators using fossil fuels, renewable fuels, or a combination of the two.

3. Based on the laws of nature, energy cannot be generated. Instead, it can be converted from one form of energy (for example, energy stored in oil or coal or received from the sun as solar energy) to another form of energy (electrical energy). However, for the sake of exposition, we will use terms like "energy generation," "electricity generation," or "renewable energy generator" as a convenient shorthand way of referring to the conversion of different forms of energy into electricity.

4. We use the term "grid-connected" to refer to SPPs that are electrically connected to a national grid or a regional grid. We use the terms "isolated" or "off-grid" to refer to SPPs that are not electrically connected to such a grid.

5. While the focus of the guide is on decisions that are traditionally referred to as regulatory decisions (for example, maximum and minimum prices, technical and commercial quality of service, and interconnection standards), this does not mean that all such regulatory decisions must necessarily be made by the national regulator. In many instances, it will be more efficient for some regulatory decisions to be formally or informally assigned to other entities, such as a rural electrification or energy agency (REA), a sustainable energy agency, or a ministry that may be providing grants to SPPs. In such situations, the national regulatory entity can provide a backstop to their decisions. Therefore, this guide describes the regulatory questions that need to be addressed, regardless of whether the answers are given by the national regulator or by some other governmental or quasi-governmental entity. In fact, two African countries, Mali and Guinea, have formally assigned traditional regulatory tasks for off-grid SPPs to their REAs (see chapter 3).

6. The boundary between regulatory and policy decisions is not always clear. For example, most observers would agree that the decision as to which pricing methodology should be used for setting feed-in tariffs for grid-connected SPPs is a policy decision because it influences the speed and cost of promoting renewable generation in a country. But it is often the case that regulators make this decision "by default" because the ministry of energy may not have acted. Therefore, for ease of exposition, all the decisions discussed in this guide will be referred to as regulatory decisions even though in some countries they may be made by a ministry or an electrification agency.

7. The single best description of the current state of electricity regulation in Sub-Saharan Africa is Kapika and Eberhard (2013). The authors provide detailed and perceptive case studies of how electricity enterprises are regulated in Ghana, Kenya, Namibia, Tanzania, Uganda, and Zambia.

8. An excellent overview of the power sector in Sub-Saharan Africa can be found in Eberhard and others (2008).

9. Our use of quotes is discussed in the later section titled "Acknowledging Controversies and Understanding Different Vocabularies."

10. Pervasive underpricing is documented in the surveys of the Africa Infrastructure Country Diagnostic (AICD). AICD found that only 10 of 21 national utilities in Sub-Saharan Africa were allowed to charge tariffs that covered their historic operating costs. And only 6 of 21 national utilities were allowed to charge tariffs that covered historic operating and capital costs. See Eberhard and others (2008, 29).

References

Barnes, Douglas F. 2007. *The Challenge of Rural Electrification: Strategies for Developing Countries*. Washington, DC: Resources for the Future.

Eberhard, Anton, Vivien Foster, Cecilia Briceño-Garmendia, Fatimata Ouedraogo, Daniel Camos, and Maria Shkaratan. 2008. "Underpowered: The State of the Power Sector in Sub-Saharan Africa." Background paper, Africa Infrastructure Country Diagnostic, World Bank, Washington, DC. https://openknowledge.worldbank.org /handle/10986/7833.

IEA (International Energy Agency). 2012. "Access to Electricity." In *World Energy Outlook* (online). http://www.worldenergyoutlook.org/resources/energydevelopment /accesstoelectricity/.

Kapika, Joseph, and Anton Eberhard. 2013. *Power Sector Reform and Regulation in Africa: Lessons from Ghana, Kenya, Namibia, Tanzania, Uganda, and Zambia*. Cape Town, South Africa: HSRC Press.

Small Power Producers, Small Power Distributors, and Electrification: Concepts and Examples

You speak a language that I understand not.

—QUEEN HERMIONE, FROM *THE WINTER'S TALE* BY WILLIAM SHAKESPEARE

Abstract

Chapter 2 defines key terms—small power producers *(SPPs),* small power distributors *(SPDs), and* electrification—*that appear throughout this guide. It describes four basic types of renewable and hybrid on- and off-grid SPPs, as well as new combinations emerging in Africa and elsewhere. Examples of operating SPPs are evaluated by their likelihood of achieving commercial viability, based on early evidence from Tanzania and other countries. (Operating examples of SPDs are discussed in more detail in chapter 10.) The chapter also discusses why grid-supplied electrification, whether obtained from an SPP or a main-grid extension, is best defined in terms of the quantity and quality of electricity supplied—rather than the traditional approach of just counting physical connections.*

What Are Small Power Producers?

In most developed and developing countries, a small power producer is described as a distributed generator (DG) or a decentralized distributed generator. *Distributed generation* (DG) is the generation of electricity by small-scale plants located near the electric loads that they serve. A *load* is a general electrical term for the amount of electricity drawn at any given moment by an appliance, machine, house, streetlight, clinic, or school. A factory will obviously have a bigger load than a household. The loads may be the DG's own, those of nearby businesses or households, or loads served by the larger grid to which the DG is connected. In developing countries, DGs are sometimes referred to as

SPPs (as in Tanzania, Sri Lanka, and Thailand) or very small power producers (VSPPs, also used in Thailand). It is usually the case that SPPs and VSPPs must satisfy both a size and a fuel or technology requirement.

For example, under the current Tanzanian regulatory rules issued by the Energy and Water Utilities Regulatory Authority (EWURA, the national electricity regulator), an SPP can be connected to a national or regional grid (a grid-connected SPP) or it may operate on an isolated mini-grid (an isolated mini-grid SPP). In either case, there are size limits. For SPPs operating on an isolated mini-grid, the SPP's maximum installed capacity must not exceed 10 megawatts (MW). In contrast, SPPs that are connected to the main or a regional grid are allowed to have an installed capacity greater than 10 MW. But when a grid-connected SPP's installed capacity exceeds 10 MW, only the power produced from up to 10 MW of the plant's total installed capacity can qualify for SPP status and benefits.[1] Thailand and Sri Lanka have adopted similar definitions for SPPs operating in their countries.[2] In countries where the total installed generation is small, such as in several West African countries, technical analyses of the grid's capacity to absorb power from DG might well suggest limiting the maximum size for SPPs to perhaps 4–5 MW. Apart from these explicit size limitations, there is the question of fuel type. In Tanzania and Thailand, all generators under 10 MW are eligible for streamlined interconnection with the grid, regardless of fuel source,[3] but in Tanzania only those using cogeneration, 100 percent renewable energy, or renewable generation with a maximum of 25 percent fossil fuel measured on an average annual basis are eligible for preferential standardized SPP tariffs.

Key Definition

Small power producers (SPPs) are independently operated, small-scale electricity-generating plants located near their customers. They may operate on isolated mini-grids or mini-grids that are connected to a larger national or regional grid. A connected mini-grid sells electricity directly to retail customers and may also buy electricity from and sell to the larger grid to which it is connected. Some SPPs connected to the national or regional grid will sell to a single customer—typically the operator of the national or regional grid. In this case, the SPP will not own or operate a mini-grid. SPPs usually must satisfy requirements related to their size and the type of fuel and technology they use.

To minimize any confusion, it is important to point out that this guide does *not* deal with regulatory systems for promoting large grid-connected renewable and cogeneration projects that sell power to a regional or national grid (*large*, in this context, means exporting more than 10 MW to the grid). While the same set of issues must be addressed for larger projects (for example, interconnection agreements, terms and conditions of a power-purchase agreement [PPA], and the level and structure of prices for sales to the grid operator), larger projects almost always require project-specific interconnection agreements and more detailed PPAs that are capable of attracting international financing. Hence, larger projects may require more than the standardized documents and rules for SPPs that are described in later chapters. Also, the guide does *not* cover the regulatory issues associated with initiatives to encourage the installation of small, household-based renewable energy systems (usually small photovoltaic [PV] systems) that sell back to local utility under a "net metering" program—although there may be considerable overlap between the regulatory requirements of these household systems and the SPPs addressed in this guide.

Renewable Energy and Cogeneration Technology

Since SPPs in many countries receive preferential tariffs only if they use renewable energy or cogeneration, it is necessary to define *renewable* and *cogeneration*. We define renewable generators as those that produce electricity using an energy source that is naturally replenished in the short term.

Renewable generation includes technologies—such as solar energy, hydropower, and wind—that do not emit carbon dioxide (CO_2), as well as biomass generation. Biomass, unlike other forms of renewable energy, may be a net emitter of CO_2. Whether biomass is a net emitter of CO_2 will depend on the fuel type (trees, crop residues, and so on), whether or not the fuel is a waste by-product, and how replanting is managed.[4] If a biomass power plant is not accompanied by measures that ensure sufficient replanting to offset CO_2 production from biomass combustion, the project will be a net producer of CO_2.

A cogeneration plant is one that simultaneously produces both electricity and useful heat. Generally this means capturing the waste heat from the electrical generation process that, in a conventional power plant, would be released through cooling towers or into rivers. By making productive use of exhaust heat, cogeneration can save considerable fuel compared with separate sources of electricity and industrial heat. While a cogeneration plant will save energy relative to two separate plants, it is sometimes the case that the primary fuel in a cogeneration plant will be a fossil fuel, such as natural gas. But in other cases, the primary fuel will be a form of renewable energy—such as bagasse in a plant that produces sugar from sugarcane.

Key Definitions

Renewable energy here refers to energy from a source that is naturally replenished in the short term. *Cogeneration* refers to an electricity generator that produces electricity, typically from a fossil fuel or a renewable fuel, and captures the exhaust heat for other useful purposes.

SPPs and Fossil Fuels

We think that the current definition of SPP used in Tanzania, which is not limited only to cogenerators or pure renewable energy generators, is appropriate for three reasons:

- First, in a straightforward reading of the term, *small power producer* refers to project size and nothing else. The term, by itself, does not imply that the SPP must use only renewable energy or a cogeneration technology. In fact, the current reality in Africa is that the number of small generators that use diesel or heavy fuel oil greatly exceeds the number of small generators that use a renewable fuel or a cogeneration technology.[5]

- Second, if a government is serious about promoting electrification (which is the paramount electricity policy goal in most Sub-Saharan African countries), regulators must be ready to process interconnection and tariff applications from all types of SPPs, not just those that use renewable generation or cogenerators. A villager who is effectively spending several U.S. dollars per kilowatt-hour ($/kWh) on batteries and a kerosene lantern wants cheaper, more reliable electricity and does not care whether it comes from renewable energy, fossil fuels, or both.

- Third, hybrid generators—generators that use a fossil fuel and one or several forms of renewable energy—are often able to supply electricity at a lower cost and for more hours than an SPP that uses just diesel, wind, or solar energy alone.[6] Therefore, *by accepting rather than prohibiting hybrid SPPs, a country is likely to achieve more success in both renewable generation and electrification than if the SPP definition were strictly limited to generators that are 100 percent renewable.* To paraphrase the words of one private developer in Africa, referring to his decision to install a combined solar-diesel isolated mini-grid: "Look, I am not doing this because it is fashionable. I am doing it because it will be cheaper and I can provide more hours of service." In our view, it would be both confusing and counterproductive for a regulator or a government energy official to announce that he wants to promote electrification by SPPs but that hybrid generators and diesel generators "need not apply."

Our recommendation is that if regulators and government policy makers are serious about expanding rural access to electricity as a goal, they should use a fuel-independent definition of SPP.[7] The key element of the definition should be the size of the generator, whether it is 10 MW or some other cutoff number. This broader definition would include the following types of SPPs:

- Renewable generators
- Cogenerators
- Generators using both renewable and fossil fuels (that is, hybrid generators)
- Generators using just fossil fuels (for example, diesel or heavy fuel oil)

*A broader definition does **not** mean that these different types of SPPs would all receive the same tariff treatment or interconnection rights from the regulator or the same level of subsidies from the government's electrification or renewable energy agency.* In fact, the opposite is likely to be true: connection rules and subsidies would vary by type of SPP. If the government's only objective for the SPP program is to promote renewable energy, then the government or regulator would want to limit the definition of an SPP to only small generators that use renewable sources. But we think that this is too narrow an approach. For most Sub-Saharan countries, the better approach is to use SPPs to promote both electrification and renewable energy.

Key Recommendation

Countries should define SPPs by their size and not their fuel. This means that small independent generators that use fossil fuel or are cogenerators or hybrid generators should all qualify as SPPs. But it is also appropriate, as described later in this guide (chapters 6 and 7), for governments and regulators to establish connection rules and subsidies that vary by type of SPP.

The Four Main Types of Grid and Off-Grid SPPs in Africa

In the United States and Europe, where the grid extends virtually everywhere, SPP systems are usually one of two types: those that sell electricity only to the grid or those that produce for self-consumption and also sell to the grid (but who may also purchase backup power from time to time). In most African countries, the starting conditions are more diverse: national grids serve some areas, isolated mini-grids serve others, and vast areas have no grid-based electricity at all. These differences create a variety of different SPP cases (table 2.1) depending on whether the project sells to the national grid[8] or to an existing or new isolated mini-grid; whether electricity is sold at wholesale[9] to the national utility, which then resells it, or whether the SPP sells at retail; and whether or not a grid-connected SPP generates for self-use or sells to other end users and therefore must purchase backup power.

We find it helpful to categorize SPPs based on three characteristics. The first is the fuel or technology used to generate electricity (renewable, cogeneration, fossil, and hybrid). The second is whether the SPP is connected to the national grid or operates on an isolated mini-grid. The third is whether the SPP is selling at retail, wholesale, or both. If one focuses on the last two characteristics, we can distinguish among four main types of SPPs in Africa.

Case 1: An Isolated SPP That Sells at Retail

As shown in table 2.1, the four cases are differentiated by whether the SPP is connected to the main grid or to an isolated mini-grid and whether the SPP sells

Table 2.1 Types of Electricity Sales Involving Small Power Producers (SPPs)

		Location of generation	
		Connected to isolated mini-grid	Connected to main grid
Nature of customers	Selling retail (directly to final customers)	Case 1	Case 3
	Selling wholesale (to utility)	Case 2	Case 4

at wholesale or at retail. The three most common pure cases are Cases 1, 4, and 2. In general, when people talk about mini-grids, they are usually talking about Case 1—a small generating plant (diesel, renewable, or hybrid) combined with a few kilometers of distribution lines and an operator that sells directly to final customers (households, public institutions, and small businesses) in one or more villages not connected to the national grid. This is the traditional definition of a mini-grid: a stand-alone, low-voltage distribution grid that is supplied with electricity from one or more small generators connected only to the isolated mini-grid. In Case 1, the SPP operates as both a producer and distributor of electricity. The mini-grid combines local generation with local distribution. In a recent International Finance Corporation (IFC) report, Case 1 was described as a "community-level mini-utility" (IFC 2012). To clarify Case 1 in more concrete terms, box 2.1 gives examples of SPPs that currently operate on isolated mini-grids in Cambodia, Mali, and Sri Lanka. On a worldwide basis, the IFC has estimated that 29 million rural households could be served on a commercial basis by isolated mini-grids.[10]

Key Definition

An isolated mini-grid is a stand-alone, low-voltage distribution grid that is supplied with electricity from one or more small generators that connect to only the isolated mini-grid.

Case 4: A Grid-Connected SPP That Sells at Wholesale to a Utility

At the other end of the spectrum is Case 4, which is an SPP that is attached to a national or regional grid. It has just one customer, usually the national utility. The SPP sells at wholesale to the national utility under what is commonly called a feed-in tariff (FIT). The SPP does not have any retail customers nor does it own or operate a distribution system. Hence, it is best described as a pure SPP. Its contribution to electrification is indirect: it provides an additional bulk supply source to a national or regional utility. In Sri Lanka there are now more than 100 Case 4 SPPs, most of which are mini-hydro producers. In Thailand there are more than 330 VSPPs online with an aggregate installed

Box 2.1 Examples of Isolated Mini-Grids (Case 1) in Three Countries

Cambodia

Cambodia has seen widespread use of isolated mini-grids because the national utility has been slow in extending the national grid into rural areas. As of December 2012 licenses had been issued to 312 private entrepreneurs, of which 288 were engaged in distribution of electricity through privately owned mini-grids. Initially, all of these rural electrification enterprises (REEs) operated as isolated mini-grids. In the past several years, 89 of the mini-grids have connected to the main grid of the national utility, sold their diesel generators, and become small power distributors (SPDs) (in Cambodia they are called "distribution REEs"). In contrast to many African countries, most of the Cambodian mini-grids, whether isolated or connected, were created without the benefit of any subsidies or grants from the government or any development agencies (that is, completely spontaneously). The typical isolated mini-grid in Cambodia has about 440 customers, an installed generating capacity of 0.15 MW, and a distribution network of about 4.4 kilometers. At present, the Cambodian isolated and connected mini-grids serve about 120,000 customers and supply approximately 42 percent of rural power in Cambodia. Since virtually all the isolated mini-grid operators use diesel-fired generators, their tariffs are high. Tariffs recently ranged from $0.40 to $1.25/kWh. Subject to approval of the regulator, most REEs were initially granted two-year licenses by the national electricity regulator, but with the possibility of receiving a longer term if the REE could demonstrate evidence of additional investments.

Mali

Within Africa, Mali has had probably more success than any other country in promoting isolated mini-grids with more than 150 in operation. Of the 60 or so private operators in Mali, most currently use small, diesel-fired generating units with high production costs. Most of these small power producers (SPPs) have received initial capital cost subsidies from AMADER (Agence Malienne pour le Développement de l'Energie Domestique et de l'Electrification Rurale, Mali's rural energy agency [REA]) to connect new customers. These capital cost subsidies have averaged about $750 per new connection. Once the connection is made, the government does not provide operating subsidies for the mini-grid operator or consumption subsidies for the operator's customers. To achieve commercial sustainability in the absence of further subsidies, the operators of these isolated mini-grids (known as PCASERS, for *Projets de Candidatures Spontanées d'Electrification Rurale*) currently charge their household customers a price of about 50 U.S. cents, which is about two to three times higher than the price charged to poor customers on the main grid under the national utility's "social tariff." This inevitably creates "tariff envy," especially in cases where an isolated mini-grid is serving a village located near another village served by the national utility. Therefore, it was not surprising that in 2011 the Malian government ordered the national utility to connect seven isolated mini-grids located close to the national grid in order to eliminate the large tariff disparity between customers served by the mini-grid and customers of the national utility. These seven mini-grids were within or very close to the designated concession area of the national utility.

box continues next page

Box 2.1 Examples of Isolated Mini-Grids (Case 1) in Three Countries *(continued)*

Sri Lanka

Between 1997 and 2011, 268 village hydro projects (VHPs) were created in Sri Lanka. These are very small, micro-hydro installations with typical average installed generating capacities of 3–50 kW, usually serving about 20–80 households. The facilities are owned and operated by community organizations known as Village Electricity Consumer Societies (VECSs). Unlike private operators in Cambodia and Mali, the owner and operator of a VECS is a community organization rather than a private businessperson.

But this does not mean that the private sector has been totally absent from the program. In fact, the opposite is true. Rather than simply relying on villages to create VHPs on their own initiative, the government sought developers to promote the VHPs and advise the community organizations. About 20 private engineering consultants and nongovernmental organizations (NGOs) received payments of about $8,000 for each successful VHP that they promoted. The developers received this payment in three installments: 40 percent upon signing a contract with the VECS, 30 percent upon verification that the generating and distribution facilities had been installed, and 30 percent after six months of operation.

Over the past several years, the grid of the national utility, Ceylon Electricity Board (CEB), has reached about 70 of these previously isolated villages. In most instances, the households closest to these new grid facilities left the VECS and became CEB customers. CEB electricity is more attractive because it generally provides better-quality supply (that is, a more reliable supply for more hours of the day). And for many VECS members, CEB electricity is also cheaper: once they are connected to the CEB grid, these households are able to buy at the subsidized national lifeline tariff of 2.5–3.0 cents/kWh (versus the typical 25 cents that they were paying as members of the VECS). Therefore, given the CEB's lower prices and greater hours of service it should not be surprising that many VECSs went out of business "when the big grid arrived."

Sources: Cambodia: Electricity Authority of Cambodia 2008, 2009, 2010, 2011; Rekhani 2011, 2012; Chanthan and Augareils 2013; Keosela 2013. Mali: Adama and Agalassou 2008; Agalassou 2011. Sri Lanka: Cabraal 2011 and authors' knowledge.

---

capacity of 1,470 MW. In contrast to Sri Lanka, most of the Thai VSPPs (about 80 percent of installed VSPP capacity) generate electricity using waste biomass and biogas.

Case 2: An Isolated SPP That Sells at Wholesale to a Utility

At first glance, Case 2 seems strange: if an SPP is operating on an isolated mini-grid that is not connected to a national or regional grid, how can it be selling wholesale to a national or regional utility? But this situation exists in many African countries. Over time, many national utilities in Africa have been forced by political pressure to construct isolated mini-grids to serve communities that were not likely to be reached by the national utility's main grid for many years. In response to this pressure, the national utility built an isolated mini-grid served by a diesel generator because this required the smallest up-front capital investment. Once the mini-grid became operational, the national utility was usually forced to sell electricity to the customers on these isolated grids at the national retail tariff, even though its production costs at these locations were likely to be several times higher than this tariff. As renewable energy has become more economical, it is now common for SPPs to approach the national utility operating an expensive "legacy" mini-grid and offer to replace some or all of the national utility's diesel-fired production with electricity that can be produced from lower-cost renewable generation. This is especially likely if the regulator sets a FIT at a level close to or equal to the national utility's relatively high, avoided costs on the isolated system. In fact, this is already happening in Tanzania, where the feed-in price for SPP wholesale sales to the national utility's mini-grids is more than twice the price that the SPP would receive for wholesale sales to the national utility on the main grid.

Combinations of Cases

It would be a mistake to assume that SPPs always fit neatly into a single category. In fact, it is common to see combinations of the four cases. For example the developer of the Andoya Hydropower project in Tanzania proposes to combine Case 1 and Case 2. The project will connect to an existing isolated mini-grid operated by the Tanzania Electric Supply Company (TANESCO) and sell power at wholesale to TANESCO so that the utility can reduce the amount of electricity that it currently generates from its isolated diesel generator (Case 2). It will also sell at retail to about 880 new customers (including 760 households) (Case 1) that currently do not have grid-based electricity. Another Tanzanian developer, the Mapembasi Hydro-Power Company, proposes to create a combination of Cases 3 and 4. It intends to sell power at wholesale to TANESCO over a new 10-km connection to the existing TANESCO main grid (Case 4) and also to make retail sales to approximately 2,000 households in 5 unelectrified villages (Case 3).

In both cases, the motive for selling to the national utility, either on the main grid or on an existing isolated mini-grid, is that it could provide a cushion of significant revenue that may make it financially feasible to sell to poor, low-consumption households at subsidized tariffs. As a buyer, the national utility would be the "anchor customer," providing a market for electricity 24 hours a day, including in the middle of the night when local usage drops to low levels. As such, the revenues from the national utility could help to ensure commercial viability as local household and commercial loads grow over time. But this would only be true if the national utility makes timely payments for the power that it purchases.

For example, in the case of one SPP that will be selling to TANESCO on an existing diesel-supplied mini-grid, it is estimated that 87 percent of the project's total revenues will come from sales to TANESCO and only 14 percent from retail sales (11 percent to commercial customers, 3 percent from households) in the project's initial years of operation. The same is true for another SPP project that proposes to sell to TANESCO on the national grid. In the early years, the project developer expects that about 95 percent of total revenues will come from sales to the national utility and only about 5 percent from sales to retail customers. For both projects, if sales were limited to just rural households and small businesses in one or more isolated villages, financial sustainability would be much more difficult to achieve. (The financial viability of different SPP cases, both pure and combined, is discussed later in this chapter.) Sales to local and commercial customers would help to mitigate the risk of nonpayment by the national utility. If the national utility fails to pay, then at least the project developer will receive some revenues from its local customers, whose risk of nonpayment can be mitigated by the use of prepayment meters.

The "anchor customer" need not always be the national utility. In fact, if the national utility is commercially insolvent and does not pay for what it purchases (or pays with a significant delay), its value as an anchor customer will be minimal to nonexistent. In this situation, it is best for the SPP to try to find some other business enterprise with a significant electrical demand and that is more likely to pay for what it purchases. For example, the owner or operator of mobile-phone towers located in rural areas could serve as an anchor customer. Its value as an anchor customer would depend on how large a base load electricity demand it can offer to the SPP provider. (Box 2.2 describes one recent initiative by an Indian company to use mobile-phone towers as anchor customers in rural India.[11]) On a worldwide basis, of the approximately 3 million mobile-phone towers in operation, about 640,000 towers are located in rural off-grid areas (GSMA 2013). At present mobile-telephone coverage in rural areas of Africa is clearly greater than that of grid-supplied electricity.

In East Africa tea plantations and mines have also been proposed as possible anchor customers for new mini-grids. There is an important difference between using mobile-phone towers versus mines or tea plantations as anchor customers. It is likely that mobile-phone towers have never been customers of the national utility, since it could not guarantee the high degree of reliability that operators need for their towers. This means that a mobile-phone tower is not a "lost load" for the national utility: it never served this load in the first place. But this may not be true of mines and plantations. If they are currently profitable customers of the national utility and the source of cross-subsidies to residential customers on lifeline tariffs, the national utility will strongly oppose losing them to a new mini-grid operator.

Purchases As Well As Sales

While the SPP cases in table 2.1 are organized by type of sales, SPPs may also need to make purchases. For example, if an SPP is connected to the main grid and is selling both at retail (Case 3) and at wholesale (Case 4), it will need to *purchase* electricity from the national utility or some other supplier in several situations:

- For backup generation, if the SPP's own generating capacity is insufficient to meet its own needs

Box 2.2 Mobile-Telephone Towers as Anchor Customers: A Recent Development in India

It has been estimated that about 150,000 of India's 400,000 mobile-telephone towers are located in off-grid areas or areas with an unreliable electricity supply from the grid. These off-grid or poorly served grid-connected towers are forced to get all or some of their electricity from small diesel generators owned and operated by a mobile-phone company or a tower company that rents out the location and a reliable supply of electricity to one or more mobile network operators. The operation of such units is a headache for most operators because producing electricity at thousands of locations is not their core business. Moreover, the cost of self-supplying electricity from a diesel generator is high. In India it has been estimated that about 40 percent of the operating expenses for a typical mobile tower are attributable to fuel and power costs. The comparable figure for Europe is about 12 percent.

An alternative business model has been developed by the Omnigrid Micropower Company (OMC).[a] OMC, founded by several former executives from India's telecommunications sector, is building small 18-kW micropower installations in rural areas that use mobile-telephone towers as their anchor customers. At each site, OMC employs about 12–15 employees (about 10 of whom are local residents). At the end of 2012, 10 micropower installations were up and running in the Indian state of Uttar Pradesh. Under the OMC business model, the company sells electricity to the tower owner or operator under a deregulated long-term power contract (that is, a contract that does not require review or approval by the state electricity regulator). OMC produces the electricity from a hybrid generating system using some combination of sun, wind, and biogas with a backup diesel generator (usually acquired by purchasing the tower owner's generator) and with some batteries to ensure a high degree of reliability. OMC's plan is to locate each micropower installation so that it can sell electricity to three to five cellular-phone towers from each generating unit. OMC estimates that it can go from "site survey" to "power on" in about 30 days because of its use of standardization and modular design.

In August 2012 OMC signed a 10-year agreement with Bharti Infratel, one of India's largest telecommunications service providers, to supply electricity to mobile-phone towers throughout the country. Bharti Infratel currently operates 33,000 towers, and 9,000 of them are off-grid or connected to a grid that provides an unreliable electricity supply. Hence, these 9,000 Bharti Infratel towers could serve as OMC's anchor customers in many rural areas throughout India. Under an enforceable long-term contract, this would provide OMC with an assured source of revenue critical for achieving "bankability" (that is, the ability to get commercial loans from a bank).

At each location, Bharti Infratel will be OMC's principal but not only customer. OMC also plans to sell electricity services to local villagers residing near the tower, using a prepaid "battery in a box" system. At the end of 2012 OMC was serving about 150,000 rural households and small businesses in Uttar Pradesh in addition to the mobile-telephone tower anchor customers. The novel feature of the OMC business model is that OMC does not build (at least initially) a traditional electricity distribution system with low-voltage lines and

box continues next page

Box 2.2 Mobile-Telephone Towers as Anchor Customers: A Recent Development in India *(continued)*

transformers to serve rural households and businesses. This saves OMC a considerable amount of money in initial capital costs. Instead of selling electricity over the traditional "online" distribution system of a mini-grid (Case 1), OMC provides electricity to its nonanchor customers from rechargeable lanterns and rechargeable battery boxes (sometimes called portable power sockets) owned by OMC and rented to its customers on a daily, weekly, or monthly basis. Once or twice a day local OMC employees deliver freshly charged battery boxes to customers. When the charged boxes are dropped off, the spent power boxes and lanterns are removed and recharged at the microgenerator site located next to the towers. With adequate roads, OMC estimates that it can serve households in villages up to 15–20 kilometers from its generating station.

At these initial locations, a single entry-level lantern with a light-emitting diode (LED) lightbulb rents for about $2/month, a savings of approximately 50 percent compared to what the customer previously paid to provide lighting with a kerosene lantern. If the customer can afford to pay more, he or she can rent a box, two lanterns, and a fan for $7/month. OMC is currently exploring the possibility of supplying customers with larger power boxes that could power irrigation pumps. OMC's customers are not required to make mandatory fixed payments or pay a security deposit, nor are they required to pay connection charges like new customers of traditional mini-grids or main grids (see chapter 5). Instead, they rent the lanterns and the boxes. In OMC's initial installations, it has been reported that more than 30 percent of potential households signed up within 45 days of initial operation. This is significantly higher than the household sign-up rate for most traditional mini-grid and main-grid extension projects. Presumably, the difference in sign-up rates is attributable to the fact that OMC's household customers do not have to pay a connection charge or security deposit.

Sources: Omnigrid Micropower Company website (http://www.omcpower.com); Raj 2012.
a. Two official policy/regulatory changes have given impetus to the OMC model. The first is a recently issued requirement of the Indian telecommunications regulator that 50 percent of rural telecom towers must be powered by renewable or hybrid energy by 2015. The second is a decision by the Central Government of India to eliminate the ability of telecom tower operators to acquire subsidized diesel oil by 2014.

- For supply to itself or to its retail customers when its own generator is offline for scheduled or unscheduled maintenance
- To restart the SPP if it had to be shut down because of a fault on the main grid or with the SPP itself
- When purchases of electricity from the national grid cost less than generating electricity from its own generator

In all of these cases, the principal (but not only) tasks for a regulator are to decide how prices will be set for both sales and purchases. A discussion of price setting for SPP sales of power (usually known as FITs) is given in chapter 7, and a discussion of price setting for SPP purchases of power (usually known as backup or supplemental power) is presented in chapter 6.

Key Observation

While SPPs can be categorized by the four types of sales described in table 2.1—based on who their customers are and whether or not they are connected to the national grid—a single SPP can itself be a combination of these cases.

Mini-Grids and SPPs: A Clarification

We have adopted the convention of using the term *isolated mini-grid* to refer to a combination of a generating unit and a distribution system that operates separately from the national or regional grids (Case 1). (See box 2.3 for a discussion of observed differences between mini- and micro-grids.) Our reason for using a single term to refer to both components (production and distribution) is that most operators of isolated distribution systems usually perform two functions: they generate electricity and then distribute the generated electricity.[12] These are also SPPs (or VSPPs) because they are small and produce electricity. Therefore, when referring to an isolated mini-grid that both generates and sells electricity, we will use the terms *SPP* and *mini-grid* more or less interchangeably. In those instances when we are referring only to the mini-grid's generation or distribution system, we will make this clear. In cases when we are referring to a generator selling electricity to the national grid, we will use the term *SPP* but not *mini-grid*.

. We will also use the term *connected mini-grid* to refer to a distribution system that is connected to, and may draw electricity from, the main grid. This describes a business arrangement in which a separate owner and operator performs commercial (metering, billing, and collections) and technical (repairs, maintenance, and replacement of distribution facilities) tasks that the main grid operator would otherwise perform. It is, in effect, a hybrid business model: two entities, usually the national utility and the connected mini-grid operator, are both involved in producing and supplying electricity to end users. This arrangement is described more fully in the discussion of small power distributors in the next section.

It is worth noting that in a technical sense, once a mini-grid is connected to the main grid, it no longer operates as a separate, electrically isolated system even if it maintains its own generation. After the connection is made, the distribution facilities of the previously isolated mini-grid become just one small component of the national grid. From an electrical perspective, the national grid and the connected mini-grid are one and the same, with electricity pulsing in both sets of wires in-phase and at the same frequency.

From the Bottom Up • http://dx.doi.org/10.1596/978-1-4648-0093-1

Box 2.3 Mini-Grids versus Micro-Grids

The term "mini-grid" is most often used to refer to isolated grids (typically ranging in power output from tens of kilowatts (kW) to tens of megawatts (MW) and serving several hundred customers) in rural areas of developing countries (Case 1 in table 2.1). Service is typically alternating current (AC), and customers can use many, if not all, of the same appliances that a customer connected to the main grid might use. In the "multi-tier framework for measuring household electricity access" provided in table 2.3, mini-grids typically provide service up to Tier 3 or 4.

The term "micro-grid" can be especially confusing, as it has very different meanings in different contexts. Sometimes it is simply a synonym for mini-grid. In other cases, particularly in developing countries, micro-grid refers to systems of very small scale, with power output ranging from hundreds of watts to a few kilowatts, and typically fewer than 150 household customers. Micro-grids, as the term is used in this context, typically provide Tier 1 or Tier 2 service (as defined in table 2.3), with electric service for lighting, TVs, fans, and charging cell phones.

While AC is the norm in both mini-grids and micro-grids, some micro-grids that combine photovoltaic panels with battery storage (usually referred to as solar micro-grids) will distribute low-voltage direct current (DC) to customers. DC has the advantage that solar panels and batteries inherently produce and store DC, and thus DC systems may have advantages in lower balance-of-system costs and higher efficiencies. DC power can be distributed with smaller poles and with thinner and cheaper wiring. Hence, DC micro-grids have been described as "skinny grids." But the disadvantages are that most appliances use AC and that low-voltage DC suffers from significant line loss. High-voltage DC has been proposed for micro-grids but poses fire and safety issues related to arcing in switches that can prevent the switches from functioning properly.

Typically micro-grids are built with smaller wires and poles than would meet the standards of the country's larger AC grid. As such, the distribution systems of micro-grids do not lend themselves to later conversion to small power distributors (SPDs) (discussed elsewhere in this chapter) that are connected to and distribute electricity produced on a larger AC network. If micro-grids are eligible to receive capital cost grants from a rural electrification agency, it seems reasonable that the amount of the grant should be lower for two reasons. The micro-grid's investment costs are likely to be lower than those of a mini-grid, and the micro-grid will probably not be able to support the same number of electricity services as an AC mini-grid. (See the discussion of tiers of electricity service later in this chapter.)

In industrialized contexts, whether in developed or developing countries, the term micro-grid can have a very different meaning. It is often used to refer to distributed generation in areas already supplied with grid electricity. The intent is to increase the use of on-site renewable generation or to improve the reliability or quality of local electric power. In this sense, a micro-grid is a local collection of electricity generation and loads, typically tens of MW or less, that is usually connected to the main grid but may be disconnected for autonomous operation to achieve very high levels of reliability (99.999 percent or more) compared to the grid (99.9 percent). This autonomous operation is sometimes referred to as "intentional islanding."

Key Definitions

Unless otherwise noted, when discussing isolated mini-grids we use the terms *SPP* and *mini-grid* interchangeably because, in this context, SPPs both generate and distribute electricity. A *connected mini-grid* is a business arrangement in which a mini-grid (whether or not it possesses its own generation source) is physically attached to a national or regional grid and sells electricity directly to retail customers.

Which Types of SPPs Are Likely to Achieve Commercial Viability? Some Early Evidence from Tanzania

Some types of SPPs will be more likely to achieve financial viability than others.[13] To make this general observation more concrete, in this section we describe some existing and proposed SPP projects in Tanzania. The country's early experience with SPPs may provide some lessons for similarly situated African countries, specifically those with a large government-owned national utility that is not recovering operating costs and whose average per capita income is low (that is, less than $2/day). In the discussion that follows, the Tanzanian SPP cases are keyed to table 2.1 and are ordered from "easier to develop" to "harder to develop."

Case 4 project: An SPP that is an agro-industrial facility and has biomass residues from its production process needs steam or process heat for its production process, and can self-supply its own electricity. The facility is connected to the main grid and is able to sell at wholesale to the national utility with a relatively small additional investment.

One example of this case is the TPC sugar estate near Mount Kilimanjaro. It installed a cogeneration plant with generation capacity of 17.5 MW in 2004–05. The cogeneration plant is powered by bagasse (shredded and crushed sugarcane) left over from sugar production. It supplies electricity and steam to the factory (which converts sugarcane into sugar), provides power to irrigation pumps in the estate's 8,000 hectares of sugarcane fields, and also supplies electricity to the homes of TPC's workers. TPC decided to build its own on-site power supply because it felt that that TANESCO's electricity supply was too unreliable, and because it had a ready source of bagasse biomass fuel supply at essentially zero cost. Prior to the installation of the cogeneration facility, TPC had experienced frequent blackouts as well as harmful variations in voltage and frequency from the TANESCO-supplied electricity. In deciding how large a generator to build on-site, TPC could have built a generator that would have met its own needs alone. But TPC decided to take the risk of building a larger generator than needed for its own consumption (17.5 MW rather than 8 MW or less) because it expected that it would be able to sell its surplus electricity to the national utility.

TPC soon discovered that the physical installation of the generator was relatively easy, but obtaining a contract from TANESCO to sell its surplus electricity was more difficult. TPC and TANESCO negotiated unsuccessfully for several years over the price that TANESCO would pay, a deadlock that was finally broken in 2010 when Tanzania's EWURA issued guidelines that specified a mandatory pricing formula for electricity sold by a main-grid-connected SPP such as TPC to TANESCO.[14] Once the SPP guidelines were issued, TPC was quickly able to sign a standardized sales contract (known as a standardized power-purchase agreement, or SPPA) that was also specified by EWURA. It was relatively easy for TPC to acquire SPP status because its electricity-generating plant was already in place and the additional investment to connect to TANESCO and operate synchronously with the TANESCO main grid was relatively small (that is, a 10-km connection line to a TANESCO substation and an investment in interconnection protection relays and additional control equipment) (TPC 2012).

An SPP such as TPC has four important advantages that are not usually available to other Case 4 SPPs.[15] First, it has an internal need for a large amount of electricity. Second, it has a need for industrial heat and steam in its production processes, which, in turn, requires an investment in a boiler that can also be used to produce electricity. Third, it has a ready source of biomass fuel (that is, the bagasse residue from producing sugar from sugarcane) that can fuel the boiler. Fourth, it is an existing commercial entity with a strong balance sheet and the ability to finance the investment through its own funds or loans from domestic or foreign banks that are already familiar with its operations and have loaned it money in the past. *Hence, unlike other potential SPP developers, it does not need to rely on project financing, which lenders consider more risky because they do not have recourse to the borrower's other assets if the project experiences financial problems.*[16] Given these advantages, it should not be surprising that agro-industrial factories are the most commonly developed type of SPP in other countries as well. For example, in Thailand 626 MW of the 1,015 MW of VSPPs online (that is, generators of 10 MW or less) in September 2011 were sugar factories burning bagasse (EPPO 2011).

Case 2 project: An SPP that sells at wholesale to an existing isolated mini-grid that is currently supplied by a diesel generator owned and operated by the national utility.

In Tanzania TANESCO owns and operates 16 isolated mini-grid systems powered by diesel and 5 powered by natural gas (World Bank and Climate Investment Funds 2013). These generators are very expensive to operate. As of June 2012, it cost about 40–45 U.S. cents/kWh to generate electricity from such diesel generators (EWURA 2012a). In addition to these high production costs, actual or perceived political constraints require that TANESCO charge all of its household customers—whether on the main grid or on one of these isolated grids—the same retail tariff. Currently TANESCO sells electricity at 3.7 cents per kWh to customers who use less than 50 kWh each month, and 14 cents for customers who use more than 50 kWh. This means that TANESCO loses at least 30 cents on every kWh sold to a customer on one of its isolated mini-grids. These heavy financial losses caused one former managing director of TANESCO to observe that the company is "financially hemorrhaging" because of its sales to customers on these diesel-powered, isolated mini-grids at the uniform national tariff.

Under Tanzania's guidelines, an SPP is allowed to charge a price of 30.3 cents for every kWh that it supplies to TANESCO on one of these isolated mini-grids in 2012. One such SPP is a proposed biomass generator located on Mafia Island, an island south of Zanzibar. The SPP's private owners are building a biomass-powered generator that will initially be supplied by the wood from old coconut trees on the island. Later the owners plan to fuel their generator with other fast-growing hardwoods. This biomass generator will produce electricity that will enable TANESCO to shut down at least one of its two diesel generators on the island. While TANESCO will still continue to lose money on every kWh that it sells at the uniform national retail tariff, it will not "bleed" as much because its input costs will decline from more than 40 cents/kWh to 23 cents/kWh (Mafia Island 2011; EWURA 2012a).

An important difference between the TPC case described earlier and the Mafia Island SPP is that the latter will not be generating for self-supply. Instead, the entire electrical output of the Mafia Island SPP will be sold to TANESCO. Moreover, the SPP will have little or no incentive to sell to new or existing retail customers on the island if it is forced to charge the same tariff as TANESCO, because, like TANESCO, it would find itself losing money on every kWh that it sells and, unlike TANESCO, it would be unable to cross-subsidize these losses from other customers. While it is legal under current Tanzanian regulatory rules to charge a retail tariff higher than TANESCO's retail tariff, it still may not be politically viable. If the developer seeks tariffs higher than TANESCO's tariffs, the SPP's customers would probably complain vehemently to political authorities that it is unfair that they should pay more for electricity than friends and relatives who live in nearby island villages that TANESCO continues to serve. Hence the existence of a non-cost-recovering retail tariff for the national utility creates a *de facto* retail price ceiling for any SPP that plans to operate near areas served by TANESCO. This, in turn, creates a strong incentive for a privately owned SPP to sell at wholesale to TANESCO and to avoid selling to any retail customers.

Key Observation

When a national utility that serves rural communities charges a non-cost-recovering retail tariff, it creates a *de facto* retail price ceiling for any new SPP that plans to operate near areas served by the utility. This, in turn, creates a strong incentive for a privately owned SPP to sell just at wholesale to the utility and to avoid selling to any retail customers.

Combined Cases 1 and 2: An SPP that sells both to TANESCO on an isolated mini-grid and to new or existing retail customers in one or more villages.

The Andoya Hydroelectric Power Company Ltd (AHEPO Ltd) is a hybrid case that would combine wholesale delivery to an existing TANESCO mini-grid and retail sales to retail customers, some new and some previously served by TANESCO. This Tanzanian developer's business plan projects that wholesale sales to TANESCO would initially provide 87 percent of the project's total annual revenues. Estimated retail sales to 882 new households in 3 villages are projected to provide only about 3 percent of total project revenues. The remaining 10 percent of revenues would come from sales to local industry, public institutions, and cell-phone companies that operate transmitting towers near the villages. Furthermore, the developer's projections were based on the assumption that he would have to charge household retail customers the same tariffs that they would pay if they were customers of TANESCO. But if allowed by the regulator, commercial retail customers (for example, stores, mills, and workshops) would pay higher tariffs than those they currently pay to TANESCO. Hence, there are two potential sources of cross-subsidies for household customers served by this project: the wholesale electricity sales to TANESCO and the retail sales to commercial customers in the village.

What is clear from this business plan is that the projected sales to TANESCO provide the financial anchor of the project. But there are two risks in using TANESCO as an anchor customer. The first risk is that the price the developer receives for sales to TANESCO will drop markedly if this isolated mini-grid becomes electrically connected to TANESCO's main grid. For example, if this had happened in 2011, under the current FIT system in Tanzania (described in chapter 7), the wholesale price that the SPP would receive would have dropped from 23 cents to an average of about 8.06 cents. Therefore, once the connection is made to the main grid, the continuing financial viability of the project will depend on growth in the number and average consumption level of household customers and growth in sales to local industries and institutions.

When the connection occurs, there will be considerable pressure on the mini-grid operator to charge all retail customers exactly the same tariffs that TANESCO charges comparable customers. And since the now-connected mini-grid operator, unlike TANESCO, will not have access to a pool of high-consumption industrial customers that can cross-subsidize its household customers, it may decide to shut down and hand over its business to TANESCO. (See chapter 10 for a more in-depth discussion of what can happen "when the big grid connects to the little grid.")

The second risk is that the SPP, whether selling to TANESCO on an existing isolated mini-grid or on the main grid, will not get paid for the energy that it has provided, or will get paid only after many months of delay. This is a significant risk for SPPs in many African countries where the SPP's anchor customer is the national utility. Since many national utilities in Africa are commercially insolvent, SPPs may find that their financial foundation is built on loose sand rather than on firm ground.

Case 1: An SPP that sells only to retail customers on a new, isolated mini-grid in one or more villages.

As noted earlier, this is the case of the pure, isolated mini-grid. Of the four basic types of SPPs, this is the hardest to create and sustain because of two major financial hurdles. The first hurdle is accumulating enough money to pay for the capital costs of the initial installation. Equity is generally scarce in rural areas.

For a proposed isolated mini-grid in Tanzania, the cost of constructing a mini-hydro-generating facility and distribution delivery system to serve about 1,800 household customers and a smaller number of public institutions and private businesses has been estimated at $1.6 million. In the past, NGOs or bilateral donors have sometimes financed the capital costs of such installations. But donors do not have deep pockets and such funding will disappear if a donor's priorities shift from the electricity sector to another sector such as health or education. And if donors are not able to supply some or all of the required capital in the form of a grant, local banks will generally be reluctant to provide loans to new business entities that do not have a financial "track record" or much equity of their own.

Even if this first hurdle of funding initial capital costs is overcome, the second financial hurdle is covering operating costs. Revenues may not cover these costs for several years, and unlike the combined cases described above, the operator of a new, isolated mini-grid will not have the financial cushion from sales to the national utility. If the cost-revenue gap is not closed in a few years, the SPP is likely to collapse.

There are several options for closing the gap:

- An operating subsidy from the government or some other outside source (Peru and India)
- Charging households a tariff higher than the retail tariff of the national utility if the national utility's retail tariff is not high enough to cover the SPP's costs (Cambodia, Senegal, and Mali)
- Charging commercial and industrial enterprises in the village a higher tariff than the tariff charged to village households (cross-subsidization)[17]
- Developing other sources of revenue and increasing the number of connected households that pay cost-recovering tariffs

The effectiveness of these measures is analyzed with actual numbers in chapter 9.

The Tanzanian electricity regulator recently proposed the second and third options above in a "second generation" set of SPP rules. In draft rules issued for public comment, the regulator proposed that:

- SPP and SPD tariffs will be allowed to exceed the TANESCO national uniform tariff if this is necessary for the SPP or SPD to recover its efficient operating and capital costs [Section 41 (5)] (EWURA 2013).
- To facilitate commercial sustainability, an SPP or SPD may propose tariffs for specific customer categories or customers within a single category, subject to the Authority's approval, that take account of the ability of these customers to pay [Section 40 (c)] (EWURA 2013, SPP rules—under consideration by EWURA at time of writing).

It remains to be seen whether isolated SPPs will be able to take advantage of this new, legally allowed pricing flexibility.

Key Observation

It is easier to implement some SPP systems than others. Based on early observations from Tanzania, we can rank four cases from easier to harder to implement: Case 4 (main-grid-connected SPP with significant internal load sells excess electricity at wholesale to the national utility), Case 2 (SPP on an isolated mini-grid sells electricity at wholesale to replace diesel-fired generation produced by the national utility), combined Cases 1 and 2 (SPP on an isolated grid sells to both the national utility and retail customers), and Case 1 (SPP on an isolated mini-grid sells just to retail customers).

What Are SPDs?

In addition to SPPs, a regulator should also be prepared to specify regulatory procedures and rules for SPDs. An SPD is an entity that buys power at wholesale (usually from the national utility) and then resells it at retail to one or more localities.[18] Most SPDs will not generate any power on their own. One exception would be a renewable SPP that initially operates an isolated mini-grid and then converts into an SPD when its mini-grid is connected to the national grid. In such cases, the operator may continue to operate its own small generator as a backup supply, to provide voltage support at the end of a long distribution line, or to make bulk sales under a specified FIT to the national grid.

SPDs are common in Asia and less so in Africa. Over the past several years, more than 200 SPDs have come into existence in Nepal.[19] The Nepalese SPDs buy at wholesale from the Nepal Electricity Authority (NEA), the government-owned national utility, and then resell this wholesale power at retail in one or more communities. A unique feature of the Nepalese SPDs is that communities provide 20 percent of the total cost of constructing distribution lines while the government contributes the remaining 80 percent. This substantially reduces communities' distribution costs.

SPDs have also been a component in the very successful electrification program in Vietnam. Making extensive use of SPDs (referred to as local distribution utilities, or LDUs, in Vietnam), the country went from 14 percent rural household electrification in 1993 to 94.5 percent in 2008. About 1,881 of Vietnam's 8,982 rural communes are served by LDUs organized as cooperatives or private companies. A typical cooperative-owned LDU supplies electricity to about 1,200–1,500 households (Van Tien 2011). In both Nepal and Vietnam, the SPDs began operating as distributors from the start, skipping the initial stage of operating as SPPs.

In contrast, the Cambodian SPDs operated first as SPPs. Over the past 10 years or so, private entrepreneurs have created more than 280 isolated rural mini-grids with the electricity typically supplied by a small, secondhand diesel-fired generator. As the national utility, Electricité du Cambodia (EDC) expanded its medium-voltage grid into rural areas served by these small entrepreneurs, licenses were granted to 82 of these operators to convert themselves from SPPs to SPDs. In these villages, EDC was willing to limit itself to serving as a bulk wholesale supplier rather than as a retail supplier. In these cases, EDC delivers the electricity to the "doorstep" of the village, and then the privately owned SPD takes over and resells the electricity within the village to new and existing retail customers. It is viewed as a "win-win-win" outcome for all parties: EDC does not have the bother and expense of selling to hundreds or thousands of small rural customers in relatively isolated villages, the private entrepreneur can continue in the electricity supply business, and rural households receive a significant reduction in price (for example, from up to $1.25/kWh for electricity supplied by a diesel-fired SPP to about 26–28 cents/kWh when the electricity is supplied by an SPD). Service hours also increase from about 4–6 hours in the evening to 24 hours per day as soon as the village is connected to the EDC medium-voltage grid. As EDC continues to expand its medium-voltage grid, it is likely that many isolated SPPs will convert themselves to connected SPDs (Rekhani 2012; Keosela 2013).

African countries could replicate the Cambodian approach, in which case many African SPDs would initially function as SPPs and then convert to SPDs when the main grid arrives. The key overarching policy and regulatory question will be: what are the regulatory rules when "the big grid arrives to connect to the little grid"? Specifically, will the existing SPP have the legal right to convert itself into an SPD when the big grid arrives, or will it be presumed that the operator of the big grid (usually the government-owned national utility) will automatically take over the operation and sale of electricity on the community's previously isolated mini-grid? And if the SPP is allowed to convert itself to an SPD, will it be required to sell to its customers at the same tariff (including lifeline or social tariff components) used by the national utility when it sells to its customers? And this raises a related question: what price should the SPD pay for the wholesale power that it purchases from its bulk supplier? In the alternative, if the SPP chooses (or is obligated) to hand over its distribution facilities and business to the main grid operator or some other entity, will there be prespecified rules to compensate the SPP for the distribution assets that it hands over to the new operator? These questions are addressed in chapter 10. At the time of this writing, these questions are under discussion in Mali, Cameroon, and Tanzania.

Key Definition

A *small power distributor* (SPD) is an entity that buys power at wholesale rates from the national utility and then sells it at retail rates to households and businesses in one or more localities. It usually will operate solely as a distributor and not have any generation of its own. A special case involves SPPs that convert to SPDs but maintain their own power supply as a backup supply source for local voltage support and for possible wholesale sales to the national or regional utility. SPDs are common in Asia, and several African countries are currently developing policy frameworks to support SPDs—among them Mali, Cameroon, and Tanzania.

Electrification: What Is It and How Can It Be Measured?

… without a definition of what we hope to achieve, it's hard to see how progress can be measured.

—EMILY HAVES, *THE ASHEN BLOG*, 2013

Electricity access is more than just poles and wires.

—UNKNOWN

Defining Electrification as Connections

Governments and donors are often inclined to define electrification as a connection to an electrical grid. The grid can be the national or regional grid or an isolated mini-grid. It is relatively easy to measure increases in grid connections. There may also be political benefits to using this definition. This was true in India, where for many years electrification was defined solely in physical terms: a connection to a grid. The basic problem with this definition is that it implies that electrification has been accomplished once there is physical access to a grid. For example, until 2004 a village was counted as electrified in official Indian government statistics "if electricity is being used within its revenue area for any purpose whatsoever" (Government of India and Ministry of Power n.d.). Under this definition, it did not matter how many households, businesses, schools, or clinics were actually receiving electricity in the village. *In fact, a village would be considered as electrified even if the only electricity supplied was to power a single light-bulb or operate one irrigation pump.* This was a politically convenient definition because it allowed national and state-level politicians to give self-congratulatory speeches about the number of villages that had been "electrified" while ignoring the fact that few if any households in the "connected" village were actually receiving electricity. The government's official electrification statistics were of little comfort to households in "electrified" villages who saw the new lines but were not offered connections to their houses or could not afford them.

In 2004 the government of India established a stricter definition of electrification, which is still in use today. Under the Electricity Act of 2003, a village is defined as electrified only when at least 10 percent of its households are supplied with electricity, and power is also supplied to public institutions such as schools, health centers, local government offices, and community centers. But there is still no specification as to the quality of electricity, as measured, for example, by the stability of voltage levels or the minimum number of hours per day it must be supplied. Combined with this somewhat stricter definition was the initiation in April 2005 of a massive program known as the Rajiv Gandhi Grameen Vidyutikaran Yojana (RGGVY), which is reported to have electrified more than 100,000 villages under the new definition by the end of 2011.[20] But even if this government statistic is accurate, it does not reveal very much about the household electrification rate—the percentage of households that have actually signed up and are receiving electricity on a grid-based connection. The post-2004 definition is an improvement over the earlier definition, but it still suffers from the basic weakness that it provides no information about the reliability, quality, and affordability of the electrical service available over the physical connection.

Defining Electrification as Needs Served

We think that a better approach is to define electrification as the supply of electricity to meet some human need. But if this definition is to be of practical use, it has to be more specific about which human needs can be satisfied. A recent World Bank report concludes that there are six basic household needs: (a) lighting, (b) entertainment and communication, (c) food preservation, (d) mechanical loads and labor saving, (e) cooking and water heating, and (f) space cooling and heating (Bhatia and others 2013). Different levels of electrification can serve some or all of these human needs. In the literature of electrification, progress in electrification has sometimes been described in terms of a ladder that goes from no electrification to preelectrification to full electrification. As one moves up the ladder, the quantity and quality of electricity supplied increases and a greater number of human needs can be satisfied.

The Traditional Electrification Ladder Approach and Its Weaknesses

Table 2.2 provides a simplified example of a ladder of electrification. At the lowest level (Step 1), batteries can supply direct current (DC) electricity that allows for lighting and radios. Moving higher up the ladder (Step 3), household or institutional PV systems can provide for task and ambient lighting, refrigeration, radios, and small TVs. Steps 1–3 involve individual consumer devices and household systems. Step 5 is a big jump because it is the first step on the ladder where consumers have access to grid-based, alternating current (AC) electricity that can be used for productive or income-generating activities, both inside and outside the household. It is also the first step in which an outside supplier provides the electricity (rather than the household self-supplying it). At the highest level (Step 7), households, businesses, and institutions connected to the main or regional grids can receive AC at a high enough voltage (usually at 110 or 220 volts) to power a full range of electrical devices, including lightbulbs, radios, TVs, refrigerators, and commercial and industrial machinery, potentially for all hours of the day.

Several problems arise in using this technology-based ladder structure to analyze different levels of electrification. *First, it implies that the normal path of electrification involves moving from one step of the ladder to the next. In fact, it is equally likely that a household may skip one or more steps.* For example, households in a village may initially be limited to using battery- or solar-charged lanterns or torches (Step 3). But if the village is finally connected to the national grid, the households may move all the way to Step 7—grid-based supply from a national or regional power supply—without ever having gone through Steps 4, 5, and 6 on the ladder.

Table 2.2 A Traditional Ladder of Electrification

	Steps	Energy source	Uses
No electrification	Step 0	Flame-based lanterns using candles and kerosene	Lighting
Preelectrification	Step 1	Battery-powered torches/ flashlights and small appliances	Lighting, mobile-telephone charging, and radios
	Step 2	Car or motorcycle batteries	Lighting, mobile-telephone charging, radios, small color and black-and-white televisions, and low-wattage appliances
	Step 3	Lanterns/torches powered by PV cells supplying electricity and solar kits supplying electricity to incandescent bulbs, CFLs or LEDs, and small appliances	Lighting, mobile-telephone charging, radios, small color and black-and-white televisions, and low-wattage appliances
	Step 4	Solar PV systems for household and community uses	Lighting, mobile-telephone charging, small color and black-and-white televisions, small refrigerators, and other low-wattage appliances
Electrification	Step 5	Isolated mini-grids using small generators (fossil, renewable, and hybrid) producing AC power for a local distribution grid	Lighting, mobile-telephone charging, color and black-and-white televisions, fans, air conditioners, refrigerators, small motors, and electric pumps
	Step 6	Grid-connected mini-grids using small generators as a backup to grid-supplied power/as a bulk supply source to the main grid	Lighting, mobile-telephone charging, color and black-and-white televisions, fans, air conditioners, refrigerators, motors, and electric pumps
	Step 7	Grid-based power supply from a national or regional grid	Lighting, mobile-telephone charging, color and black-and-white televisions, fans, air conditioners, refrigerators, motors, and electric pumps

Note: AC = alternating current; CFL = compact fluorescent light; LED = light-emitting diode; PV = photovoltaic.

Second, the steps in the ladder are defined in terms of a specific technology or energy source (Column 3). But technologies will change over time and new technologies may arise. This suggests that a better approach is to define levels of electrification in terms of the outcomes—the human needs that can be satisfied—rather than the means (that is, particular technologies or energy sources) by which these outcomes can be achieved.

Third, the ladder in table 2.2 does not differentiate the quality of electricity supplied by different projects that use the same technology. Instead it seems to suggest that once the technology is in place, all projects using that technology will achieve the same level of service. This is not true. Consider Step 5—the case of an isolated mini-grid. This is a general descriptive term that ignores the fact that there can be considerable variation in the quality and quantity of electricity services provided by different mini-grids. For example, a hybrid mini-grid consisting of solar, diesel generator, and battery components may be able to provide an electricity supply for 24 hours a day, and households connected to this mini-grid can be comfortable buying and operating a small refrigerator. But in households connected to a mini-grid supplied by a diesel generator that is operated only 4–5 hours each night, it would not make sense to buy a refrigerator because the food stored would spoil during the 18 hours when electricity is not available. While both the hybrid and diesel generators supply mini-grids, they provide very different levels of electricity service. *Hence, the basic weakness of the electrification ladder in table 2.2 is that it defines levels of electrification in terms of technology and physical equipment, rather than the quality and quantity of electricity that is provided.*

Measuring Electrification by Its Attributes

A recent World Bank staff report (Bhatia and others 2013) proposes a different methodology for measuring electrification. It defines electrification in terms of the quality and quantity of electrical service supplied.[21] The basic premise of the report is that several key attributes of the service provided directly affect how the electricity can be used to satisfy human needs. The report concludes that seven basic attributes determine the "quality" or "usability" of electricity supply for a household:

- Quantity of supply *(measured in terms of maximum potential household load)*
- Duration of supply *(measured in hours supplied over a 24-hour period)*
- Evening supply *(hours supplied during the evening)*
- Affordability of supply *(cost of a stipulated consumption package as a percentage of income)*
- Legality of connection *(a legal or illegal supply)*
- Quality of supply *(evidence that electricity was not supplied at targeted voltage levels)*
- Reliability of supply *(frequency of unscheduled interruptions)*[22]

Table 2.3 shows the different tiers of electricity service, as defined by these key supply attributes. Hence, the tiers represent different levels of usability. As one moves to higher tiers with better attributes (for example, more hours of service, more reliability), additional electricity services can be supplied and it becomes possible for households to satisfy an increasing number of the six human needs listed earlier. For example, a solar home system could supply the eight specific uses listed under Tier 2. But if one moves up to Tier 3 by establishing a larger solar home system or mini-grid for the community (Case 1 in table 2.1), four additional uses of electricity (air cooling, food processing, rice cooking, and washing machine) could become available to households served by that mini-grid.

At first glance, tables 2.2 and 2.3 seem very similar: Table 2.2 refers to steps on a ladder and table 2.3 refers to tiers of a matrix. But there is a big difference between the two. Table 2.2 measures electrification in terms of inputs or technologies, whereas table 2.3 measures electrification in terms of the quantity and quality of electricity that can be supplied. In other words, table 2.2 is keyed to

Table 2.3 Multi-Tier Framework for Measuring Household Electricity Access

	Tier 0	Tier 1	Tier 2	Tier 3	Tier 4	Tier 5
Attributes of access						
Quantity (peak available capacity in watts)	—	>1 W	>20 W	>200 W	>2,000 W	>2,000 W
Duration of supply	—	>4 hrs	>4 hrs	>8 hrs	>16 hrs	>22 hrs
Evening supply	—	>2 hrs	>2 hrs	>2 hrs	4 hrs	4 hrs
Affordability (of a standard consumption package)	—	—	Affordable	Affordable	Affordable	Affordable
Legality	—	—	—	Legal	Legal	Legal
Quality (voltage)	—	—	—	Adequate	Adequate	Adequate
Feasible applications (and indicative wattage)						
		Radio (1) Task lighting (1) Phone charging (1)	All those in Tier 1 plus: General lighting (18) Air circulation (15) Television (20) Computing (70) Printing (45)	All those in Tier 2 plus: Air cooling (240) Food processing (200) Rice cooking (400) Washing machine (500)	All those in Tier 3 plus: Water pump (500) Refrigeration (300) Ironing (1,100) Microwave (1,100) Water heating (1,500)	All those in Tier 4 plus: Air conditioning (1,100) Space heating (1,500) Electric cooking (1,100)
Possible electricity supply technologies						
	Dry cell	—	—	—	—	—
	Solar lantern	Solar lantern	—	—	—	—
	Rechargeable batteries	Rechargeable batteries	Rechargeable batteries	—	—	—
	Home system	Home system	Home system	Home system	Home system	Home system
	Mini-grid/grid	Mini-grid/grid	Mini-grid/grid	Mini-grid/grid	Mini-grid/grid	Mini-grid/grid

Source: Bhatia and others 2013.
Note: Applications may not actually be used. — = not available.

inputs in the form of technologies that can produce electricity, whereas table 2.3 is keyed to outputs in terms of the quantity and quality of the electricity that can be supplied. These outputs determine which of the six basic household needs can be satisfied.

Another way to think about table 2.3 is that it describes the "supply" side, the level of electrification that will be available to consumers in a community. *This does not mean that villagers will actually use the available services.* The actual demand or use of these services will depend on family income levels and the relative prices of substitute services from sources other than electricity.

For example, Tier 3 makes it possible for a household to install a washing machine, but it would not make sense for even wealthy households in a village to purchase a washing machine when they could employ someone to do the washing at lower cost.

To go from theory to practice, Bhatia and others (2013) have created a questionnaire to measure both the level of electricity supply that is available and whether households are actually using it. Since the goal is to collect unbiased information on what happens at the household level, the questionnaire would be administered to households rather than to suppliers. From the questionnaire responses, two indices would be created: one would measure the available quantity and quality of electricity supply (the supply side); the other would measure how that supply is actually being used (the demand side). Whether calculated at the village, provincial, or country level, these two indices would provide accurate measures of electrification in terms of human needs that are being or could be satisfied.

Key Recommendation

In its simplest form, electrification should be defined as the supply of electricity to meet some human need rather than simply the installation of physical connections or the use of a particular technology. The six key household human needs are (a) lighting, (b) entertainment and communication, (c) food preservation, (d) mechanical loads and labor saving, (e) cooking and water heating, and (f) space cooling and heating. Electrification should be measured in terms of whether the quantity and quality of electricity supplied is adequate to satisfy one or more of these basic human needs.

Measuring Electrification: From Theory to Practice

We think that the analytical framework and indices proposed by Bhatia and others (2013) are clearly superior to the village and household connection statistics that many governments and donors currently use. But if this new framework is going to be more than just an interesting analytical concept, some way to administer the questionnaire on a periodic basis in many countries must be established. This is easier said than done. One way would be for the United Nations to sponsor the questionnaire as part of its recently announced "2014–2024 Decade of Sustainable Energy for All."[23] But this would take time and money to accomplish. In the meantime, in Africa it should be possible to take advantage of the fact that the 26 new electricity regulatory commissions and 15 new REAs would find the questionnaire useful in meeting their own statutory obligations (see the next sections). If these two types of government entities could be persuaded to test the questionnaire, and if it produces useful data for satisfying their legal mandates, then it might be easier to make the case for a coordinated global effort to measure progress in achieving the United Nations' goal of universal energy access by 2030.

Electrification and the Regulator's Need to Monitor Quality of Service

It is not widely recognized that there is substantial overlap between the need to create more accurate measures of electrification and the existing statutory responsibilities of most electricity regulators to measure the quality of service provided by entities that they regulate. For example, in Africa virtually every one of the new electricity regulators has legally mandated responsibilities regarding quality of service. Most African regulatory statutes require the regulator to set minimum quality-of-service standards and then to monitor and enforce these standards.[24]

Regulators in Africa and elsewhere have generally interpreted quality of service as having three components (see chapter 9 and appendix C). The first is *quality of product*, which is measured by the technical parameters of supply (such as whether frequency and voltage are at or near required technical levels). The second is *quality of supply*, which is usually measured by the availability and continuity of supply. For example, how many hours of the day does the SPP operator provide electricity? How frequent are unexpected blackouts and how long do outages last (at each occurrence and in total over the course of a year)? The third component is the *quality of commercial service*. This refers to the quality of service provided in the operator's commercial interactions with customers. For example, how long does it take to get a basic connection or a reconnection after a disconnection? Does the supplier provide accurate meter readings and bills?

In the case of the first two components of quality of service—quality of product and of supply—there is a significant overlap between the monitoring responsibilities of electricity regulators and the usability attributes that Bhatia and others (2013) recommend for measuring tiers of electrification. For example, the regulator's quality of product (stability and level of frequency and voltage) overlaps with Bhatia and others' (2013) proposed quality attribute of electrification. And the regulator's quality of supply (frequency and duration of unplanned outages) corresponds to the reliability attribute (see table 2.3). Hence, this seems to suggest that African regulators could, in the normal course of performing their statutory duties, also monitor a country's progress on electrification. The recent experience in Tanzania provides some insights as to how this might be done.

Monitoring Quality of Service by the Tanzanian Regulator

In July 2009 EWURA issued a "Customer Service Charter," a document that described the obligations of the national utility, TANESCO, in the electricity service it provides to main-grid-connected household customers. This charter established specific quantitative minimum standards for service to households. For example, the charter created targets for advance notification of planned outages and reporting requirements on unplanned outages. But EWURA, like most regulators, discovered that it was relatively easy to specify quality of service standards for a national utility but more difficult to implement effective monitoring of these standards. To date, EWURA has used two main approaches to monitoring TANESCO's performance. The first approach involves quarterly self-reporting by TANESCO. But an independent audit conducted in 2011 found that many of the numbers reported by the national utility were not credible. The second approach involved a survey that EWURA commissioned in 2011 of 2,001 randomly selected household customers of TANESCO in different parts of the country. Some of the survey's 68 questions related to quality of supply and quality of product. EWURA hopes to repeat these surveys of TANESCO's customers on a regular basis. If this is done, we think that future surveys could, with some small adjustments, be designed to also provide information on whether TANESCO, mostly a main-grid supplier, is effectively providing the Tier 5 service level that would be expected of a main-grid operator.

Measuring Connections and Service on Mini-Grids: The Role of the Rural Energy Agency

Tanzania's Rural Energy Agency (REA) is better positioned than the regulator to monitor the electrification performance of isolated and connected mini-grids. Like equivalent agencies in most other African countries, Tanzania's REA provides grants to mini-grid developers that propose to connect new rural customers. The grants are intended to lower initial connection costs. To ensure that the money has been used as promised, the REA conducts an independent audit to confirm that the connections have actually been made before releasing the grant money. Intending to go a step further, Uganda's REA plans to perform a follow-up audit to ensure that the newly connected customers are actually receiving electricity several months after the connection is made.

Presumably, most REAs would also want to ensure that physical connections have led to continuing electricity service at whatever quality standards were specified in their grant agreement with the mini-grid developer. For example, is the developer providing Tier 3 service when it promised to provide Tier 4 service? In this context, Bhatia and others' (2013) questionnaire could serve a dual purpose: it would allow the REA to verify that service commitments made in the grant agreement are being fulfilled; and it would also provide ground-level information on the quantity and quality of electrical service that mini-grid operators provide. And since most African REAs also subsidize the installation of solar home systems, an REA could use the same questionnaire to determine the level of electricity service received by households that have installed such systems.

Key Observation

Electricity regulatory commissions are typically required to monitor the quality of electricity supply of the entities that they regulate. REAs are usually required to validate the installation of new connections that have received grants from the agency. In the course of performing these legally mandated functions, both entities are well positioned to develop accurate ground-level data on the status of electrification in their countries.

Notes

1. The installed capacity of a grid-connected SPP may be considerably larger than 10 MW. For example, the Mitkarasin Sugar Company, a bagasse and rice husk plant in Thailand, has an installed capacity of 39 MW. But it is only eligible for VSPP prices on the electricity produced from 10 MW of its total installed capacity and actually exports only 8 MW (the rest is used inside the sugar company). The same convention was adopted in Tanzania. The TPC sugar plant, a biomass cogeneration plant located near Mount Kilimanjaro, has an installed capacity of 18 MW, but only the electricity produced from 10 MW is eligible for the SPP FIT for grid-connected SPPs. The remaining electricity is used inside the sugar factory or to power irrigation for the sugarcane fields.

2. In Thailand there are two sets of regulations with similar names: SPP regulations apply to generators that export 10–90 MW; VSPP regulations apply to generators that export up to 10 MW. In this guide when we refer to Thailand we will be making reference to the Thai VSPP regulations as they are most similar to the Tanzanian and Sri Lankan SPP regulations.

3. EWURA (2012b) defines an SPP as: "an entity generating electricity using renewable energy, fossil fuels, a cogeneration technology, or some hybrid system combining fuel sources . . . and either sells the generated power at wholesale to a Distribution Network Operator (DNO) or sells at retail directly to end customers or some combination of the two. An SPP may have an installed capacity greater than 10 MW but may only export power outside of its premises not exceeding 10 MW." A distribution network operator is responsible for the operation of a distribution network serving 10,000 customers or more.

4. While some forms of biomass energy arguably produce no net carbon (because carbon dioxide released is reabsorbed by sustainably managed fuel crops, or because the fuel is a waste product that would have been burned anyway), some generation of electricity from biomass combustion can contribute to global greenhouse production through burning wood or other crops that are not replenished. Carbon neutrality will depend on the type of biomass used and its net land-use effect (Searchinger and others 2009).

5. Small fossil fuel generators, typically fired by diesel or fuel oil, are common throughout Africa. These small generating plants typically fall into two categories. The first category consists of backup generators operated at plants or commercial installations in the many countries where the grid supply is unreliable. The second category consists of generators operating in isolated communities that do not have access to grid-supplied electricity. It has been estimated that 4,000 MW, or approximately 6 percent of Sub-Saharan Africa's installed generating capacity, is of the first type. There are no good estimates of this second type because many of them operate without government approvals (Foster and Steinbuks 2008).

6. The cost and service advantages of hybrid generating systems for both isolated and connected mini-grids are discussed more fully in appendix A.

7. At the time of this writing (September 2013), EWURA, the Tanzanian electricity regulator, has proposed widening its definition of SPP so that it is no longer limited to cogenerators and renewable generators exclusively. EWURA has proposed allowing hybrid generators that use a fossil fuel in addition to renewable energy, as long as the fossil fuel generates no more than 25 percent of the total production calculated on an annual basis.

8. Since the buying utility in most African countries is the national utility, we will adopt the convention of using the term *national utility* to refer to the buying utility for grid-based SPPs.

9. Wholesale simply means that the electricity that is sold by the SPP will be resold; retail means that the electricity is sold to a final user (that is, it will not be resold). A bulk sale of electricity means the sale of a large quantity of electricity. The two terms, *bulk* and *wholesale,* are sometimes incorrectly used as synonyms. In fact, a bulk sale of electricity can be either a wholesale sale (for example, an SPP selling to the national utility) or a retail sale (for example, the national utility selling to an industrial customer who will not be reselling the electricity).

10. The IFC estimate is based on an estimate of unelectrified rural households located in relatively densely populated villages distant from the main grid with monthly expenditures on lighting and nongrid sources of electricity equal to $8.50 or higher per month (IFC 2012, 148–49). The IFC estimate is based on the concept of an "addressable market," which it defines as "the number of households that could afford to pay the full commercial price of a service (based on current spending levels for traditional energy), if it was offered by an efficient company, earning a commercial return on capital but not constrained by lack of finance or excessive regulatory restrictions." In this guide, we examine how to design and implement "nonexcessive" regulatory requirements so that the IFC estimate could actually be achieved. We are not aware of any similar estimates for Case 4 or the hybrid cases shown in table 2.1.

11. Other companies and organizations are also pursuing village-level electrification in India with micro-grids that use mobile-phone towers as anchor customers. These include Gram Power (http://www.grampower.com) and the SPEED Program funded by the Rockefeller Foundation (Jhirad and Lewis 2012).

12. Any isolated mini-grid system will have three physical components or subsystems: production, distribution, and users' subsystems. Production refers to the generators, an energy control system, and batteries and converters if there is a need to convert DC power to AC power. Distribution refers to the low- or medium-voltage lines needed to bring electricity to the users. The user subsystem includes meters, internal wiring, and grounding (ARE 2011). Isolated mini-grids typically operate at low voltages of around 11 kilovolts (kV) or less.

13. This section draws from a variety of sources. It benefitted greatly from discussions with Mr. Krishnan Raghunathan, who provided technical assistance on many of Tanzania's early SPP projects. It also draws from discussions with developers and their advisers and the SPP applications for provisional licenses and tariffs filed with EWURA.

14. A detailed description of this pricing formula that creates a "feed-in tariff" is given in chapter 7.

15. For example, another type of Case 4 SPP (a grid-connected SPP that sells at wholesale to a utility) would be a mini-hydro generator that proposes to sell exclusively at wholesale to the national grid, which has no on-site production needs and requires project financing. This has been the dominant SPP model in Sri Lanka. It has also been proposed by several developers in Tanzania, but it remains to be seen whether it will be commercially viable under the existing FITs in Tanzania.

16. In fact, the first few SPP projects were developed by existing agricultural enterprises that all relied on balance sheet financing.

17. In effect, the mini-grid would be seeking the right to cross-subsidize the low tariffs charged to low-consumption households with the higher tariffs charged to

commercial and industrial installations. Such cross-subsidies exist in many African countries. For example, in Tanzania, industrial and commercial users pay 14 cents/kWh while low-consumption households (that is, less than 50 kWh/month) pay 4 cents/kWh. Hence, industrial consumers pay higher tariffs despite the fact that the cost of serving an industrial customer is less than the cost of serving a small household.

18. Another term that overlaps with SPD is *energy services company* or ESCO. But the terms are not exactly the same. Most SPDs limit themselves to distributing grid-quality AC electricity purchased from another source over a local low-voltage network. In contrast, an ESCO may provide this service but may also supply other energy services such as selling and maintaining solar home systems, selling energy-efficient lightbulbs and appliances, and offering charging services for mobile phones and other appliances. And if the energy comes from renewable sources, the ESCO may refer to itself as a renewable energy services company or RESCO.

19. Most of the Nepalese SPDs are local cooperatives. From early reports, the SPDs are much more successful than the Nepal Electricity Authority (NEA) at reducing commercial losses, providing technical service, and achieving high collection levels (Mahato 2010).

20. These statistics are reported on the Ministry of Power's website (http://www.powermin.nic.in/bharatnirman/bharatnirman.asp). All of these villages were electrified through extensions of the central grid.

21. This proposed approach has been largely adopted by the World Bank and International Energy Agency (World Bank and IEA 2013) for use in measuring progress in the Sustainable Energy Access for All initiative.

22. A complete discussion of how each attribute is measured can be found in Bhatia and others (2013).

23. See http://www.un.org/News/Press/docs/2012/ga11333.doc.htm.

24. See EWURA (2008, sections 28–30) for the quality-of-service responsibilities of the Tanzanian electricity regulator.

References

Adama, Sissoko, and Alassane Agalassou. 2008. "Mali's Rural Electrification Fund." Presentation at the Sustainable Development Week, Washington, DC, February. http://siteresources.worldbank.org/INTENERGY2/Resources/presentation8.pdf.

Agalassou, Alassane. 2011. Personal communication. March.

ARE (Alliance for Rural Electrification). 2011. "Hybrid Mini-Grids for Rural Electrification: Lessons Learned." Brussels. http://www.ruralelec.org/fileadmin/DATA/Documents/06_Publications/Position_papers/ARE_Mini-grids_-_Full_version.pdf.

Bhatia, Mikul, and Heather Adair Rohani. 2013. "Defining and Measuring Access to Energy." Presentation at Learning Days, SDN Forum, World Bank, March 7.

Bhatia, Mikul, Nicolina Angelou, Elisa Portale, Ruchi Soni, Mary Wilcox, and Drew Corbyn. 2013. *Defining and Measuring Access to Energy for Socio-Economic Development.* Washington, DC: World Bank and Energy Sector Management Assistance Program (ESMAP).

Cabraal, Anil. 2011. "Empowering Communities: Lessons from Village Hydro Development in Sri Lanka." Presentation, World Bank, Washington, DC, May 25.

Chanthan, Ky, and Pascal Augareils. 2013. "Potential for Increasing the Role of Renewables in Mekong Power Supply: Cambodia." Presentation at the CPWF Mekong Basin Development Challenge, Hanoi, Vietnam, February 20.

Electricity Authority of Cambodia. 2008. *Report on Power Sector of the Kingdom of Cambodia*. 2009 edition. Compiled by Electricity Authority of Cambodia from Data for the Year 2008 Received from Licensees. Annual Report of the Electricity Authority of Cambodia, Phnom Penh. http://www.eac.gov.kh/pdf/reports/Annual%20report%20 2008.en.pdf.

———. 2009. *Report on Power Sector of the Kingdom of Cambodia*. 2010 edition. Compiled by Electricity Authority of Cambodia from Data for the Year 2009 Received from Licensees. Annual Report of the Electricity Authority of Cambodia, Phnom Penh. http://www.eac.gov.kh/pdf/reports/Annual%20report%202009.en.pdf.

———. 2010. *Report on Power Sector of the Kingdom of Cambodia*. 2011 edition. Compiled by Electricity Authority of Cambodia from Data for the Year 2010 Received from Licensees. Annual Report of the Electricity Authority of Cambodia, Phnom Penh. http://www.eac.gov.kh/pdf/reports/Annual%20Report%20 2010%20En_final.pdf.

———. 2011. *Report on Power Sector of the Kingdom of Cambodia*. 2012 edition. Compiled by Electricity Authority of Cambodia from Data for the Year 2011 Received from Licensees. Annual Report of the Electricity Authority of Cambodia, Phnom Penh. http://www.eac.gov.kh/pdf/reports/Annual%20Report%202011En_%20Final2.pdf.

EPPO (Energy Policy and Planning Office), and Ministry of Energy. 2009. "Regulations for the Purchase of Power from Very Small Power Producers (for Generation Using Renewable Energy)." http://www.eppo.go.th/power/vspp-eng/Regulations%20 -VSPP%20Renew-10%20MW-eng.pdf.

———. 2011. "Status of Purchase of Electricity from VSPP September 2011 (PEA)." http://www.eppo.go.th/power/data/index.html.

EWURA (Energy and Water Utilities Regulatory Authority). 2008. "Standardized Tariff Methodology for the Sale of Electricity to the Main Grid in Tanzania under Standardized Small Power Purchase Agreements." http://www.ewura.go.tz/pdf /public%20notices/SPP%20Tariff%20Methodology.pdf.

———. 2012a. "Detailed Tariff Calculations for Year 2012 for the Sale of Electricity to the Mini-Grids in Tanzania under Standardized Small Power Purchase Agreements in Tanzania." http://www.ewura.go.tz/pdf/SPPT/2012/2012%20SPPT%20Calculation %20for%20Mini-Grid.pdf.

———. 2012b. "The Electricity (Development of Small Power Projects) Rules." Proposed for Public Consultation. Dar es Salaam, Tanzania.

———. 2013. "The Electricity (Development of Small Power Projects) Rules." Proposed for Public Consultation, Dar es Salaam, Tanzania. ewura.go.tz/sppselectricity.html.

Foster, Vivien, and Jevgenijs Steinbuks. 2008. "Paying the Price for Unreliable Power Supplies: In-House Generation of Electricity by Firms in Africa." Working Paper, Africa Infrastructure Country Diagnostic, World Bank, Washington, DC. http://www .infrastructureafrica.org/system/files/WP2_Owngeneration_2.pdf.

Government of India, and Ministry of Power. n.d. "Definition of Electrified Villages." http://rggvy.gov.in/rggvy/rggvyportal/def_elect_vill.htm.

GSMA. 2013. "Mobile for Development Programmes/Mobile for Development." http:// www.gsma.com/mobilefordevelopment/programmes.

IFC (International Finance Corporation). 2012. *From Gap to Opportunity: Business Models for Scaling Up Energy Access*. Washington, DC: IFC. http://www1.ifc.org/wps/wcm/connect/ca9c22004b5d0f098d82cfbbd578891b/EnergyAccessReport.pdf?MOD=AJPERES.

Jhirad, David, and Jessica Lewis. 2012. "SPEED-Smart Power for Environmentally Sound Economic Development." Presentation to the United Nations Workshop on Facilitating Energy Access and Security: Role of Mini/Micro Grids.

Keosela, Loeung. 2013. "Status of Power Sector in Cambodia." Presentation to the Renewable Energy Workshop, Chiang Mai, Thailand, January.

Mafia Island. 2011. Personal communication. September.

Mahato, Rubeena. 2010. "Power Sharing, Nepali Style." *Nepali Times*, July 23.

Raj, Anil. 2012. "The Micropower Opportunity: Paving the Way for Rural Electrification." Presentation, World Bank, November 15.

Rekhani, Badri. 2011. Personal communication. November.

———. 2012. Personal communication. March.

Searchinger, Timothy D., Steven P. Hamburg, Jerry Melillo, William Chameides, Petr Havlik, Daniel M. Kammen, Gene E. Likens, Ruben N. Lubowski, Michael Obersteiner, Michael Oppenheimer, G. Philip Robertson, William H. Schlesinger, and G. David Tilman. "Fixing a Critical Climate Accounting Error." 2009. *Science* 326: 527–28.

TPC. 2012. Personal communication. February.

Van Tien, Hung. 2011. Personal communication. June.

World Bank and Climate Investment Funds. 2013. "Scaling Up Renewable Energy Program in Low Income Countries." https://www.climateinvestmentfunds.org/cif/srep.

World Bank and IEA (International Energy Agency). 2013. *Sustainable Energy for All: Global Tracking Framework*. Washington, DC: World Bank and IEA. http://documents.worldbank.org/curated/en/2013/05/17765643/global-tracking-framework-vol-3-3-main-report.

CHAPTER 3

The Regulation of Small Power Producers and Mini-Grids: An Overview

There you have it—reforms on unprepared ground, and copied from foreign institutions as well—nothing but harm!

—From *The Brothers Karamazov*, book 11, chapter 9, by Fyodor Dostoevsky

Regulation can provide a fertile ground. But regulation does not make a market.

—IFC official, World Bank Group workshop, January 30, 2012

The less we have to do with government, the happier we are.

—Indian micropower developer, November 2012

Abstract

Chapter 3 describes the technical, commercial, economic, and process decisions that regulators face, and provides examples of how each type of decision can affect small power producers (SPPs) and mini-grids. It examines the concept of "light-handed regulation," and explores its application to SPPs and how it can produce both positive and negative consequences. The chapter also considers the threshold question of when to regulate and when not to regulate. In cases where SPP regulation is necessary, the chapter examines whether regulation should be performed by the national electricity regulator, a rural electrification agency (REA), or the local community.

What Is Regulation?

Regulation implies government control of an enterprise. Economic regulation is usually imposed when an entity has monopoly power, whether because it is a natural monopoly or it has been given a legal monopoly or both (Breyer 1982, 15–16). When a government regulates an enterprise, it imposes direct or indirect controls on the enterprise's decisions or actions. Almost any regulatory decision or process will affect an enterprise's revenues or costs. The three universal tasks of national electricity regulators that oversee traditional monopoly sellers are:

- Setting maximum and minimum prices
- Establishing minimum quality-of-service standards
- Specifying entry and exit conditions (usually through licenses, permits, or concessions) (Brown and others 2006, chapter 1)

Regulating SPPs is different from regulating a traditional monopoly utility. If a grid-connected SPP is selling to a national utility (Case 4 from the previous chapter), the SPP will not have any monopoly power as a seller. In fact, the opposite will be true. The SPP will be a small entity selling to a monopoly buyer that is likely to have many other supply options. In many instances, the national utility may not be a willing buyer for several reasons: first, it may believe that it is being forced to pay too much for the SPP's power; second, it may not want the administrative bother of buying power from many small producers; and third, it may want to remain the sole producer/supplier of electricity in the country. In any of these situations, if an SPP program is going to be successful, the regulator will find that it needs to regulate the actions of the buyer (that is, the national utility) in addition to its traditional responsibility of regulating the actions of the seller (the SPP).

Key Definition

The three principal tasks of an electricity sector regulator are setting maximum and minimum electricity prices, establishing minimum quality-of-service standards, and specifying entry and exit conditions through licenses, permits, and concessions.

Three Types of Regulatory Decisions That Affect SPPs

Electricity regulators are usually described as economic regulators. But this is only partially true. Most electricity regulators also make technical and process decisions that affect SPPs.

- A *technical decision* is usually an engineering decision. For example, a regulator must decide on the technical standards in the interconnection agreement that provide for safe and robust electrical connections between the national utility and a grid-connected SPP. A regulator will also have to decide on safety standards for both grid-connected and isolated SPPs that serve retail customers. For example, if an SPP is serving retail customers in a rural village, the regulator will need to specify the minimum clearance between the ground and the wires on distribution poles. While the content of these rules is technical, the effects of the regulator's rules are both technical and economic. For example, Thailand requires only a few standard relays for interconnection of small induction generators to the national grid, whereas regulators in other countries may require more extensive, and expensive, protection equipment.

- *An economic or commercial decision* typically affects which party is responsible for paying for what, or sets the price that some entity may be allowed to charge for either the wholesale or retail sale of electricity. For example, a regulator usually decides who will pay for the cost of the interconnection between the SPP and the national grid operator. Of all decisions affecting grid-connected SPPs, the one decision that gets the most attention in numerous books and articles is the price that a grid-connected SPP receives for the power that it sells to the national or regional utility (usually referred to as the feed-in tariff, FIT). This is not surprising, because the level and stability of this price are critical for the SPP's economic viability. But as we will see in the sections that follow, there are less-visible (but equally important) regulatory decisions that will also affect an SPP's economic viability.

- A *process decision* is one that specifies the process by which the regulator's technical and economic decisions are made and enforced. For example, does the regulator consult with some or all stakeholders before making a technical or economic decision affecting SPPs? Is the consultation conducted publicly or privately? Must the regulator meet deadlines, either established by law or self-imposed, for making key decisions? What happens if the regulator fails to meet the deadlines? What information must a potential SPP supply in applying for a provisional or final license? In addition to the regulator's process decisions, the regulator may also specify the process by which an SPP and buying utility interact with each other. For example, the regulator may specify the number of days that a national utility is given to respond to a request for interconnection by an SPP.

Table 3.1 gives some specific examples of these different types of regulatory decisions.

The Importance of Regulatory Process

Of the three types of regulatory decisions, technical and economic decisions usually get the most attention because they tend to be more visible and to have an obvious impact. For example, it is clear that few, if any, main-grid-connected SPPs will be created if the price that they will be paid for electricity sold to the national utility is set below the SPPs' costs of supply. But even if the regulator sets a price that ensures economic viability for SPPs, the regulatory system may still fail if the specified decision-making process involves too many steps, if government entities ignore one another's responsibilities, or if the regulator

Table 3.1 Examples of Different Types of Regulatory Decisions

Technical regulations

A technical decision is usually an engineering decision. Technical decisions include:
- Regulations to provide for safe and robust electrical connections between the national utility and a grid-connected small power producer (SPP)
- Distribution system safety standards for both grid-connected and isolated SPPs
- Technical standards for allowable voltage and frequency variations and total harmonic distortion (THD) variations
- Required relays for generators of different sizes and types

Commercial regulations

A commercial or economic decision affects which party can receive a license or permit and who is responsible for paying for what, or the price that can be charged for either the wholesale or retail sale of electricity. Some examples include:
- Information and approvals that must be provided to obtain a license or permit
- Sharing arrangement for the cost of the interconnection between a separate or mini-grid-based SPP and the national grid operator
- Price that a grid-connected SPP receives for the power that it sells to the national or regional utility (that is, the feed-in tariff, FIT)
- Price charged to the mini-grid for backup power because of planned or unplanned maintenance of its system

Process regulations

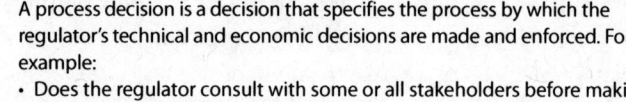

A process decision is a decision that specifies the process by which the regulator's technical and economic decisions are made and enforced. For example:
- Does the regulator consult with some or all stakeholders before making a technical or economic decision?
- Is the consultation conducted publicly or privately?
- Does the regulator make its decisions before, after, or at the same time as nonsector regulatory decisions (for example, incorporation, acquisition of land titles, tax registration, environmental approvals, and so on)?
- How many days does the utility have to respond to a request for interconnection by an SPP?
- Does the regulator delegate (either formally or informally) some decisions to other government bodies such as the rural energy agency?

fails to enforce its decisions in a timely manner. As one SPP developer observed, "[b]y the time the regulator gets around to enforcing his decision, I will be bankrupt." Or as another SPP investor noted, "I cannot sustain endless negotiations." So an effective regulatory system for SPPs requires both fair and efficient technical and economic decisions as well as timely processes for making and enforcing decisions. (The most important initial regulatory process for SPPs is the application and review process for licenses and permits. The licensing approaches used in several countries are discussed in chapter 4.) The bottom line is that an SPP regulatory system will be successful only if the decision-making processes are quick and are not overly expensive for SPPs to comply with and for regulators to administer.

Key Observation

Electricity regulators make three types of decisions: technical, economic, and process. Process decisions are the hidden underside of SPP regulation. An SPP regulatory system will be successful only if its decision-making processes are quick and not overly expensive for both SPPs and those who purchase from SPPs.

Light-Handed Regulation: When It Works and When It Doesn't

The Cost of Regulation

Regulatory rules will typically specify maximum or minimum prices, minimum service standards, information that must be supplied to acquire a license or permit, and procedures that must be followed to satisfy these regulatory requirements. Complying with regulatory rules costs time and money. This is true regardless of whether the regulated enterprise is privately, publicly, or community owned. For SPPs, regulators need to be especially conscious of the costs of regulation because many SPPs operate on the "razor's edge" of commercial viability. This is especially true of new SPPs that intend to serve isolated communities (Case 1: isolated SPP that sells at retail). Their costs are high because of the need to transport equipment and supplies over long distances, and their revenues are low because many of their customers can afford to buy only small quantities of electricity. Unnecessary regulation, even though well intentioned, can easily destroy the commercial viability of these SPPs. SPPs that propose to serve isolated mini-grids are not likely to develop unless the regulator makes a conscious effort to create a light-handed regulatory system for such SPPs.

What Is Light-Handed Regulation?

In practice, light-handed regulation usually implies that:

- The amount of information required by the regulator is minimized.
- The number of separate regulatory processes and decisions are as few as possible.
- Documents are standardized.
- Related decisions made by other government or community bodies are communicated to and utilized by the regulator.

Table 3.2 gives some real-world examples of light-handed regulation applied to SPPs.

Clearly, it makes sense to employ light-handed regulation when it can achieve some or all the goals that a heavier and more intrusive form of regulation would achieve. One might reasonably ask: why would anyone want to adopt "heavy-handed" regulation when some form of light-handed regulation

Table 3.2 Examples of Light-Handed Regulation for SPPs

Characteristics of light-handed regulation	Example
Minimize the amount of information provided to the regulator.	For very small power producers (VSPPs) (installed capacity of 100 kilowatts [kW] or less), the VSPP need not make a retail tariff filing with the Energy and Water Utilities Regulatory Authority (EWURA). But if the EWURA receives complaints about the tariffs, it reserves the right to review the VSPP's tariffs using a publicly available cost-of-service model employed for the larger small power producers (SPPs) (Tanzania).
	In setting feed-in tariffs (FITs), the regulator does not require individual cost-of-service studies for each SPP (Tanzania, Sri Lanka, and Kenya), but instead sets generic-technology-based tariffs or tariffs based on estimates of the buying utility's avoided costs.
Minimize the number of separate regulatory processes and decisions.	Licenses are not required for SPP projects less than 1 megawatt (MW) (Tanzania).
Use standardized documents or similar documents created by other agencies, and make documents available on the Web.	Standardized power-purchase agreements (PPAs) and standardized application forms are used for interconnection to a national or regional utility (Tanzania, Thailand, and Sri Lanka).
	A standardized template for prefeasibility studies is used when SPPs that wish to sell to the national utility apply for provisional approvals (Sri Lanka).
	A standardized model electricity supply agreement is preapproved by the regulator for villages served by the private operator of an isolated mini-grid with an SPP (Cambodia).
Rely on related decisions by other government agencies or community bodies.	The regulator gives considerable weight to the rural energy agency's (REA)'s approval of an SPP business plan when the regulator reviews license applications (Tanzania).
	The regulator gives considerable weight to the renewable energy agency's issuance of an energy permit when it makes its decision as to whether it will issue a generation license (Sri Lanka) (pre-2011).

could achieve some or all of the same outcomes at a lower cost for all parties? But this does not imply that light-handed regulation should be blindly adopted in all situations involving SPPs. In other words, light-handed regulation should not be a mantra applied without thinking through the consequences.

Key Observation

Regulation creates costs for SPPs that can reduce or even destroy the commercial viability of SPPs, particularly SPPs that serve isolated communities using mini- and micro-grids. Therefore, regulators should make a conscious effort to create a "light-handed" regulatory system for such SPPs.

When Light-Handed Regulation Backfires

Light-handed regulation has had undesirable consequences in both Sri Lanka (pre-2011) and Nepal, where it was used for SPPs that wished to sell solely at wholesale to the national utility on its main grid (Case 4: a grid-connected SPP that sells at wholesale to a utility). Both countries created "first-come, first-served" application and review systems that allowed SPPs to "lock in" an exclusive position in the queue with minimal effort. In the case of Nepal, the regulatory threshold was set low in four ways that are instructive for Case 4 SPPs:

- Low application fees
- Easy (that is, lengthy) deadlines
- No required prefeasibility studies, or acceptance of copied prefeasibility studies
- No serious monitoring of milestones in project development between deadlines

As a consequence of this light-handed approach, Nepal has been overwhelmed with applications for projects between 1 and 25 megawatts (MW) that propose to sell hydropower at wholesale to the national utility. It is quite clear that this light-handed regulatory system has led to an avalanche of applications from speculators who were simply seeking a high place in the queue with no real intention of developing their proposed project. This has had the effect of clogging up the regulatory approval system with phantom projects and has caused delays in processing serious projects. Nepalese law requires that applications for survey licenses be processed in 21 days, but in fact the average processing time is about 6–7 months. It has been reported that more than 1,500 projects have received "survey licenses" (the equivalent of a provisional approval), which give the license holder an exclusive five-year right to develop a project. To date, only about 10–20 such projects have actually been developed, so the present system is clearly not functional. Recently, the Nepalese government has been presented with several proposals for tightening deadlines and raising the quantity and quality of information required to obtain a survey license.

Sri Lanka had a similar (though not as extreme) experience. Between 1996 and 2008, Sri Lanka had a loose, utility-managed approval process for SPPs that wished to sell at wholesale to the national utility (Case 4). As in Nepal, the regulatory system in Sri Lanka led to a backlog of paper projects that clogged up the system but produced little in the way of additional electricity. Hence, in both countries an overly light-handed regulatory system hurt rather than helped in the development of SPPs.

This experience suggests that light-handed regulation may not be equally appropriate for all types of SPPs. As a regulatory strategy, we think that it has substantial merit for regulating new isolated mini-grids that propose to serve new rural customers at retail (Case 1: an isolated SPP that sells at retail). In contrast, light-handed regulation can easily backfire if it is applied to SPPs that want to sell only at wholesale to the national utility and face little or no pressure to develop their proposed project within a specified time (Case 4).

To Regulate or Deregulate? A Specific Example

While there are examples of regulators operating in a light-handed way, this is not the norm. *When in doubt, regulators tend to regulate.* This is often the least-risky political strategy because a regulator can always justify a decision by claiming it was just doing its job. But unnecessary regulation can do more harm than good.

Some modern regulatory statutes will give regulators the discretion not to regulate (that is, regulatory forbearance) **or** to regulate in various ways.[1] Regulators should take advantage of this flexibility. This is especially true when an electricity regulator is presented with a new technology or a new business model for delivering electricity. In this situation, the first question the regulator should always ask is: *Should this entity be regulated, deregulated, or regulated in a different way?*

Why regulate? The question is best answered by going back to the basics: what is the principal economic reason for regulating in the first place? In the literature, economic regulation is usually justified on the grounds that an enterprise should be regulated if it has monopoly power.[2] This means that the enterprise does not have any actual or potential competition because some or all of its customers are captive customers who have no viable alternatives, either through self-supply or from other suppliers. In this situation, the presumption is that if the regulator does not control the supplier's monopoly power, it will be able to charge captive customers unreasonably high prices and earn high profits. This is considered both inefficient and unfair.[3]

A Specific Example

To make the discussion less abstract, let us consider the case of an emerging business model for delivering electricity to rural areas that the Omnigrid Micropower Company (OMC) (http://www.omcpower.com) is pioneering in India. OMC is one of several companies in India that propose to sell electricity to mobile-phone tower owners or operators under a long-term power sales agreement using hybrid generation (for example, solar and diesel). A key element of the business model is that the enterprise will also provide energy services to surrounding villages by renting rechargeable battery boxes, lanterns, and appliances to households and businesses on a daily, weekly, or monthly basis. (See box 2.2 for a fuller description of the business model as it has developed in India.) Several companies are reportedly considering introducing a similar business model in rural Africa. If this happens, African electricity regulators will need to decide whether or not to regulate these companies. And if there is regulation, what should be regulated?

It is important to remember that regulation is not an "all-or-nothing" proposition. Regulation is multidimensional. In the case of the OMC business model, the

African electricity regulator will need to decide whether the following elements of regulation are necessary:

- *Tariff regulation*, which entails approving (a) the prices that the enterprise proposes to charge for the sale of electricity to tower owners or operators, and (b) the leasing charges for daily, weekly, or monthly rental of precharged battery boxes, lanterns, and other appliances to households and small businesses
- *Licensing*, which entails requiring the enterprise to obtain a license or permit to operate
- *Safety regulation*, which entails establishing and enforcing safety rules for some or all of the enterprise's operations

Tariff Regulation

An OMC-type enterprise will earn revenues from two principal commercial transactions: (a) the sale of electricity under a long-term contract to the mobile-phone tower owner/operator, and (b) the short-term leasing of battery boxes, lanterns, and other electrical appliances to village households and businesses. Let us consider whether either of the transactions needs to be regulated.

Before the arrival of this new business, the tower owner or operator self-supplied electricity to the tower from its own on-site diesel generator. Presumably, a tower owner/operator would enter into a long-term supply agreement with this new enterprise *only* if the new business is willing to supply the same or a higher quality of electricity at a lower price. Therefore, it would be inaccurate to say that the tower owner or operator is a captive customer with no alternative at the time that the contract is signed. The owner/operator could continue to self-supply but instead chooses to sign a contract with the new provider because it is a better deal.

The situation is different with village households. Two characteristics are relevant to the tariff regulation decision in their case. First, OMC is not selling electricity to households but rather leasing electrical appliances on a daily, weekly, or monthly agreement with households. Most electricity regulatory statutes limit regulation to the sale of electricity. So from a legal perspective, the leasing of charged electrical devices and appliances may not constitute a "sale of electricity" under many statutes. Second, from an economic perspective, there are several competing businesses. In any village, at least several businesses or individuals are able to offer battery and mobile-phone charging services. Presumably, households would choose an OMC-type supplier only if it offers a better deal— the charged appliances are delivered directly to the household's doorstep, the prices are lower, and the quality of energy service (for example, light from lanterns using light-emitting diodes) is better. If this supplier tried to raise its prices, its customers would not be "locked in," because they have not signed any long-term leasing agreements. Within a day, a week, or a month, they can always return to the local businesses that had previously offered battery charging. In other words, the new supplier faces competition.

From the Bottom Up • http://dx.doi.org/10.1596/978-1-4648-0093-1

Licensing

A license is a government-granted right to conduct a specified business. When a government agency issues a license, it is based on a determination that a particular entity has the technical and financial capacity to conduct a certain type of business. Licenses are often justified as a form of consumer protection. The usual rationale is that it is more efficient for the government to conduct "due diligence" on the competence of the entity through a licensing process than for individual consumers to do so.

Most modern regulatory statutes in Africa and elsewhere exempt SPPs from applying for licenses if they have an installed capacity of 1 MW or less. The usual justification for this exemption is that governments do not have the resources to perform the due diligence needed to license potentially hundreds of small entities throughout the country. In effect, this is an administrative rationale for exempting SPPs from applying for a license.

Apart from this administrative justification, we think that there are three other compelling reasons for exempting this type of operation from the need to apply for a license. The first is that small consumers face little or no risk if they buy services from the new supplier, as long as the new supplier does not require a connection fee or a security deposit for the rental of its batteries and lanterns. The second is that a new supplier cannot block other entities from providing the same service. It does not have a government-granted exclusive franchise to serve a specified geographic area that would be a legal barrier to potential competitors. Nor does it control physical or legal access to distribution wires. If it fails to perform as promised, the disappointed consumer does not lose money and can easily turn to another supplier because there are no significant barriers to setting up a competing business. The third is that governments do not normally issue licenses (apart from a general business license) for the right to provide the commercial service of charging batteries and mobile phones. So why should the new entrant be required to have such a license if its competitors do not?

Some might argue that even if there is no rationale for the government to issue a license to protect consumers, the government may still have a need to know which villages are receiving this type of service when planning for grid expansion. So the regulator might impose a requirement that the enterprise register its business for information purposes. But *registration is different from licensing*. Registration is a form of self-reporting that does not require the approval of the regulator. Instead, registration allows the regulator and other government agencies to know that the enterprise is out there and is providing a type of electrical or energy supply service.

Safety Regulation

Battery boxes and battery-operated appliances usually operate at low voltages, so they present a much lower risk of electrical shock than the 120 or 230 volts used in typical household wiring. But the batteries still have some safety risks: short circuits of battery terminals can release currents of hundreds or thousands of amperes; melting metal conductors can create the risk of burns; and leakage of electrolytic or toxic metals can be harmful to humans. In Europe and the United States, household electrical appliances would normally be tested by a government agency responsible for consumer product safety rather than an electricity regulator. Since such agencies do not exist in many African countries, we would recommend that one of the international appliance-testing laboratories such as the Underwriters Laboratories (UL) test battery boxes for safety.

In addition, the low- or medium-voltage line that runs from the hybrid generator to the site of the mobile-telephone tower raises potential safety concerns. Hence, it seems reasonable that this part of the entity's operation should be subject to whatever regulations (for example, height of the poles, need for fencing, and so on) the electricity regulator or another government agency has established for the safe operation of generators and low-voltage lines.

In summary, our recommendation for this new type of enterprise for Africa is:

- No tariff regulation
- Registration of the business for information purposes but no requirement for regulatory approval of a license or permit
- Certification of the safety of the battery boxes by a credible international safety-testing laboratory as well as application of safety regulations that would apply to all other electricity suppliers that operate generators and medium- and low-voltage electrical facilities

---

Key Recommendation

Regulators have a tendency to overregulate. When confronted with an enterprise that proposes a new technology or business model, the threshold question for every regulator is: should this activity be regulated, deregulated, or regulated in a different way? If the enterprise does not have a monopoly power because its customers have substitutes, then tariff regulation may be harmful. But even if tariffs are deregulated, there may still be a need for requiring registration (but not licensing) and safety regulation.

Who Should Regulate SPPs and Mini-Grids?

In this guide, for ease of exposition, we assume that most regulatory decisions relating to SPPs are performed by a separate, designated national electricity regulatory entity. Separate electricity or energy regulators now exist in more than 20 Sub-Saharan African countries (Camos and others 2008). These entities were created in the expectation that separate and independent regulatory bodies could make better (that is, technically more competent and less politicized) decisions than a government ministry. *But regulation of SPPs need not always be performed by a separate national electricity regulatory entity.* For example, in Sri Lanka and Thailand—two countries that have had considerable success in promoting both

From the Bottom Up • http://dx.doi.org/10.1596/978-1-4648-0093-1

grid and off-grid SPPs—the initial regulatory systems for SPPs were developed by government ministries or the cabinet because no separate electricity regulator existed at the time the decisions needed to be made. It is generally agreed that both regulatory systems have been quite successful in creating effective SPP programs. Now that separate new electricity regulatory entities have been created in both countries, some future SPP regulatory decisions will be made by the regulators rather than by ministries. It is too early to say whether the transfer of SPP regulatory decisions to the new national regulators will continue to produce good outcomes.

Regulation by Rural Electrification Agencies

Even when a national electricity regulator does exist, it may be more efficient for the regulator to delegate to other bodies, either formally or informally, regulatory decisions affecting SPPs.[4] For example, more than 15 rural electrification agencies (REAs) have been created in Africa in the past few years (Mostert 2008). Almost all of them give grants to SPPs to lower their initial capital costs. In processing the applications for these grants, the REAs perform a review of the SPP's business plan, which usually involves a detailed review of projected costs and revenues. *The reality is that the "business plan" review of an REA is very similar to a traditional "cost of service" review that a regulator would undertake.* The purpose of the REA business plan review is to ensure that the SPP's revenues are high enough so that the SPP will be financially viable but not so high as to allow the SPP to earn monopoly profits at the expense of its customers. In addition, most REAs have a legal mandate to maximize the number of new households that will receive electricity. Hence, the REAs have to take a close look at the affordability of the tariffs that the SPPs propose to charge. It is clearly not in the interest of REAs to give a grant to an entity that will not be commercially sustainable. Therefore, most REAs are already acting like quasi-regulators when they balance commercial viability against the affordability of the electricity service that will be provided by those who receive grants.

Given the fact that an REA has the same basic concerns as a regulator—ensuring the commercial viability of the SPP, protecting the SPP's customers from monopoly profits, and establishing minimum quality-of-service standards—it will often be more efficient (by eliminating duplication) for the national electricity regulator to formally delegate, if legally permissible, the setting of tariffs and quality-of-service standards to the REA. Or if the national law does not permit formal delegation, the regulator could state that it will give considerable weight to the business plan reviews conducted by REAs when it issues licenses or permits or sets maximum tariffs and minimum quality-of-service standards.[5] Under either approach, the regulator will have to reserve the right to review the REA's regulatory decisions and make adjustments if the regulator finds that the REA is doing an inadequate job.

Another approach is to legally assign or transfer to the REA all regulatory responsibilities over SPPs that operate on isolated mini-grids (Cases 1 and 2). In fact, this is the regulatory approach that has been taken in Guinea and Mali.

In both countries, the REAs—AMADER (Agence Malienne pour le Développement de l'Energie Domestique et de l'Electrification Rurale, Mali's rural energy agency) in Mali and BERD (Bureau d'Électrification Rurale Décentralisée, Bureau for Decentralized Rural Electrification) in Guinea—are legally responsible for traditional regulatory responsibilities (that is, price setting and minimum quality-of-service standards) for all off-grid SPPs. This approach is "cleaner" and presumably easier to implement than the indirect approach of assigning regulatory responsibilities over isolated mini-grids to the national electricity regulator and then encouraging the national regulator to delegate or give considerable deference to the decisions of the national REA.

But once the isolated mini-grid is connected to the national grid (see chapter 10), this approach—dividing regulatory responsibilities between the national regulator and the REA—will become much harder to manage. Therefore, when the "big grid" connects to the "little grid," a regulatory handoff would make sense. This is effectively a hybrid regulatory arrangement: it explicitly assigns regulatory responsibilities over off-grid SPPs to the country's REA for the first 5–7 years of the SPP's existence, or until the isolated SPP connects to the main grid, or whichever comes first. Once any of these events occur, regulation of the SPP is handed over to the national regulator.

If an REA is directly or indirectly assigned regulatory responsibilities over SPPs that operate an isolated mini-grid, how should it exercise its regulatory responsibilities? We think that it would be relatively easy for an REA to perform regulatory functions over SPPs if it does three things. First, the REA should explicitly require information about proposed tariff levels and structures in the business plans required for grant applications. Mali's REA, AMADER, has imposed this requirement and uses this information to set allowed tariffs and overall revenue levels. Second, the REA should explicitly incorporate its regulatory decisions in the formal grant agreements that it signs with SPPs. Third, the elements of the grant agreement should be publicized in the villages that will benefit from the proposed electrification.

Traditionally, such grant agreements are like contracts: they specify rights and obligations of the grant recipient as the *quid pro quo* for receiving a capital cost grant.[6] The AMADER grant documents include provisions that specify maximum tariffs and minimum construction and service standards. If a country has no REA, these same provisions would normally be included in licenses and permits issued by the national electricity regulator. Hence, in countries where there is a functioning REA and a national electricity regulator, such provisions could be included in both the REA's grant agreement with the SPP and by reference in any license or permit issued by the regulator to the SPP.

This would be, in effect, a form of "regulation by contract," in which the regulatory contract is embedded in the REA grant agreement (Bakovic, Tenenbaum, and Woolf 2003, 13–14). This is not a radical proposal. The United States, Costa Rica, and Bangladesh have all adopted the approach of assigning explicit regulatory responsibilities over rural electrification cooperatives to grant-providing agencies (Barnes 2007). In these three countries, the agency that gave grants or soft loans was a rural cooperative agency, but we see no reason why the same approach could not be used by an REA giving grants to any entity, regardless of whether it is a cooperative, a community organization, or a private operator.

Key Recommendation

Rural electrification agencies (REAs) typically require the submission of a business plan in deciding whether to give grants to SPPs. The business plan review that the REA undertakes is very similar to a traditional "cost of service" review that would be performed by a regulator. Therefore, the regulator should consider formally or informally delegating regulatory responsibilities to REAs, especially for SPPs operating on isolated mini-grids. The details of the price and quality-of-service regulation can be incorporated in the REA grant-giving document. But the regulator should always reserve the right to take back regulatory responsibilities if it finds that the REA is doing an inadequate job.

Community Regulation of Privately Owned SPPs

Another regulatory option is to let communities perform regulatory functions. This is especially relevant in the case of a private operator who proposes to build, own, and operate an isolated mini-grid that will sell to households and businesses in one or more isolated villages (Case 1: isolated SPP that sells at retail). Most private operators clearly recognize that they will be able to construct and operate mini-grid systems only if there is "buy-in" or acceptance from the villages that will be served. While the private operators may be legally required to get a license or permit from the national regulator, the document will be of little or no value unless the local government and villagers also support the project.

Village-level support can be given formally or informally. One formal option, used successfully in Cambodia, would encourage a private operator to sign an electricity service agreement with designated representatives of the village (that is, a village electricity committee or a local governmental unit). An example of one such contract in Cambodia is a 15-year electricity service agreement specifying the rights and responsibilities of the village (Smau Khney) and the private operator (Mahé and Chanthan 2005, annex 2).[7] Like a grant agreement between an REA and a mini-grid operator, such an agreement can also specify traditional regulatory parameters such as maximum prices and hours of required service.[8] In this way, it can serve as a form of "regulation by contract" at the local level. Since most villages are not likely to have the knowledge or expertise to develop such contracts on their own, the regulator or REA could develop a model version of such a contract. In Cambodia the electricity service agreement between the private operator and the village electricity committee also required that the private operator provide a small annual budget of about $200 to support the operation of the committee.

Those who argue in favor of such an agreement maintain that it is critical to get explicit buy-in from any village that will be served. Without this, villagers will likely complain that they were not consulted and that they had little or no input in deciding the key terms and conditions of a government-granted, multiyear monopoly for electricity in their village. Such a complaint was heard in a recently completed survey of nine villages in Mali. In Badinko, a village served by a private operator who had received grants from AMADER (the REA), the members of the village electrification committee stated that they needed "more information regarding the terms of services, contracts, responsibilities of AMADER, private operator and users" (Rodriguez and Janik, forthcoming). The advantage of an electricity service agreement between the village and the potential private operator is that these issues could be discussed and agreed upon in advance.

But from a developer's perspective, a disadvantage is that *requiring* a formal signed agreement creates yet another hurdle for potential private operators. Developers will argue that villagers will make impossible demands, such as 24-hour service at the same prices charged by the national utility in the capital. Developers have also expressed concerns that village electrification committees may experience delays in forming, set decision-making standards that are unworkable (for example, requiring unanimous approval), and get easily side-tracked by disputes that may have nothing to do with the merits of the mini-grid project.

Our recommendation is a compromise solution with two key elements: first, the regulator or REA would make available a model service agreement between a developer and a village; and second, the developer would provide evidence that the agreement was publicized in the village. If the developer is receiving grant money, this evidence would need to be shown to the REA; if applying for a license, the evidence would be shown to the regulator. We would not, however, require that there be a signed agreement in all situations because that could lead to major delays. If the villagers believe that the agreement is not fair, they would be given an opportunity to explain why to the regulator if a license is awarded or to the REA if a grant agreement is signed.

If the village and the private operator do sign a model electricity service agreement, the regulator's role (whether performed by an REA or a national electricity regulator) could be reduced or eliminated. Specifically, the regulator's role could be limited to incorporating the electricity service agreement by reference in any license or permit that the regulator is required to issue or if there is a grant agreement between the REA and developer. And if this is done, the regulator's or REA's role could be limited to serving as a mediator or arbiter of disputes over implementation of the supply agreement.

This type of decentralized regulatory arrangement has three potential advantages:

- First, the village government is likely to feel it has *more ownership* because it directly negotiated the contract with the private developer. Consequently, the village government or electricity committee will feel responsible for ensuring that the developer complies with the terms and conditions of the supply contract. This is quite different from relying on a national regulator or REA in a distant capital to administer a piece of paper called a "license" or "grant agreement," the terms and conditions of which may be unknown to the village and largely beyond its control. And if no license is required, as is the case in Tanzania if the plant is 1 MW or less, then this agreement between the village and developer would be a village-level substitute for such a license.

- Second, the village electricity committee can *assist in monitoring* compliance with quality-of-service standards if established in a license or grant agreement. It is easy for a national regulator to issue quality-of-service standards for decentralized energy service providers. But it is often difficult and expensive for a national regulator to monitor whether the providers in distant and isolated villages are actually complying with the standards. If a village electricity committee or government is actively involved, it can act as the regulator's "eyes and ears" at the local level.
- Third, it reduces the likelihood of *corruption*. Corruption is less likely to occur when regulation is openly shared with a village governmental body. The overriding incentive for the village is to get results: a reliable electricity supply for new and existing customers. If there are delays or requests for bribes, this will raise the cost of electricity to the village. Therefore, the village committee has a strong incentive to take timely action in a way that a national regulator would not.

Key Recommendation

For new isolated mini-grids, the regulator should encourage informed involvement by the local community. This does not require formal delegation of regulatory responsibilities to a community body. A compromise solution would have two elements: first, the regulator or REA would make available a model service agreement between the developer and the village; and second, the developer must show that the agreement was publicized and discussed in the village as a condition for receiving grant money or a license or permit.

Self-Regulation by Community Organizations or Local Governments

Some isolated mini-grid systems may be owned by community organizations or local governments. In such situations, it is common for the local government to be granted the explicit legal authority to regulate its own SPP. In effect, the mini-grid is allowed to self-regulate. It sets its own tariffs and is not required to go to the regulator or any other national government agency for approval of these tariffs. For example, in Peru, the national rural electrification regulations explicitly exempt small, municipally owned SPPs from any tariff regulation by OSINERGMIN (Organismo Supervisor de la Inversión en Energía y Minería, the country's state energy and mining investment regulator).

The rationale for self-regulation is that the SPP's owners are also its customers. Hence, there is no incentive for the owners to charge high monopoly prices. To do so would simply mean transferring money from one pocket to another. But the problem with self-regulation by communities, cooperatives, and local governments is that they often set prices too low rather than too high. Without some outside pressure or control, there is a high likelihood that a community- or local-government-owned, isolated SPP will charge prices at a level that does not cover operating costs and rarely covers depreciation for future equipment replacement.[9] If a community-owned SPP is to be sustainable, it must have an incentive to charge prices high enough to operate as a commercially viable entity without creating an elaborate and costly regulatory system. In chapter 9, we examine what these incentives, whether positive or negative, might be.

Key Observation

Several different entities can regulate SPPs in support of or as a substitute for a national utility regulator. These include REAs, communities or community organizations, and local governments. Such regulation is more likely to be effective if there is a binding contract between the SPP and the regulating entity (that is, regulation by contract). But self-regulation of a community-owned SPP tends to lead to prices that are set too low without countervailing pressures or constraints (see chapter 9).

Notes

1. For example, Section 23 (3) of Tanzania's Electricity Law states that the "authority may prescribe maximum tariffs of a generic nature or simplified tariff methodologies, applicable to licensees or persons exempted under section 18." Section 18 refers to generators with less than 1 MW of installed capacity at one site in rural areas or distribution entities serving off-grid systems with less than 1 MW of maximum (that is, peak) demand. Section 26 of Rwanda's 2011 Electricity Law states that: "With regard to rural electrification license, the regulatory agency shall establish a simplified license in order to expedite licensing for rural electrification projects. Such licenses shall be granted to those operating in rural areas."

2. See Breyer (1982, 15) and Kahn (1988, 11–12). Kahn's two-volume *Economics of Regulation* is widely recognized as the seminal modern work on regulatory theory and practice. While Kahn is acknowledged as the leading U.S. scholar on regulation, he was more than just a renowned academic. He also served as a practicing regulator of electricity in New York and as a deregulator of airline travel for the entire United States.

3. Economists focus on efficiency, and politicians focus on fairness. For an economist, the harm caused by a monopolist is that it will charge prices that are not socially optimal. For a politician, the economist's concept of optimal and nonoptimal pricing is too abstract. Instead, politicians are concerned about what will affect votes for them in the next election. And in most developing countries, the hot-button issue raised by households (and voters) in isolated villages served by a mini-grid operator is that they are paying higher prices for electricity than households served by the national utility.

It is these price differentials rather than profit levels that get the attention of local members of parliament.

4. For two surveys that reach different conclusions on the effectiveness of REAs and rural electrification funds (REFs), see Mostert (2008) and Matly (2010).

5. The presumption in many legal systems is that a government entity that has been assigned a responsibility does not have the legal right to formally reassign this responsibility to any other entity. In common law systems, this legal doctrine is referred to as *delegatus non potest delegare* (that is, what is delegated by the legislature cannot be redelegated to another entity unless the law specifically allows for redelegation).

6. For an example of a well-designed combined grant and concession agreement between an REA and a private mini-grid operator, see AMADER (n.d.), "Concession Contract" and "Specifications Annexed to Concession Order."

7. The electricity service agreement in Cambodia specified technical engineering requirements (for example, grounding, types of poles, distance between poles and cables), location of meters, responsibility for meters that were intentionally broken or tampered with, number of new customers to be connected within a specified period of time, duration of service on weekdays and weekends, subsidized tariffs for poor households, a local system for handling complaints and funding of subsidies for poor customers, and the administrative expenses of the village electricity committee.

8. In fact, prices were not specified in this agreement because it was thought that this would infringe on the legal responsibilities of the Electricity Authority of Cambodia (EAC), the national electricity regulator.

9. This is explicitly recognized in Nepal's Community Electricity Distribution Bylaws. Section 10a mandates that electricity distribution entities (that is, small power distributors, SPDs) set aside 10 percent of their monthly sales revenues for repairs and maintenance (NEA 2003).

References

AMADER (Agence Malienne pour le Développement de l'Energie Domestique et de l'Electrification Rurale, Malian Agency for the Development of Household Energy and Rural Electrification). n.d. "Concession Contract." Unofficial English translation. http://ppp.worldbank.org/public-private-partnership.

———. n.d. "Specifications Annexed to Concession Order." Unofficial English translation. http://ppp.worldbank.org/public-private-partnership.

Bakovic, Tonci, Bernard Tenenbaum, and Fiona Woolf. 2003. "Regulation by Contract: A New Way to Privatize Electricity Distribution?" Working Paper 14, World Bank, Washington, DC. http://rru.worldbank.org/Documents/PapersLinks/2552.pdf.

Barnes, Douglas F. 2007. *The Challenge of Rural Electrification: Strategies for Developing Countries.* Washington, DC: Resources for the Future Press.

Breyer, Stephen G. 1982. *Regulation and Its Reform.* Cambridge, MA: Harvard University Press. http://site.ebrary.com/id/10313884.

Brown, Ashley C., Jon Stern, Bernard Tenenbaum, and Defne Gencer. 2006. *Handbook for Evaluating Infrastructure Regulatory Systems.* Washington, DC: World Bank.

Camos, Daniel, Maria Shkaratan, Fatimata Ouedraogo, Cecilia Briceño-Garmendia, Vivien Foster, and Anton Eberhard. 2008. "Underpowered: The State of the Power Sector in Sub-Saharan Africa." Background Paper, Africa Infrastructure Country

Diagnostic, World Bank, Washington, DC. https://openknowledge.worldbank.org /handle/10986/7833.

Kahn, Alfred E. 1988. *The Economics of Regulation: Principles and Institutions.* Vol. 1. Cambridge, MA: MIT Press.

Mahé, Jean Pierre, and Ky Chanthan. 2005. *Rehabilitation of a Rural Electricity System.* GRET and Kosan Engineering, Phnom Penh, Cambodia. http://www.gret.org/wp -content/uploads/07409.pdf.

Matly, Michael. 2010. "Best Practice of Rural Electrification Funds in Africa." Review Paper, ICTS-NTUA and SOFRECO, Clichy, France.

Mostert, Wolfgang. 2008. *Review of Experiences with Rural Electrification Agencies: Lessons for Africa.* Draft Report, European Union Energy Initiative—Partnership Dialogue Facility. http://www.mostert.dk/pdf/Experiences%20with%20Rural%20Electrification %20Agencies.pdf.

NEA (Nepal Electricity Authority). 2003. "Nepal Electricity Authority Community Electricity Distribution Bye Laws, 2060." http://www.nea.org.np/images/supportive _docs/Community%20Electricity%20Distribution%20Bylaw.pdf.

Rodriguez, Sebastian, and Vanessa Lopes Janik. Forthcoming. "Case Studies on Gender and Electrification from Mali." World Bank, Washington, DC.

CHAPTER 4

Regulatory Processes and Approvals: Who Approves What, When, and How?

It's not like baking a cake where you follow a recipe. No. We are all different. But we can take certain things, certain key lessons, and apply those lessons and see how they work in our environment.

—Former trade minister of Colombia commenting on regulation (World Bank 2009, viii)

Where regulation is burdensome, success tends to depend on whom you know rather than on what you can do.

—World Bank (2009, vii)

By action and inaction, choices are made.

—Robert Samuelson, *Washington Post*, December 2011

Abstract

In this chapter, we examine the regulatory approvals that a small power producer (SPP) needs to begin operations, and the processes by which these approvals are obtained. We compare Sri Lanka's current approvals and processes with those used or proposed in Tanzania and Kenya. The focus will be on approvals and processes for SPPs that wish to connect to the national or a regional grid and sell to the national utility or some other buyer that is legally obligated to purchase electricity from the SPP (Case 4 in chapter 2: a grid-connected SPP that sells at wholesale to a utility). We recommend six key characteristics of a workable and efficient regulatory system for approving applications for this type of SPP.

Regulatory processes and approvals are the hidden underside of regulation. The regulatory processes that affect small power producers get much less attention than the more visible substantive decisions, such as the level and structure of feed-in tariffs (FITs) or the specific provisions of standardized power-purchase agreements (PPAs). But the reality is that regulatory processes—who approves what, when, and how—are just as important to a potential SPP as the regulator's substantive economic and technical decisions. Even if a regulator's proposed economic and technical rules are acceptable to all parties, an SPP may never operate if the procedures for obtaining regulatory approvals are unclear, take too long (because there are too many steps or the sequence of steps is unclear), or cost too much money. SPPs are vulnerable to both the direct (for example, application fees) and indirect (for example, time) costs of regulation, because most SPPs operate on the "razor's edge" of financial viability. Therefore, it is important to consider how to create an efficient and effective approval process.

Sub-Saharan Africa has a bad reputation for its regulation of businesses. In the World Bank's annual worldwide survey of general business regulation, Sub-Saharan African countries have consistently ranked near the bottom for ease of getting regulatory approvals. In 2009 the average ranking of 38 surveyed Sub-Saharan countries was 138 out of 181 (World Bank 2008, 1). When compared to other developing countries, the existing regulatory procedures for starting and operating a small business in most African countries involve too many steps and take too long. If Africa's poor regulatory practices are repeated in its regulation of SPPs, the SPP programs, no matter how well intentioned, will fail.

Key Observation

Regulatory processes—who approves what, when, and how—are just as important to a potential SPP as the regulator's substantive economic and technical decisions.

The Key Approvals Required: Electricity Sector–Specific versus General Approvals

Government approvals for SPPs fall into two general categories: approvals that are specific to the electricity sector and approvals required for any new business enterprise (that is, non-sector-specific approvals). Among the electricity sector approvals, the approval that usually gets the most attention, apart from the approval of tariffs, is the license or permit approval. A license can be thought of as an admission ticket. It is the regulator's formal permission to conduct a specified electricity business activity such as generation, transmission, distribution, supply, or combinations of these activities. When a regulator grants a license, the license specifies rights and responsibilities for the recipient. In the case of SPPs, the license will typically give an SPP the right to install a generating plant of a specified maximum size at a particular location and the right to sell electricity to specified retail and wholesale customers for a specified period of time under one or more tariffs that will need to be approved by the regulator.

From the Bottom Up • http://dx.doi.org/10.1596/978-1-4648-0093-1

If a license or permit were the only approval required of SPPs, the start-up process for SPPs would be relatively simple and the up-front transaction costs would be minimal. But, in fact, SPPs must also obtain numerous other approvals from many government entities (both national and local) before they can start operations. Nonsector or general business approvals vary from country to country, but the core set of such approvals usually includes the following:

Right to Operate a Business[1]
- Registering as a business
- Obtaining construction or building permits
- Registering the property
- Obtaining a license to operate a factory
- Registering as an entity that will pay taxes
- Registering for tax concessions (for example, exemption from import duties or a five-year holiday from paying profit taxes)

Land and Natural Resource Rights
- Proof of ownership or usage rights to land
- Approval of the right to use a specified amount of water or other natural resource at a particular location

Environmental Approvals
- Completion of an environmental review at whatever level is specified by the national environmental agency
- Review and approval by the river and/or irrigation authority
- Statements from the relevant government agency that the project is not in a protected area (for example, a national park, a wildlife preserve, a protected coastal region, or a protected cultural or archeological area)

In addition to these business and environmental approvals, most SPPs will need an approval from one or more local governmental authorities for the right to conduct business in that locality.

Key Observation

Governments usually require SPPs to obtain both sector and nonsector approvals before starting their operations. These approvals include (but are not limited to) a license to conduct a certain type of electricity business, the general right to operate as a business, land and natural resource rights, environmental approvals, and permission from local authorities to conduct business at a particular locality.

In this chapter, we will address several questions:

- How should the electricity-sector-specific approvals be coordinated with general business and environmental approvals? Should they be done in parallel or sequentially? If performed sequentially, what should be the sequence of approvals?
- How should the electricity-sector-specific approval process be structured if several separate government agencies have legal responsibilities for the sector (for example, a ministry of energy, a national electricity regulator, and a renewable energy agency)?
- In addition to government and regulatory approvals, what decisions should be made by the grid operator and buyer of SPP power (who are often the same entity)?

A Successful Example: The Regulatory Process in Sri Lanka

To make this discussion less abstract, we focus on the current regulatory processes for obtaining approvals to build and operate a grid-connected SPP in Sri Lanka.[2] Why Sri Lanka? We think that Sri Lanka merits closer attention because the country has a good track record in approving SPPs, the regulatory approval process has been reformed several times, and the current process that was initiated in April 2008 is transparent and well documented. Since the inception of Sri Lanka's SPP program in 1996, it has had considerable success on the ground, as measured by SPPs that are actually producing electricity. By the end of 2011, 102 SPPs (each less than 10 megawatts, MW) owned and operated by the private sector, with a total capacity of 243 MW, had received all the necessary sector and nonsector approvals and were selling electricity to the national utility, the Ceylon Electricity Board (CEB). Of the 102 operating SPPs, 92 were small hydro plants, 3 were wind, and 1 was a biomass plant using rice husk. The remaining six plants included one SPP operating on waste heat, another one using sustainably grown biomass, and four solar photovoltaic (PV) SPPs.

The pre-2008 system for SPP approval (that is, the old system) in Sri Lanka was a loose, utility-operated review system with very little direct involvement by government entities. Its first step was the issuance of a letter of intent (LOI) by the CEB. The three widely recognized shortcomings of the old system were that (a) the CEB was the *de facto* decision maker for allocating the country's renewable energy resources (even though it had no legal authority to do so), (b) extensions to the end date of LOIs were granted by the CEB even to projects clearly making very little progress toward construction and commercial operation, and (c) the CEB had little interest and ability to push agencies outside the electricity sector to complete their reviews and make a decision.

In 2007 the government, under pressure from SPP developers, decided that there was need for a new government agency to take a more active role in promoting renewable energy by SPPs and larger producers, and that the existing review and approval process had to be rationalized to produce decisions in less time. Given the very significant changes made in the new SPP review and approval system, it is worth taking a closer look at it to see if there are any lessons to be learned for other countries. *Our focus will be on the current regulatory approval process for small, main-grid-connected hydro projects.* This type of SPP is especially relevant for a number of East African countries (Tanzania, Uganda, Kenya, and Rwanda) that have numerous sites on which small hydro facilities could be developed.

Overview of the Approval Process

In Sri Lanka the Sustainable Energy Authority (SEA) rather than the Public Utilities Commission of Sri Lanka (PUCSL) (the electricity regulator) takes the lead in conducting the sector-specific review and approval process for SPPs. The SEA review and approval process has three major steps: (a) resource verification, (b) provisional approval, and (c) an energy permit. The key features of the second and third steps are summarized in figure 4.1.

The first step, *resource verification*, is not a formal approval step. Instead, it is limited to the SEA staff telling the developer if another entity has already filed an application or set up an operating facility at the proposed location. It can be thought of as a limited initial screening to make sure that the proposed site is potentially available. In contrast, the next two steps, the issuance of a *provisional approval* and an *energy permit*, are formal steps performed by the SEA with assistance from other government entities. The PUCSL is required to undertake a separate review in deciding whether to issue a generating license to the SPP. Since 2011 the PUCSL review and license decision is made after the SEA's award of the provisional approval but before its issuance of an energy permit.

From a legal perspective, Sri Lanka has a dual-authorization system for proposed mini-hydro generators: an approval for the exploitation of a natural resource (the energy permit) and an approval for the right to generate and sell electricity (the generation license). Within this dual-authorization system, the energy permit is recognized as the key approval. Hence, in contrast to Tanzania, the Sri Lankan electricity regulator plays a subsidiary role to the renewable energy agency in deciding if a proposed SPP will be allowed to come into existence.[3]

Step 1: Resource Verification

This step is encouraged but not required. SPP developers are encouraged to consult the SEA staff to see if another project has already applied to use the same proposed location. When the inquiry is made, the SEA staff checks the database of projects that have received either provisional approvals or energy permits to determine whether another developer has already applied for this site. If the site has not been sought by another developer, the developer is urged to submit a formal application for a provisional approval within three months of making this

Figure 4.1 Energy Permit Application Process in Sri Lanka

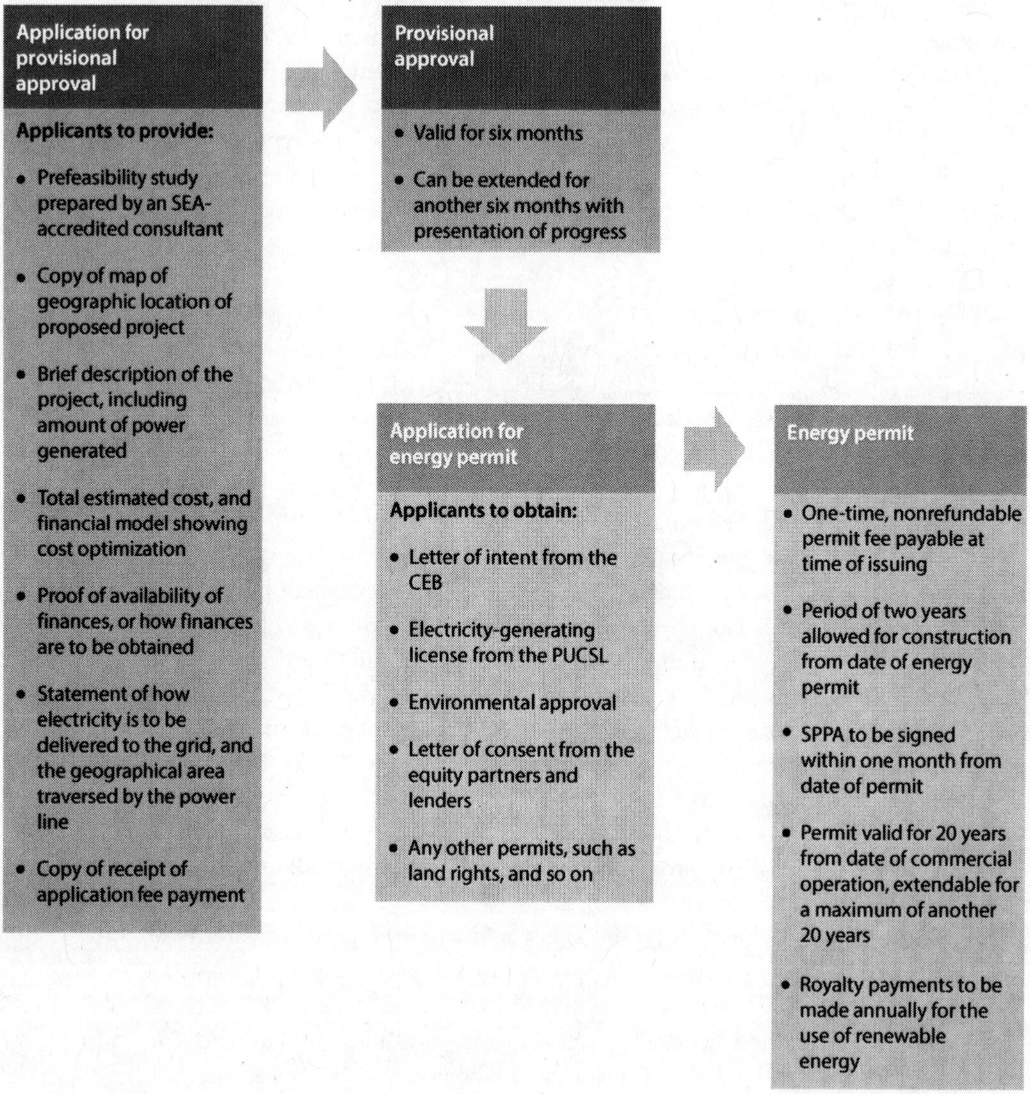

Application for provisional approval	Provisional approval
Applicants to provide:	• Valid for six months
• Prefeasibility study prepared by an SEA-accredited consultant	• Can be extended for another six months with presentation of progress
• Copy of map of geographic location of proposed project	
• Brief description of the project, including amount of power generated	
• Total estimated cost, and financial model showing cost optimization	
• Proof of availability of finances, or how finances are to be obtained	
• Statement of how electricity is to be delivered to the grid, and the geographical area traversed by the power line	
• Copy of receipt of application fee payment	

Application for energy permit	Energy permit
Applicants to obtain:	• One-time, nonrefundable permit fee payable at time of issuing
• Letter of intent from the CEB	• Period of two years allowed for construction from date of energy permit
• Electricity-generating license from the PUCSL	• SPPA to be signed within one month from date of permit
• Environmental approval	• Permit valid for 20 years from date of commercial operation, extendable for a maximum of another 20 years
• Letter of consent from the equity partners and lenders	
• Any other permits, such as land rights, and so on	• Royalty payments to be made annually for the use of renewable energy

Source: Adapted from the Sri Lankan SEA 2011.
Note: CEB = Ceylon Electricity Board; PUCSL = Public Utilities Commission of Sri Lanka; SEA = Sustainable Energy Authority; SPPA = standardized power-purchase agreement.

initial inquiry. Encouraging this cost-free preapplication inquiry avoids duplicate applications for the same site, which would waste the time and resources of the developer, the SEA staff, and other government agencies.

Step 2: Provisional Approval
Project Approval Committee
An SPP must apply to the SEA for a provisional approval. Based on the SEA staff review, the director-general makes a recommendation to a Project Approval

Committee (PAC), which is the final decision-making authority to accept or reject an application within eight weeks of the application being filed. Membership in the PAC is specified by law and comprises 10 members representing the heads of various government entities or their designees. In addition to these members, the district secretary, the highest-level civil servant in the district where the SPP will be located, is also invited to the PAC meeting. The PAC meets in the capital, Colombo, every month to consider several applications for various sites for provisional approval.

Eligibility

Sri Lanka's Sustainable Energy Law provides that any entity—whether an individual, a company, or a cooperative—can apply for a provisional approval at any time. The SEA encourages the application to be made in the name of a "special purpose company," because it is relatively easy to change the ownership arrangements that underlie a special purpose company. A special purpose company stays the same but the owners of the company may change over time, which minimizes the need to obtain new approvals every time there is a change in ownership. Also, under Sri Lanka's investment regulations only companies are eligible for tax concessions, such as a customs duty waiver for all power plant equipment and a tax holiday on profits for five years.

Fees

The application fee for a provisional approval is keyed to the size of the project. A proposed 1 MW plant would pay an application fee of SL Rs 100,000 ($757) and the fee for each additional megawatt is SL Rs 50,000 ($378). Hence, a 10 MW SPP's application fee would be SL Rs 550,000 ($4,166). Applications are processed on a first-come, first-served basis. To avoid any dispute over who applied first, applicants are issued an electronic token at the SEA office when they hand in their application.

Prefeasibility Study

To discourage purely speculative applications that tie down a desirable site but never get developed, the SEA's rules "raise the bar" by requiring that the application be accompanied by a prefeasibility study performed by a consultant who has been accredited by the SEA. To ensure that the prefeasibility study is serious, the SEA provides checklists of the information that must be provided. The SEA's checklists vary by technology: for example, small hydro, biomass-grown, wind, waste (agricultural, industrial, and municipal), and waste heat. If the prefeasibility study is incomplete, the application will be rejected by the SEA and the applicant will lose its place in the queue until it returns with a complete application. At present about 10 individuals or consultants have been accredited to perform prefeasibility studies for a mini-hydro SPP, which currently costs around $3,000.

What is reviewed? In reviewing the application for a provisional approval, the PAC looks for evidence that the proposed SPP would not infringe on the rights of other operating SPPs or SPPs that have been granted provisional approvals or an energy permit. In addition, the PAC determines whether the proposed project would interfere with an existing or planned project of the CEB and whether it has a reasonable likelihood of getting other regulatory approvals (for example, an environmental approval). The representatives from the other agencies that are represented on the PAC help the SEA to make an early and informed evaluation. For example, if a developer proposes to build his facility in a wildlife reserve, it is almost certain that the application would be rejected. The applicant does not need to have these other approvals in hand when applying for a provisional approval but he will need all the approvals for an energy permit. In addition, the SEA's rules state that an applicant may apply "irrespective of whether the person holds any rights to the resource or land rights."

Temporary exclusivity. If the PAC grants a provisional approval, the approval gives the applicant the exclusive right to develop the specified site for a six-month period with a possible six-month extension. The SEA's letter granting provisional approval also notifies the developer of the documents and approvals that must be completed or obtained to apply for an energy permit. During the 6- or 12-month period, the applicant must file quarterly progress reports with the SEA using a prespecified reporting format.

Reapplications
If the applicant fails to apply for an energy permit at the end of one year (assuming that he has received a six-month extension), the law requires that the provisional approval be cancelled. If the applicant wants to continue to pursue the project, it must reapply for a new provisional approval. The SEA's decision on whether to grant a new provisional approval requires that the applicant show that it has made significant progress in project preparation. The SEA's system for determining significant progress is based on a transparent marking scheme, wherein an applicant must achieve a score of at least 40 out of 100 points (SEA 2011, 7).

Exclusivity: Rationale, Rights, and Responsibilities
In Sri Lanka when the PAC grants a provisional approval, the applicant has the exclusive right to develop a particular SPP project at a specified location for up to one year with the right to reapply for an extension at the end of one year. Both Kenya and Tanzania offer much longer periods of exclusivity. In Kenya consultants to the Ministry of Energy have proposed that an applicant who receives approval for its expression of interest (which is essentially an application for a provisional approval) by a committee convened by the Ministry of Energy is "entitled to a three (3) year exclusivity period to further

assess and develop the project" with the possibility of an additional three-month extension at the end of the three years.[4] In Tanzania provisional licenses awarded by the electricity regulator (Energy and Water Utilities Regulatory Authority, EWURA) are valid for three years.[5] To the best of our knowledge, the longest period of exclusivity is given in Nepal. If an SPP receives a survey license (which is akin to a provisional approval), the developer has five years before it need apply for final approval (Government of Nepal 1992, Preamble, Paragraph 5).

The rationale for granting provisional approvals (or similar approvals in other countries) is that no applicant is going to make the considerable investment of time and money to develop a project if there is a risk of someone else coming in and building the proposed project after the initial developer has completed all the preparatory work. Therefore, the provisional license is, in effect, a government granted "temporary monopoly" to create an incentive for developers to develop a particular project at a specified location. If the developer is successful in obtaining all the required approvals for the project, the temporary monopoly can be converted into a longer-term (for example, 20-year) monopoly with the award of the final approval: an energy permit and generation license (Sri Lanka), or a generation license (Kenya and Tanzania).

A government will only want to give a provisional approval to serious proposals that are likely to be developed in the near term as the *quid pro quo* for granting exclusive rights for a specified period of time. It is not in the national interest to "tie up" sites that are unlikely to be developed or whose development will be delayed. To achieve this outcome, most regulatory systems will contain both a carrot and a stick. The carrot is the period of temporary exclusivity; the stick is that the provisional approval can be taken away if certain milestones are not met during or at the end of the exclusivity period. For the stick to be credible, however, the developer must believe that the provisional approval will be taken away if he fails to perform.

Among the three countries analyzed here (Tanzania, Kenya, and Sri Lanka), Sri Lanka probably has the best regulatory system for eliciting serious applications for provisional approvals. Three features of the current Sri Lankan system lead to this outcome. First, obtaining a provisional approval requires more than just racing to the SEA's offices and filing numerous applications to get a place in the queue. An SPP must pay a nontrivial application fee and prepare a serious prefeasibility study costing several thousand dollars for the chance of getting a provisional approval. For each renewable energy technology, the SEA has developed a detailed checklist of information that must be included in the prefeasibility study. If the application and prefeasibility study are not complete, the application will be turned down. Second, Sri Lanka's Sustainable Energy Law mandates that the provisional approval is good for only one year and at the end of that year, the applicant must reapply for a new one. Third, if the provisional approval is removed after one year and the applicant reapplies for a new one, the objective criteria for obtaining a new one are specified in detail. The applicant has to achieve a minimum score on these criteria that measure its progress in moving the project forward, such as:

• Obtaining nonelectricity sector approvals and environmental clearances
• Undertaking project development actions (acquisition of land rights, feasibility studies, land surveys)
• Measuring installed capacity and electricity production
• Obtaining financing and technical support

In contrast, the process in Tanzania is less clear. The developer has three years and must file quarterly reports with the electricity regulator who has granted the provisional license, but it is uncertain as to what constitutes acceptable progress during this term. Kenya, like Tanzania, gives developers three years to bring an SPP project to operation once it receives approval for its expression of interest application from the Feed-In Tariff Committee convened and chaired by the Ministry of Energy. Kenya and Tanzania are thus similar in that they both require quarterly progress reports, but in Kenya it has been proposed that the Feed-In Tariff Committee evaluate the quarterly progress reports and assign point values for each achieved development task or approval. The project thus runs the risk of losing its provisional approval if it fails to show an improvement of at least 15 points between each quarterly report (ECA and Ramboll 2012). Since this proposal is yet to be officially adopted by the ministry, it is uncertain whether it will actually be implemented.

What Information Is Required and Reviewed in the Application for Provisional Approvals?

Table 4.1 is a checklist of the information that an applicant must submit to the SEA while applying for a provisional approval. *Note that the application must include a prefeasibility study.* The content of the study varies somewhat by the type of technology proposed for the SPP project. The SEA gives specific guidance on prefeasibility studies for five different renewable technologies or fuels (small hydro, wind, biomass-grown, waste [agricultural, industrial, or municipal], and waste heat).

Table 4.1 Information Required of an SPP Applying for Provisional Approval in Sri Lanka

1. Prefeasibility study prepared by an SEA-accredited consultant with a one-page summary
2. Copy of the map of the geographic location of the proposed project
3. Brief description of the project, including the amount of energy that is expected to be generated
4. Total estimated cost and financial model showing the optimization criteria adopted
5. Proof of availability of financing or how financing will be obtained
6. Statement of how electricity will be delivered to the national grid and geographic area to be traversed by the power line
7. Copy of the SEA receipt for the application fee

Source: SEA 2011, 4.
Note: SEA = Sustainable Energy Authority; SPP = small power producer.

As noted earlier, in this first-round screening, the PAC, convened by the SEA, examines only some of the information provided by the applicant. The principal focus of the review is whether the proposed project: (a) would interfere with the operation of an existing SPP project or a hydro project that is operating or planned by the CEB, (b) would infringe on the development rights of another SPP that has received a provisional approval or an energy permit, (c) is located in an environmentally sensitive area (for example, a wildlife preserve or a protected coastal area), or (d) would have access to available capacity in the distribution substation that would be needed to transmit the SPP's electricity to the CEB's main grid.

Sizing. In addition to these considerations, it appears that the SEA does some review of the proposed installed capacity and the likely energy output of the proposed hydro facility. As discussed later in box F.1 in appendix F, the existence of a resale market for energy permits has created an incentive for some developers to overstate the expected installed capacity and energy output of projects whose permits they wish to resell. There is some review of these estimates by the SEA staff. But it is unclear whether the SEA staff has enough information to detect anything other than gross exaggerations of the two parameters.

Financial and economic viability. Another key issue is how deeply the PAC should review project economics and to what extent it should do so at this first stage or at a later regulatory stage. There are two potential economic reviews. The first is to decide whether there will be sufficient equity and debt financing to supply the capital needed to fund the construction of the facility. The current application form for provisional approval states that the applicant must provide "proof of availability of adequate finances or the manner in which the required finances for the project are to be obtained" (SEA 2011, 5). Presumably, this requirement could be satisfied with a simple statement: "I will get 50 percent from Bank ABC, 20 percent from my brother, and the balance of 30 percent from Mr. X." So it is not clear what added value is obtained from a PAC review of such a general statement. Moreover, if the original developer is allowed to resell some or all of its ownership interests to other parties before construction begins (see below), the information provided will be inaccurate as soon as the ownership interests are sold. Given this low threshold of proof of financing (effectively nothing more than a "here is what I plan to do" statement) and the fact that ownership and financing options may change markedly during the development process, it is not obvious that the PAC's review of possible financing arrangements serves any useful regulatory purpose.

The second economic issue is whether the project itself will be economically viable. In other words, will revenues exceed costs (including financing costs)? If the original developer or future owners guess wrong on future project revenues and cost, they will be the only ones that are hurt by faulty estimates. Their faulty estimates will not lead to higher FITs or higher retail tariffs for the CEB's captive customers since neither is directly affected by an individual SPP's costs. And since any individual SPP project will be small relative to the CEB's supply needs, the CEB is unlikely to be hurt if an individual SPP commercially fails. So it is unclear if there is a public interest in having the PAC review the financial projections of a proposed SPP project.

Transmission availability. In the application for provisional approval, the SPP is required to provide information on how it proposes to deliver its electricity to the CEB and the geographic areas that will be traversed by any new proposed line. This allows the CEB representative to give a quick and preliminary opinion as to whether the proposed interconnection and delivery arrangement is feasible. This review is beneficial as it provides early feedback from the CEB as to whether the proposed interconnection is viable. If it is not viable, it provides an opportunity to look at alternative routings or possible reinforcements that would make the project viable. In Kenya it has been proposed that this feedback from the national utility be formalized in a preliminary, one-page Grid Connection Opinion that would be issued by the Kenya Power and Lighting Company (KPLC), the grid operator and buyer of the SPP's electricity, at the same time that the SPP receives an approval of its expression of interest application (which is the functional equivalent of an application for a provisional approval) (ECA and Ramboll 2012, 48). In Sri Lanka the CEB provides the PAC with a preliminary opinion on transmission availability but it does not appear that this information is automatically shared with the project developer.

Key Recommendation

A regulator or other government agency should grant a provisional license or similar document that provides a period of temporary exclusivity to potential SPP developers at a particular site. It should be clear that the provisional license will expire if the project fails to achieve one or more milestones within a specified time period.

Step 3: Energy Permits

The application for an energy permit is the final step in the SEA approval process. To apply for an energy permit, the applicant must demonstrate that it has obtained the following permits and approvals:

- Electricity generating license from the public utilities commission
- Letter of intent from the CEB (the single buyer) that it is willing and able to purchase electricity from the applicant
- Environmental approval
- Letters of consent from equity partners and lenders
- Any other required approvals or consents from institutions, agencies, and persons

In addition to these items, the applicant must also submit a comprehensive feasibility study prepared by one of the SEA-accredited consultants. This is a more comprehensive version of the prefeasibility study that was prepared to get a provisional approval. The cost of such a study for a 10 MW hydro plant is likely to be in the range of $75,000–125,000. An applicant who is awarded an energy permit must pay a one-time fee that is keyed to the generating capacity of the proposed facility. For plants under 10 MW in installed capacity, the permit fee is SL Rs 500,000/MW ($3,787/MW). A 10 MW plant would have to pay a permit fee of about $37,870.

Once an applicant receives the approval, the SEA refers to the applicant as a "developer." After receiving an SEA permit, the developer is required to obtain a signed standardized PPA (SPPA) with the CEB within one month. The SEA also requires that the plant become operational (that is, start commercial operations) within two years after receiving the energy permit. If the developer finds that it is unable to meet the commercial operation date, it can submit an explanation to the SEA as to why it needs more time. If the SEA's board of management finds merit in the developer's explanation, it may grant more time. If the developer can demonstrate significant progress over the two-year period and if there are no new conflicts between the project and other projects, the SEA may extend the construction period beyond two years.

When the SEA grants an energy permit, it gives the developer the right to use the designated natural resource for a 20-year period with the possibility of an additional 20-year extension. The PPA received from the CEB and the generation license obtained from the PUCSL both have 20-year terms. But the starting dates are usually slightly different so there are differences in the end dates of the three documents (the generation license, energy permit, and PPA). There are ongoing discussions on how to adjust the different end dates so that they will be the same for all three documents.

Should Resale of Provisional Approvals, Final Permits, or Licenses Be Allowed?

While there is no formal reporting system that documents resales of provisional permits and licenses in either Sri Lanka or Thailand, there is considerable anecdotal evidence that resales are common. Typically, these are resales of some or all of the ownership interests in a possible SPP project that has received some government approval and would give the new owner(s) exclusive rights to develop for a specified period of time. In Sri Lanka it has been reported that the resale price of ownership interests in mini-hydro projects has ranged from the equivalent of $10,000/MW to $40,000/MW. Similar resales have been reported in Thailand. One Thai photovoltaic (PV) developer reported that individuals or companies have paid as much as 5 million baht ($300,000) per MW of installed capacity for signed PPAs for solar PV systems (and twice this if sold with an accompanying land title). Whenever there is a resale, it provides the new owner with the legal right to sell electricity to the national utility (or some other obligated buyer) at a prespecified price or pricing formula. Whatever approval is being resold may sell at a very high price because it comes with the right to a guaranteed future stream of revenues once the SPP starts making sales to the national utility.

It is not surprising that reports of resales at high prices have generated considerable controversy in both countries. The debate over resales is really a debate over speculation. Those who oppose resales make arguments that are usually variants of the following statements:

> These resellers are just a bunch of squatters and speculators. Why should they be allowed to make a lot of money just because they acquired a piece of paper from the national utility or some government agency? We don't need speculators; we need real developers. What makes it worse is that these resellers don't have any real incentive to develop the project. They just clog up the system and waste everyone's time. And I would not be surprised if they paid a bribe to some official at the utility or at the government agency to get one or more favored positions in the queue. All of this is very distasteful and it gives renewable energy a bad reputation.

Let us take a closer look at the two principal criticisms.

Criticism 1: The speculators just sit on these projects. The projects never get developed or are developed only after significant delays. Speculators tie up valuable limited capacity at the buying utility's substations and distribution feeders that could be used for real rather than paper projects.

One obvious response to this criticism would be to establish tighter deadlines and enforce them. This was done in Sri Lanka in 2008 when the country switched from a loose utility-managed system to a much tighter government-managed system. As described earlier, once an SPP receives a provisional approval in Sri Lanka, by law the developer is given six months to acquire an energy permit with the possibility of a single six-month extension. If the developer fails, he must reapply for a provisional approval, which will be granted only if he receives a score of 40 points or higher against the objective criteria designed to measure the project's progress against certain specified milestones (SEA 2011, 7).[6]

While it is tempting to recommend a similar system for African countries that are starting SPP programs, it is important to remember that the current Sri Lankan system was put into place after 11 years of operating under a much looser system. Hence, the current Sri Lankan system has benefited from considerable in-country experience in learning how long it takes to complete different studies and to obtain different approvals. With the benefit of this on-the-ground experience, the time needed in Sri Lanka to conduct studies and obtain approvals has been reduced over time. In addition, Sri Lankan officials have had the benefit of real-world evidence that could be used to make informed decisions for tightening deadlines. But what might take three months in Sri Lanka could initially take a year or more in an African country. The danger of adopting the tight Sri Lankan deadlines in an African country would be that, if enforced, they may kill off an SPP program while it is still in its infancy.

Ultimately, it is a judgment call as to how tight initial deadlines should be. The best strategy for African countries just starting an SPP program is to establish easier deadlines in the early years of the program and as experience is gained, tighten them. It is also recommended that even these looser deadlines be combined with a monitoring system such as now exists in Sri Lanka and as proposed in Kenya. A monitoring system can provide valuable information on where the processing roadblocks are and real-world evidence as to when deadlines could be tightened. It also provides objective information on which projects are not being developed.

Two specific features of the proposed Kenyan monitoring system are worth considering. First, it requires documented progress from one six-monthly progress report to the next, and the developer must show an improvement of at least 15 points against prespecified project progress assessment criteria in each new progress report (ECA and Ramboll 2012, 24). This means that a speculator will not have the luxury of just "sitting" on a project. Second, the three-year deadline between receiving approval for the expression of interest (the functional equivalent of a provisional approval) and the issuance of final approval stays fixed regardless of any change in ownership arrangements. Both features will help a government to achieve its objective of projects being developed sooner rather than later.

Criticism 2: Consumers should not be forced to pay for the high profits that speculators will earn just because they were able to get the first slot in the queue. This is not fair!

It is easy to blame "speculators" for high profits when they resell their place in the approval queue, but they are able to do so only because the regulatory entity or some other government agency set FITs at too high a level. This happened in Thailand in 2006 when the Energy Policy and Planning Office (EPPO) in the Ministry of Energy set a FIT for grid-connected, PV generators as the sum of the buying utility's avoided cost plus a premium of 8 baht/kilowatt-hour (kWh) (26 U.S. cents), which produced a total price of about 36 cents/kWh. This high price created the equivalent of a solar "gold rush" as both speculators and real developers rushed to claim a place in the queue that would give them the legal right to receive these high FITs. The hundreds of applications that were filed and the many reported resales provided concrete, after-the-fact evidence that the FIT had been set too high. In all fairness to the Thai Ministry of Energy, the premium of 8 baht/kWh was not outrageously high at the time. The industry was asking for 12 baht/kWh and complained vociferously at 8 baht/kWh. Solar module prices have dropped precipitously, however, from about $4.50/watt in 2006 (Dunnison 2011) to $0.70/watt or less in 2012 (A-E-S Europe GmbH 2012).

Even when numerous applications and resales at high prices show that a government has set too high a FIT, government officials, like most people, do not like to publicly admit to mistakes. Instead, they will strongly condemn resellers and call them "speculators" or "social parasites." This is to draw attention away from the fact that any higher retail prices that consumers are forced to pay for SPP-produced electricity are the result of the government's own pricing decision.

Key Recommendation

Regulators should allow for the resale of provisional licenses as long as the new owners comply with the originally established deadlines. The price for resales should not be regulated. If high resale prices are observed, the government should lower the FITs offered for new grid-connected SPPs.

In our view, a government's overriding interest should be to get good SPP projects online and producing electricity. A government has tools available to do this. For example, if the government imposes serious time-bound approvals, the initial developer will either do the project himself or sell off development rights to someone else who can bring the project to fruition. It should not matter to the government whether it is the first or second owner who brings the project across the finish line. In addition to imposing major deadlines, the government can also establish credible intermediate milestones and monitor progress against these milestones (as has been proposed in Kenya). By doing so, it creates both positive and negative incentives for the initial owner. The positive incentive is that resale will take place at a higher price if the initial developer is able to perform the required studies and get approvals and land rights. The negative incentive is that the initial developer will lose its place in the queue if it fails to meet required milestones and deadlines, and the resale value will disappear completely.

As an alternative or as a complement to the monitoring of milestones, the approving authority could require that the applicant post a performance bond when it receives a provisional approval. If the SPP fails to meet the deadline (for example, commercial operation date), it would lose the money in the performance bond. As noted earlier, performance bonds are now required in Thailand for projects of 100 kW or larger in capacity. When an SPP signs a PPA with the purchasing utility, it must post a bond of about $6/kW of proposed installed capacity. It loses this money if it fails to complete the project by the commercial operations date specified in the PPA (or achieve approved extensions). In tracking new applications for very small power producers (VSPPs), a clear drop in applications can be seen at around the time (June 2010) that the bid bond was introduced, but it is difficult to determine whether this indicates causation or is simply a coincidence (figure 4.2).[7]

One downside of the current bond requirement in Thailand is that it requires a cash outlay early on in project development, when cash is scarce. Another is that the money is forfeited to a buying utility. So if the buying utility does not want to purchase from an SPP, it has an additional financial incentive to slow down the connection process. If performance bonds are used, the better approach would be for the bond money to be paid to an electrification or renewal energy fund rather than to the purchasing utility.

Figure 4.2 Megawatts of Renewable Energy under the Thai VSPP Program in Various Stages of Development

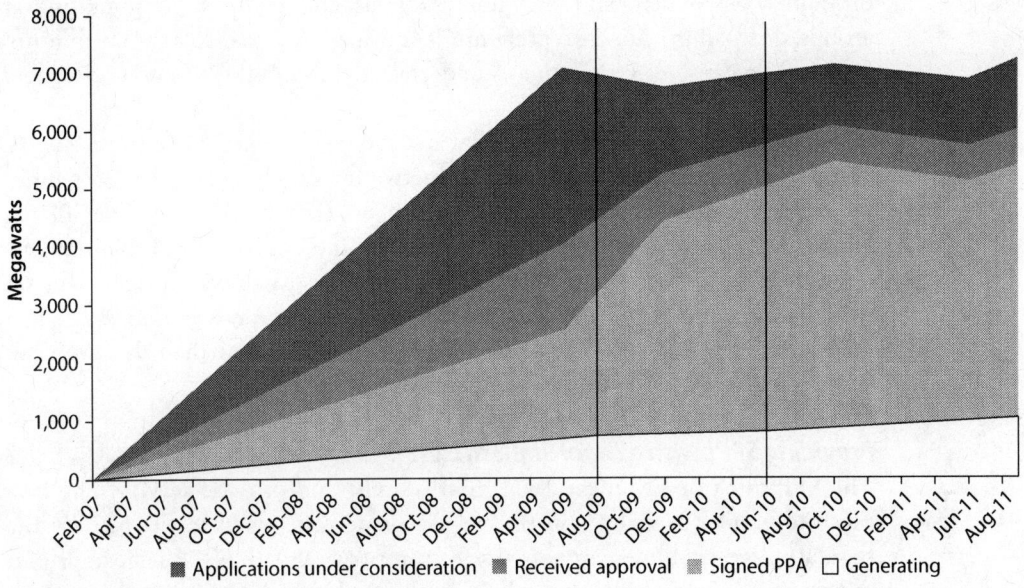

Source: Tongsopit and Greacen 2012.
Note: PPA = power-purchase agreement; VSPP = very small power producer.

Key Recommendation

The approval process that grants a developer the right to construct and operate a grid-connected SPP should:

- Grant only time-bound approvals and permits.
- Establish credible intermediate milestones that the developer must achieve and/or require the developer to post a performance bond tied to specific milestones when it receives provisional approval.
- Use the proceeds of the forfeited bid bonds to fund a general electrification fund rather than give it to the buying utility.
- Implement a credible process to monitor the developer's progress.
- Lower FITs if high prices are observed in a resale market.

Licensing: Does the SPP Have a Legal Right to Exist?

Recall that most Sub-Saharan African countries are pursuing the scale-up of grid-based access through two parallel tracks: a centralized track and a decentralized track. The centralized track or the "top-down" approach to electrification is usually undertaken by one or two national entities such as the state-owned national utility, the rural electrification agency (REA), or the ministry of energy. It is mostly accomplished through extension of the existing high- and medium-voltage grid. In contrast, the decentralized track promotes electrification through the actions of numerous owners and operators such as cooperatives, community-user groups, or small private entrepreneurs. It is usually referred to as a "bottom-up" approach because electrification is undertaken through the actions of a variety of nongovernmental entities.

From a regulatory perspective, a key question is: *how can a licensing regime be designed to accommodate both tracks?* If the two tracks are going to be compatible, the licensing regime must accomplish two things. First, it must grant the SPP the legal right to exist in the first place. Second, in the case of an SPP that initially operates in an isolated mode, the regulator must specify the SPP's rights when it finally gets connected to the main grid. In this chapter, we will address the first issue—the right to exist. The second issue—what happens when the two grids come together—will be examined in chapter 10.

Providing SPPs with a Legal Right to Exist

The SPP must be given the legal right to exist and provide service. The best place to grant this right is in the license or permit issued to the SPP by the regulator, renewable energy agency, or ministry.[8] But the SPP license or permit, by itself, does not provide the complete solution if the SPP's license is in conflict with the license granted to the national utility. This is not a problem in countries where the national utility's service area is designated as a "band" within 100–200 meters around existing distribution facilities. Anything outside this band is available to be served by SPPs. This is the licensing or concession approach taken in most of Latin America and Mali, but not in many African and Asian countries. In these countries, a state-owned utility is typically granted an exclusive nationwide service area. Presumably, the state-owned utility was granted an indefinite or very lengthy exclusive monopoly license because it was expected that there would never be a need for any other electricity suppliers.[9]

Indonesia provides an example of this approach. Until 2009, PLN, the state-owned national utility, had an exclusive legal monopoly to be the sole supplier to end-use customers throughout the county. But the expectation that PLN would succeed in electrifying the entire country was never achieved. When faced with a uniform national tariff that did not cover its costs, especially on the country's outer islands, PLN adopted the strategy of minimizing financial losses rather than maximizing electrification (World Bank 2005, xi). So even though PLN was actually serving only about 67 percent of the population in 2004, it was illegal for other entities to sell electricity to retail customers in a geographic area that was not served by PLN.

This changed in 2009. Under a new electricity law, PLN no longer has a legal monopoly to supply and distribute electricity to retail customers throughout the country. The right to serve end-use customers was extended to regionally owned companies, private business entities, cooperatives, and nongovernmental organizations (NGOs). But the law still grants PLN the "right of first priority" for the

Regulatory Processes and Approvals: Who Approves What, When, and How?

103

provision of electricity, and other entities can become electricity suppliers only if PLN waives this right. In the three years since the law went into effect, it appears that very few SPPs have taken advantage of the fact that the legal barrier to their existence has been removed. It is unclear whether this is attributable to PLN refusing to waive its priority or the inability of SPP developers to obtain grants or loans.

A similar situation existed in the Philippines. Prior to 2001 the national utility had an exclusive franchise to serve the entire country. But as with Indonesia, the national utility was unable or unwilling to serve many rural areas. In 2001 the Electric Power Industry Reform Act (EPIRA) gave the Energy Regulatory Commission the right to grant permission to qualified third parties (QTPs) to supply power in franchise areas where the incumbent was not supplying power. Hence, the specific goal of the EPIRA was to allow SPPs to operate mini-grids to serve otherwise uncovered areas, especially in the unserved portions of the franchise areas of more than 100 electricity cooperatives. But it took until 2006 for the regulator to issue rules to implement this provision of the EPIRA and, at the time of this writing, very few parties have succeeded in negotiating the regulatory red tape to become a legal QTP. For example, it took PowerSource Philippines in Palawan five years to gain approval for its tariff and service contract after the electricity cooperative decided not to exercise its first right of refusal. So while the legal right exists on paper, the initiative appears to have failed on the ground because the regulatory process is too complicated.

How Can a Legal Handoff to SPPs Be Accomplished?
If SPPs and mini-grids are going to be a viable option in countries where a national utility's service area covers the entire country, there has to be a workable mechanism for a legal handoff of the obligation to serve from the national utility to the SPP operator. This can be accomplished in several ways.

One way would be to require the national utility to submit a business plan to the minister of energy or the regulator that indicates the geographic areas that the national utility plans to serve over the near and mid term. Before 2009 this approach was used in Indonesia, but it did not work very well. The plans submitted by PLN tended to be overly optimistic, and after the plans were filed, PLN's actual provided service and expansion were usually below projections.[10] It is possible that the projections would have been more realistic if PLN had been required to submit the plans with a performance bond that it would have forfeited if it failed to meet its projections.

Another way would be to allow mini-grid developers to make a direct request to the regulator or other national or local decision-making body for an operating license in an area that is not currently being served by the national utility. The national utility would be given the opportunity to object to the developer's proposal, but the objection would have to be accompanied by a written commitment and a performance bond to serve the currently unserved area within a specified period of time. The regulator would then make a decision within 30 or 45 days as to which of the two requests to serve would be granted. If a license is granted to the mini-grid applicant, the license would presumably have to be guaranteed for a minimum period of time and with prespecified options that would be made available to the mini-grid operator when the main grid finally arrives (see chapter 10). This approach allows new entrants to take the initiative instead of waiting for some action by the national utility.

Key Recommendation

SPPs must be given the legal right to exist. In cases where the government has granted a nationwide service area monopoly to the national utility, regulators and policy makers must provide clear language allowing SPPs to exist within that service area, and establish a clear and simple regulatory process for SPPs to apply for this legal right to exist. One way to accomplish this is to require that the national utility periodically submit a near- to medium-term business plan indicating the geographic areas it intends to serve. The information in this plan would be made publically available. Another way is to allow SPP developers to request an operating license directly from the regulator, while allowing the national or regional utility to object to the request—but only if it provides a written commitment and performance bond to serve the area sought by the SPP within a specified time.

Recommended Characteristics of a Good Regulatory Review and Approval System

Different countries use different procedures and institutions to make the final decision as to whether an SPP will be allowed to operate in a country's electricity sector. In Sri Lanka the "go, no-go" decision is made by the SEA based on the recommendation of a PAC consisting of representatives of various government and electricity sector entities. The Sri Lankan regulator's licensing decision is almost automatic once the SEA decides to grant a provisional approval and the buyer (CEB) has granted an LOI to the SPP. In Tanzania the key decisions for an SPP—the issuance of a provisional and a final license—are made solely by the national electricity regulator. (At present, Tanzania does not have a renewable or sustainable energy agency.) Hence, the two countries use two very different institutional models for granting an SPP the approval to operate in the electricity sector. In Sri Lanka, the key regulatory and approval decisions are made by the agency that is charged with promoting renewable energy (that is, a promoting agency), whereas in Tanzania the key decisions are made by the national electricity regulatory agency (that is, a gatekeeper agency). Both approaches seem to be effective.

We believe that no single institutional arrangement is equally appropriate at all times and in all countries. But this does not mean that any and all institutional arrangements will be equally effective. In our view, a good review and approval process for SPPs should include the following six key features:

1. The Review and Approval Process Is Transparent and the Decision-Making Criteria Are Known to Applicants

In Sri Lanka, the SEA has published a guide, available on its website (http://www.energy.gov.lk) that describes the review and approval process in considerable detail. It describes the sequence of required steps and includes copies of applications, checklists, and sample approval documents. (See figure 4.1 for an overview of what is required at each of the principal steps.) The review and approval process is similarly well documented in Tanzania. Guidelines, rules, and sample documents for SPPs are all available on the website of the national electricity regulator (http://www.ewura.go.tz).

Both Sri Lanka's guide and Tanzania's guidelines go beyond simply describing the recommended sequence of steps. Both documents provide information on the actual criteria that will be used to make a decision at each step of the process. The goal is to shine a bright light on what is often a "black box" of government decision making. In addition, efforts have been made to minimize uncertainty about next steps in the overall process. For example, the SEA letter that grants provisional approval also provides the applicant with a list of specific documents and approvals that the SEA will require to move to the next step, the issuance of an energy permit. Based on the experience in Sri Lanka, Tanzania, and elsewhere, we have summarized in box 4.1 the key requirements to operationalize transparency in an SPP regulatory review and approval process.

2. The Role of the Buying Utility Is Limited to Technical Decisions[11]

In Tanzania the new proposed rules state that the buying utility, the Tanzania Electric Supply Company (TANESCO), is allowed to accept or reject a request for interconnection or sale of power from an SPP "solely" on the basis of "(a) its determination of the ability of the local electrical network to accept power from a power plant of the proposed type, size and power export capacity at the proposed location; and (b) its determination as to whether the proposed project conflicts with other on-going private or DNO [distribution network operator] projects" (EWURA 2013, section 9.2).

TANESCO is prohibited from making its connection decision based on its judgment of the commercial feasibility of the proposed SPP project. The rationale for limiting a buying utility's decision-making power (and not giving it a more open-ended review authority over a proposed SPP project) is that it may favor its own generation or believe that it is being forced to pay too much for the SPP-generated electricity and turn down the project by finding an excuse. In other words, it may not always be a willing buyer. In instances when an SPP believes that its proposal has been unfairly dealt with, it should have the legal right to appeal to the regulator.

In contrast to Tanzania, the new approval system in Thailand, put in place after the solar PV PPA "gold rush," requires screening by a managing committee

Box 4.1 What Are the Characteristics of a Transparent Review and Approval Process for SPPs?

Transparency is desirable for at least two reasons. First, it reduces the likelihood of corruption. Second, government officials and potential developers will usually make better decisions if they have more information. Government officials will have to spend less time reviewing flawed applications and repeating explanations of basic procedures to new developers. Developers, in turn, will make better investment decisions if they understand the "rules of the game." Providing accurate and timely information has not been the norm for many African governments. As one Tanzanian developer observed: "Good information is often one of the scarcest commodities."

What are the minimum requirements for a transparent small power producer (SPP) review and approval process? We recommend that the following items be posted and regularly updated on the website of the government agency that has the lead role:

- Up-to-date copies of standard documents such as:
  - Applicable tariffs, including the tariff that the SPP will receive for electricity sold to the national utility, as well as the structure and amount of backup or standby power tariffs that might be applicable
  - Application form for initial government approvals (Sri Lanka—provisional approval; Tanzania—provisional license; Kenya—expression of interest)
  - Application form for permanent government approvals (Sri Lanka—the energy permit and the license to generate and supply; Tanzania—the generation and supply license)
  - Progress report forms
  - Progress assessment criteria
  - Technical guidelines for interconnection and operations
  - Technical guidelines for distribution equipment and operations
  - Standardized power-purchase agreements (PPAs) for wholesale sales to the main- or mini-grid operator
  - Requests for an SPP interconnection to the main grid or a mini-grid
  - Initial utility evaluation of the SPP's request for interconnection
  - Typical costs of equipment needed to connect to the buying utility
  - Grid codes that include utility processes or procedures related to SPPs
- A list of SPPs showing size, location (using GIS coordinates), technology, and generating capacity that have received provisional and final approvals, and the expiration dates of the approvals
- A list of all non-electricity-sector approvals that SPPs of different technologies will need in order to receive final sector approval
- A list or map of areas that are likely to be "off limits" to SPPs
- The best available information on the likely expansion of the national and regional grids, capacities of substations, and capacity already allocated to developers and applicants

box continues next page

Box 4.1 What Are the Characteristics of a Transparent Review and Approval Process for SPPs?
(continued)

Even when the process and individual decision-making steps are transparent, this does not mean that all information submitted by applicants should be publicly available. For example, the prefeasibility studies submitted by applicants to the SEA in Sri Lanka are deemed confidential, since a developer will not go to the expense of undertaking the due diligence required for such studies if the information can easily be acquired by a competing applicant at little or no cost.

(comprising representatives from buying utilities), the EPPO, and the ministry of energy. And, unlike Tanzania, the review is not limited to whether there is sufficient capacity at its distribution substation (or how much it would cost to expand that capacity to accommodate the proposed SPP). It appears that the managing committee in Thailand has been given *de facto* ability to screen the project for readiness across several dimensions: land, financing, technology, and permits from other agencies. So if a utility does not want to buy from an SPP or VSPP, this new review system provides opportunities for significant influence in denying the project on nontechnical grounds (Tongsopit and Greacen 2012).

But even if the buying utility's review is legally limited to a technical one, this does not mean that the review will be necessarily fully objective. As shown in box 4.2, technical reviews can provide the buying utility with considerable discretion in setting connection and operational requirements for individual SPPs. If the utility with legally mandated buying obligations does not wish to comply, it can impose high-cost technical obligations on the SPP. If an SPP believes that the buying utility's technical review has been unfair, it should have the right to appeal the proposed technical requirements to the electricity regulator.

3. The Sequence of General Business Approvals and Electricity Sector Approvals Are Clear and Logical

Should General or Electricity-Specific Approvals Come First?

As a general rule, we think that it is best that *general business and environmental approvals/consents should come before any final electricity sector approval* such as an energy permit (Sri Lanka), generation and sales license (Tanzania), or generation permit or license (Kenya). In addition, we recommend that the SPP be required to demonstrate that it has received approval to use the natural resource that will be used to generate electricity before seeking any electricity regulatory approvals. Government agencies that issue general business documents (for example, incorporation, tax registry, and factory registrations) are sometimes reluctant to do so if the applicant has no official status in the electricity sector. Therefore, we recommend that the regulator, the renewable energy agency, or the appropriate energy minister issue some form of provisional approval or temporary license while the SPP is seeking general business approvals.[12]

Box 4.2　Excessively Stringent Technical Requirements

Utilities have very legitimate safety and reliability concerns related to the interconnection of distributed generators. In some cases, however, utility insistence on expensive technical requirements—when simpler and cheaper options are possible—might act as a barrier (intentional or not) to interconnection by SPPs. Some examples we have seen include:

Requiring a direct transfer trip. A direct transfer trip is an optical fiber link used to trip a remote relay, also known as intertripping. Because a transfer trip requires installation of a fiber-optic cable between the substation and the distributed generator, it can be very expensive. Much cheaper options for tripping relays include over current, over/under voltage, over/under frequency, zero sequence over-voltage, and others familiar to protection engineers that use the characteristics of the electrical signal at the connection point to determine whether a fault or other condition exists on the line that warrants a relay trip.

Requiring excessive upgrades to the substation, distribution line, or other utility assets. In many countries the cost of upgrading utility substations or distribution lines is borne by distributed generators that wish to connect to these facilities. Utilities can use this as a pretext to require the SPP to pay for more line or substation capacity than is strictly necessary for interconnection.

Imposing requirements for supervisory control and data acquisition (SCADA) systems. SCADA systems are widely used by utilities to monitor power plants and exercise control over how the plants operate. When full SCADA connectivity is required for a major power plant, the developer is typically required to pay for the total costs of such links and facilities. A full SCADA is generally not needed, since most SPPs are less than 10 MW. When an SPP is connected to a larger national grid, it is very unlikely that the SPP will have any operational effect on the national grid, which is sufficiently large that a SCADA connection is not needed. But as the number of SPPs increase, it may make sense to establish cell-phone carrier connections that will allow the central dispatcher to remotely determine the level of current production for all SPPs on the system.

Non-Electricity-Sector Approvals and Consents

We have no general recommendation as to the optimal sequence for general, non-sector-specific approvals (also referred to as ancillary approvals). For the sake of reducing delays, it would clearly be better if these approvals could be obtained in parallel, since sequential decisions take longer.[13] For each individual approval, the number of required steps should also be kept to a minimum. It is beyond the scope of this guide to suggest specific ways to reduce the time required to obtain general business approvals. Detailed suggestions on how to streamline general business regulatory processes can be found in the World Bank's annual *Doing Business* publication.

Electricity Sector Reviews and Approvals

Electricity sector reviews and approvals are made by both the buying utility and the lead government agency (which could be the national electricity regulator, a ministerial committee, or a renewable energy agency). The buying utility must initially decide if it is physically capable of connecting to the SPP (on existing or expanded facilities), and then if it can receive the electricity produced by the SPP. The initial decision for the lead government agency is whether the SPP should be given some provisional approval to start the development process. It is not obvious that the utility's connection review must always precede the government's provisional approval or vice versa.

Request for an Interconnection by an SPP
In both Sri Lanka and Tanzania, the initial request for interconnection by an SPP is described as a request for a letter of intent (LOI). To avoid any confusion, it would probably be more accurate to describe it as a request for a letter of intent to interconnect and to purchase electricity by the buying utility. In the request for an LOI, the SPP developer is required to specify the technical details of its proposed generating facility. In Tanzania, once the utility receives a complete application, it has 30 days in which to respond. It can accept or deny the request; if it denies the request, it must give written reasons for the denial.

If the request for an LOI is approved and the SPP indicates that it wishes to continue the process, the next mandated step is an engineering assessment (EA). The purpose of the assessment is to produce a detailed estimate of the cost of making the connection. In Tanzania the SPP has the option of obtaining the assessment from the buying utility or hiring its own engineering consultant to make the assessment using technical standards and guidelines established by the buying utility. The second option will generally be more appealing to SPPs in many Sub-Saharan African countries because the national utility may not have the money to pay for an EA or the national utility may have a slow and cumbersome procurement process (often mandated by a public procurement law) that usually produces higher prices than if a private company were to seek the same services.

The Standardized Power-Purchase Agreement (SPPA)
In Tanzania once the national utility issues an LOI, it has 90 days in which to sign an SPPA with the SPP. This does not mean that the electricity will immediately flow upon signing the SPPA. If additional facilities are required, they need to be built. And even if no additional facilities are required, there is still a need to do interconnection protection and testing. Finally, meters must be installed and tested. But from the perspective of an SPP, once the SPPA is signed, the developer can show that the utility is obligated to purchase its electricity at the FITs specified by the regulator. The SPPA is important to close the financing arrangements with investors and lenders to the project.

Final License and Contingent Final License
Once an SPP has an LOI and an SPPA, we would recommend that the approval process be structured so that the SPP can immediately apply for a final license or its equivalent. This is the approach taken in Sri Lanka and recommended in

Kenya, and is common to any PPA, not only for SPPs. In contrast, in Tanzania the final license is not issued until the regulator tests the SPP's equipment to determine that it can safely and reliably generate electricity, that is, just before the SPP's commercial operation date. The better approach would be to grant the final license at an earlier time—specifically, as soon as the SPP has received an LOI and an SPPA. But this could be characterized as a "contingent final license," as its legal effectiveness would be contingent on the SPP demonstrating to the regulator that it has obtained financing and passed the reliability and safety tests required for commercial operation.

This does not mean that everything could be a candidate for a contingency. For example, basic business and local government approvals should *not* be included as contingencies in a final SPP generation and sales license. Instead, they should be completed and documented before the electricity regulator or other government agency is asked to act on the application for a final license. But other electricity sector approvals or tests (for which there is a reasonable probability that they will be obtained or passed) could be added as contingencies to the final license.

The strategic advantage of granting the final license at an earlier time is that it facilitates the acquisition of financing. In general, it is much easier for an SPP to get financing if it can show that it has an SPPA and final license in hand. Construction will still have to be completed and testing will have to be performed, but potential equity investors and lenders will draw comfort from the fact that no more major approvals will need to be obtained.

A contingent final license is different from a provisional license. A contingent final license indicates that a final license will be issued subject to the completion of construction and specified final engineering tests. In contrast, a provisional license does not give any certainty about the issuance of a final license. Instead, it provides a period of temporary exclusivity that allows a potential developer to undertake economic, legal, and technical studies to determine whether the proposed project is likely to be viable.

4. The Regulator, in Its Final Approval for a License or Tariffs, Makes Use of the Information and Decisions of Other Government Bodies

In the consultants' proposal for Kenya, the requirement that the electricity regulator rely on the determinations of the Ministry of Energy's Feed-In-Tariff Committee is made explicit. The current recommendation is that: "EOI [Expression of Interest]-approved applicants with a signed PPA for a project of 10 MW [or less] will be *automatically issued* [emphasis added] with the Energy Generation Permit or License as long as the procedures in the Act, the Regulations and other statutory approvals have been adhered to and the relevant fees paid" (ECA and Ramboll 2012).

In Tanzania, in recently proposed "second-generation" SPP rules, the regulator stated that: "an applicant may, while submitting its application for tariff approval to [EWURA—the national regulator] submit the proposed tariffs that are calculated based on REA project evaluation spreadsheets" (EWURA 2013, 20).

While the proposed rule does not say how EWURA will use this information, it seems to imply that EWURA will give considerable weight to the decisions made by the REA.

We think that it makes sense for the regulator, to the extent permissible by law, to rely on the decisions made by an REA for several reasons:

- The REA or rural electrification fund (REF) is almost always more knowledgeable than the regulator about the specific technical and commercial operations of the SPP, especially in the case of SPPs that operate isolated mini-grids.
- The agency or fund will have a better appreciation of the cost implications of imposing different regulatory requirements.
- If the regulator decides to undertake traditional regulatory tasks, it will often simply be repeating many of the determinations already made by the REA.
- The two principal sources of income for typical SPPs that operate isolated mini-grids are subsidies and tariffs. Therefore, the REA's decisions on subsidies and the regulator's decisions on tariffs need to be closely coordinated.[14]

5. The Review and Approval Process Incentivizes National and Local Government Agencies to Make Timely Decisions and to Insulate These Decisions from Political Determinations

When the Sri Lankan government consulted electricity sector stakeholders in 2007 on possible changes to the existing SPP approval process, *the biggest single complaint from SPPs was the difficulty of getting timely reviews and approvals from national and local government agencies outside the energy sector.* The SPPs said that their applications were often ignored or lost in government agencies that seemed to have little or no interest in processing their requests. In response to this complaint, the government created, under the new SEA law, the PAC that was described earlier. The PAC has helped reduce the SPPs' complaints of non-electricity-sector applications getting lost or ignored in government bureaucracies. Several features of the PAC have led to this improvement. First, the membership and operations of the PAC are established by law rather than regulation. In Sri Lanka, as in most countries, putting the requirements into law carries more weight than if requirements are specified in lower-level regulations or administrative arrangements. Second, the 12 members of the PAC include representatives from inside and outside the electricity sector.[15] The organizations represented on the PAC still retain their individual review and approval responsibilities. Third, the PAC meets monthly and is required to make a decision on applications for provisional approvals within eight weeks after receiving the applications. Fourth, the district secretary, the highest-level civil servant in the district where the project will be located, is invited to attend the meeting.

When an SPP applies for provisional approval to the PAC, the government organizations represented on the PAC are not obligated to have completed processing applications or reviews required by their individual statutes. Instead, the PAC members are simply asked whether there is anything in the SPP application that is likely to be patently inconsistent with their separate statutory responsibilities. For example, if an SPP seeks to locate in a national park, it is highly likely that its application for a provisional approval will be turned down. Hence the PAC's review of the application for a provisional approval can be thought of as an initial screening.

The fact that an SPP receives a provisional approval from the PAC is no guarantee that it will ultimately receive the individual regulatory approvals required from each government agency represented on the PAC. The PAC process was carefully designed to respect the separate legal responsibilities of each of the PAC members. But it clearly raises the visibility of individual SPP projects within all the government agencies that have to give approvals, and creates informal pressure on different government agencies to deal with the application in a timely way. The representatives from each government agency have to share updates on the status of applications at the monthly PAC meetings, and it would be embarrassing to report that an application has been "lost" or a local office has not made any progress on the application—and the PAC representative of that agency does not know why. Another advantage of the PAC process is that representatives of the different government agencies know that once a month they will have the opportunity to talk directly to one another, which reduces misunderstandings and is almost always faster than communicating through memos.

6. The Government or Lead Agency Commits to Making an Outside Review of the Existing Review and Approval System Every Two to Three Years to Obtain Objective Recommendations for How It Can Be Improved

No administrative or regulatory system is perfect for all times and all places. But once a system is in place, it becomes difficult to make changes because there will always be private entities and public agencies who benefit from maintaining the status quo. Therefore, it is good strategy to make a publicly announced commitment to reviewing the operation of the regulatory system every two to three years of operation. This was done in Tanzania, where EWURA, the national regulator, recently proposed a set of "second-generation" modifications to both its administrative processes and substantive rules for SPPs (see box 4.3). The proposed changes were the outcome of public consultation among all stakeholders, who were given the opportunity to review and critique them in two workshops and in filed written comments. Similarly, the new SPP review and approval system in Sri Lanka was the result of a dual requirement: (a) the need for a lead agency to promote and regulate renewable energy development, and (b) the need for one agency to coordinate with other regulatory agencies. The requirement was fulfilled in the drafting of the new SEA law, developed with significant stakeholder participation.

Ideally, the review process should be initiated with an independent and objective evaluation. This is because stakeholders—who know the system best—will inevitably be influenced by individual commercial interests. Similarly, government officials will have a strong incentive to provide comments designed to protect their agency's turf. One way around these issues is to conduct (or at least to initiate) a review by bringing in outside experts or consultants who have no

Box 4.3 "Second-Generation" Changes to the Rules Governing Tanzania's SPPs

After public consultations with sector stakeholders in 2011 and 2012, the Energy and Water Utilities Regulatory Authority (EWURA) proposed the following key changes to its existing rules:

Provisional licenses. Small power producers (SPPs) will be offered the option of a provisional license that would provide them with the rights of temporary exclusivity for three years to conduct preparatory activities such as assessments, studies, financial arrangements, land acquisitions, construction, and other activities leading up to an application for a final license.

SPP working group. The composition of a broad-based SPP working group with advisory rather than decision-making responsibilities (for example, in calculating updated values for feed-in tariffs, FITs) is clearly outlined. If the working group is unable to reach a consensus recommendation, minority views are to be accommodated.

Very small power producers (VSPPs). Licenses are optional for VSPPs (generators up to 100 kilowatts), but registration and periodic reports are mandatory. VSPP retail tariffs do not require prior approval, but EWURA reserves the right to review them upon receiving a complaint, using the same tariff-setting methodology specified for SPPs.

Hybrids. A hybrid generator is eligible for the same benefits as a full renewable SPP (including renewable energy FITs), if the electricity generated by the hybrid SPP uses no more than 25 percent of a fossil fuel or some other nonrenewable source on an annual basis.

Small power distributor (SPD). An SPD category (with specific rights and privileges) has been created for entities that purchase electricity from a bulk power supplier and then resell it to end users at retail.

When the big grid connects to the little grid. Where the main grid connects to a previously isolated mini-grid, the SPP is given the right to sell electricity to the main-grid operator, purchase some or all of its electricity needs from the main-grid operator, or undertake a combination of the two.

Retail tariffs. There are more detailed principles for minimum and maximum allowed retail tariffs. The SPP may charge tariffs that account for different customers' ability to pay (that is, that allow for cross-subsidies). SPPs and SPDs are also allowed to charge retail tariffs that exceed the national uniform tariff. New rules for the pricing of backup power have been established.

Source: EWURA 2013.

---

ties to sector stakeholders. Unless carefully chosen, however, such outsiders may lack detailed understanding of the country context. Also, their analysis and recommendations may get ignored and their reports may end up gathering dust. To avoid this outcome, they should issue recommendations in a formal "consultation document" that the regulator or some other government entity is required to issue for public comment.

A different approach that has been tried in Africa involves a peer review by counterparts from other African countries who are assisted by knowledgeable consultants. The most successful example of this approach has been the peer review of the general performance of six African electricity regulatory commissions in Ghana, Kenya, Tanzania, Uganda, Namibia, and Zambia. These commissions agreed to have their performance evaluated by fellow African electricity regulators with assistance from regulatory experts from the University of Cape Town's Management Programme in Infrastructure Reform and Regulation (Kapika and Eberhard 2013).

Each evaluation consisted of a one-week, on-site visit by the chief executive officers of several neighboring electricity regulatory agencies, accompanied by regulatory experts from the University of Cape Town. The review team interviewed the local regulators as well as those whom they regulate, along with government officials at ministries with direct or indirect responsibilities in the electricity sector. At the end of the week, the outside review team gave a briefing on its preliminary evaluation and recommendations, followed up by a full written report that was made available to sector stakeholders several weeks later. The early indications are that this approach has been quite successful, which probably reflects the fact that regulators will pay more attention to recommendations that come from fellow regulators who have to "walk the same walk." Plans are under way to extend this initiative to other countries in southern Africa. A similar system could be adopted for reviewing the national regulatory and policy systems that apply to SPPs in particular, or to national electrification and renewable energy programs in general.

Notes

1. Our focus in this chapter will be on the regulatory processes and approvals that are specific to the electricity sector. We will *not* address reforms to general business regulation—the approvals that would apply to any business, regardless of the sector in which the business operates. The best single worldwide review of "good" and "bad" general business regulatory practices can be found in the World Bank's annual Doing Business survey (available at http://www.doingbusiness.org).

2. The discussion of the Sri Lankan regulatory system in this section is based on the Sri Lanka Sustainable Energy Authority guide (SEA 2011) and one of this guide's author's (Tilak Siyambalapitiya) extensive working experience with Sri Lankan SPPs.

3. For a clear description of Sri Lanka's current review and approval process for SPPs, see SEA (2011). In addition to its detailed description of the process, the guide contains copies of the actual documents used by the SEA including: the application form for provisional approval, provisional approval, energy permit, checklist of prefeasibility study contents (by renewable technology), and summary of prefeasibility study.

4. See *Feed-In Tariff Policy: Application and Implementation Guidelines* (ECA and Ramboll 2012). This report was widely circulated among stakeholders in the Kenyan electricity sector and was discussed at open workshops, though only some of the consultants' recommendations have been formally adopted by the Ministry of Energy. Appendices to the report contain copies of the standardized application (called the Feed-In-Tariff Project Application Form) and other reporting documents similar to those found in Sri Lanka's SEA (2011) Guide.

5. In Tanzania the issue is complicated by the fact that the regulator is not the only entity that offers exclusivity to SPP developers. The regulator's rules also permit TANESCO, the national utility, to offer exclusivity. For example, if an SPP developer receives

a "letter of intent" from TANESCO, it has an exclusive right to the site for a period of 12 months with the possibility of two additional six-month extensions. And if the developer signs a PPA with TANESCO, this gives the developer the exclusive right to develop an SPP at that site for two years (EWURA 2013). What is unclear is how these buyer-granted periods of exclusivity are coordinated with periods of exclusivity granted by the regulator when it issues a provisional license.

6. The scoring measures progress in completing feasibility studies, acquiring access to land resources, and obtaining statutory approvals and environmental clearances.

7. The $6 dollars/kW would be equivalent to a performance bond of $6,000/MW. Some have argued that this is too small a financial penalty when the overall cost per MW is likely to be at least $1 million.

8. See Ehrhardt and Burdon (1999) for a discussion of when "free entry" (that is, no regulatory approval for entry) would be preferable.

9. The legal presumption of an exclusive monopoly to serve an entire country was recently rejected in a decision of the Supreme Court in Jamaica. The court found that while the minister of energy had the power to issue an all-island license to the Jamaica public service company, the minister did not have the authority to issue an exclusive license. In Jamaica the statute does not specifically mention "exclusive" licenses, so the court concluded that the minister had exceeded his legal authority in issuing an exclusive license. It is not clear that this decision would be relevant in countries where the minister or regulator has the explicit authority to issue exclusive licenses or an exclusive license is granted by statute. See Supreme Court of Judicature of Jamaica (2012).

10. A variant of this approach is expected to be proposed by the Tanzanian electricity regulator. In its "second-generation" SPP rules, EWURA proposes the following requirement of the national utility: "The DNO shall, on or before the 1st of January of each year, issue a document indicating the names of the villages and districts to which the DNO intends to expand its distribution system to serve new customers in the coming 12 months, 24 months, and 36 months" (Article 58). The rationale for the proposed rule is to give all potential developers of isolated mini-grids equal access to regularly updated information about TANESCO's proposed expansion of its high- and medium-voltage grids. DNO stands for "distribution network operator." At present, TANESCO is the only DNO in the country.

11. This implicitly assumes that the buying utility does not have to make any major capital investments to connect to an SPP, which is the norm in most countries. The cost of interconnections is usually borne by the SPP. Nepal is an exception. In Nepal it appears that the NEA, the national utility, must pay for the capital costs of investments that are needed to connect an SPP to its system. In such a situation, it seems reasonable to allow the national utility to perform an economic review of a proposed SPP project. For example, it would be unfair and uneconomic to force the national utility to build and pay for a 75-kilometer line to a proposed small 500 kW SPP.

12. It is important to eliminate obvious inconsistencies between different regulatory approvals. One early SPP in Tanzania could not get a business license because the Business Registrations and Licensing Agency (BRELA) said that it would not issue a business license unless the SPP first obtained a generation license by EWURA. This created a problem because EWURA would only issue a full generation license when the SPP was clearly ready to generate electricity. Fortunately, the problem was resolved when BRELA decided that it would accept a "provisional generation license" issued by EWURA as evidence that the SPP had the legal possibility of operating as

an SPP. And since EWURA did not require a "business license" as a prerequisite for issuing a provisional license (though a business license was a prerequisite for receiving a full generation license), the potential conflict in the requirements of EWURA and BRELA was resolved.

13. Similarly, some electricity sector approvals could take place at the same time as non-electricity-sector reviews and approvals. For example, an SPP's request for interconnection could take place at the same time that the SPP is seeking approvals such as business registration, tax certificates, and local governmental approvals.

14. The issue of delegation of regulatory functions—whether formal or informal, temporary or permanent—is discussed more fully elsewhere (Reiche, Tenenbaum, and Torres de Mästle 2006, 22–25). A regulator's ability to delegate will depend on how the underlying laws are written. If the law prohibits the regulator from delegating its regulatory responsibilities, most regulators will still have the legal right to defer or give substantial weight to the decisions of other government bodies.

15. The PAC comprises representatives from 12 organizations. These include the Central Environmental Authority, the Forest Conservation Department, the Wildlife Conservation Department, the Irrigation Department, the CEB, the Land Commissioner, the Mahaweli Authority (a river basin development agency), the Board of Investment of Sri Lanka, the SEA, the Coast Conservation Department, the division in which the project will be implemented, and the provincial council of the area where the project will be implemented.

References

A-E-S Europe GmbH. 2012. "Price Trend PV Modules: Price Trend Photovoltaic Modules—Updated Weekly." *Clean Energy Investment*, October 12. http://www.europe-solar.de/catalog/index.php?main_page=page_3.

Dunnison, David. 2011. "U.S. Residential PV Market Driven by More than Price." *D-bits*, April 5. http://d-bits.com/residential-pv-price-sensitivity/.

ECA (Economic Consulting Associates), and Ramboll Management Consulting. 2012. *Feed-In Tariff Policy: Application and Implementation Guidelines*. Government of Kenya, Ministry of Energy.

Ehrhardt, David, and Rebecca Burdon. 1999. *Free Entry in Infrastructure*. Report prepared for the World Bank, London Economics International, LLC, Boston, MA. http://www.castalia-advisors.com/files/1857.pdf.

EWURA (Energy and Water Utilities Regulatory Authority). 2013. *The Electricity (Development of Small Power Projects) Rules*. Proposed for Public Consultation. http://www.ewura.go.tz/sppselectricity.html.

Government of Nepal. 1992. *Electricity Act, 2049 (1992): An Act Made for the Management and Development of Electricity*. http://www.propublic.org/tai/download/Electricity%20Act%201992.pdf.

Kapika, Joseph, and Anton Eberhard. 2013. *Power Sector Reform and Regulation in Africa: Lessons from Kenya, Tanzania, Uganda, Zambia, Namibia and Ghana*. Cape Town, South Africa: HSRC Press.

Reiche, Kilian, Bernard Tenenbaum, and Clemencia Torres de Mästle. 2006. "Electrification and Regulation: Principles and a Model Law." Energy and Mining Sector Board

Discussion Paper, World Bank, Washington, DC. http://siteresources.worldbank.org/EXTENERGY/Resources/336805-1156971270190/EnergyElecRegulationFinal.pdf.

Samuelson, Robert J. 2011. "Why We Need to Fix Social Security, and Other Year-End Reflections." *Washington Post* (blogs), December 28. http://www.washingtonpost.com/blogs/post-partisan/post/why-we-need-to-fix-social-security-and-other-year-end-reflections/2011/12/28/gIQAiJyUMP_blog.html.

SEA (Sri Lanka Sustainable Energy Authority). 2011. *On-Grid Renewable Energy Development: A Guide to the Project Approval Process for On-Grid Renewable Energy Project Development*. http://www.energy.gov.lk/pdf/guideline/Grid_Renewable.pdf.

Supreme Court of Judicature of Jamaica. 2012. *Meadows vs. Blaine et al.*, JMSC Civ 110 (Civil Division 2012).

Tongsopit, Sopitsuda, and Chris Greacen. 2012. *Thailand's Renewable Energy Policy: FiTs and Opportunities for International Support*. Palang Thai, May 31. http://www.palangthai.org/docs/ThailandFiTtongsopit&greacen.pdf.

World Bank. 2005. *Electricity for All: Options for Increasing Access in Indonesia*. Washington, DC: Energy and Mining Unit, Infrastructure Department, East Asia and Pacific Region, World Bank. http://siteresources.worldbank.org/INTINDONESIA/Resources/Publication/280016-1106130305439/Electricity-for-All-Options-for-Increasing-Access-in-Indonesia.pdf.

———. 2008. *Doing Business 2009: Comparing Regulation in 181 Economies*. Washington, DC: World Bank. http://www.doingbusiness.org/~/media/GIAWB/Doing%20Business/Documents/Annual-Reports/English/DB09-FullReport.pdf.

———. 2009. *Doing Business 2010: Comparing Regulation in 183 Economies*. Washington, DC: World Bank.

CHAPTER 5

The Regulatory Treatment of Subsidies, Carbon Credits, and Advance Payments

Show me the money!

—CHARACTER IN *JERRY MAGUIRE*, A 1996 MOVIE

You cannot develop a long-term sustainable strategy based on subsidies and grants.

—GENERAL ELECTRIC EXECUTIVE, WORLD BANK ENERGY DAY, 2012

... like most principles, it is more easily expressed in abstract than satisfied in practice.

—U.S. DOE (2007, 8-5)

Abstract

In chapter 5 we describe different subsidies that small power producers (SPPs) and small power distributors (SPDs) can receive, and ways to close an initial equity gap. We also explore the regulatory implications of capital cost subsidies and tariff cross-subsidies, explain how SPPs can earn carbon credits and how regulators should handle carbon credit revenues. Finally, we examine the regulatory issues that arise when receiving advance payments to provide equity financing.

Types and Sources of Subsidies Available to SPPs and Their Customers

A broad definition of a *subsidy* is cash or other transfer of something of value to an economic agent, whether it is a producer or consumer. Subsidies can be targeted at both SPPs and their customers. The subsidies provided to SPPs are usually referred to as producer or supply-side subsidies. Producer subsidies can benefit SPPs by lowering their costs or increasing their revenues.[1]

Subsidies provided to SPP customers are known as consumer or demand-side subsidies.[2]

The fact that a producer subsidy is targeted at an SPP operator does not mean that the benefits of the subsidy will stay with the SPP operator. SPP customers may also benefit from producer subsidies. For example, if a subsidy lowers an SPP's costs, the SPP may lower the tariffs charged to its customers. Or if it provides critical additional revenue that ensures the SPP's commercial viability, rural households will benefit by getting access to grid electricity that otherwise would not be available until the main grid arrives.

The two most common consumer subsidies are connection subsidies and consumption subsidies. A connection subsidy is a one-time grant that allows a household, business, or public institution to connect to an SPP system. A consumption subsidy (sometimes described as a quantity-based subsidy) is an ongoing subsidy that reduces a customer's cost of consuming electricity by reducing the customer's tariff. (Among countries that have set up explicit programs to subsidize both rural customers' connections and consumption, Peru has an especially clear and well-run program; see box 5.3, in a later section, for a description of the Peruvian program.)

Subsidies can be further distinguished by source. In other words, who funds the subsidy? As shown in table 5.1, subsidies received by SPPs usually come from one of four sources: national or subnational governments, external donors, other electricity consumers who are not in the SPP's service area, or other customers

Table 5.1 Types and Sources of Supply Subsidies Available to SPPs and SPDs

Type	Source
Subsidies that increase revenues	
Feed-in tariffs with premiums	Government/donors/buying utility's customers
External operating subsidies	Government/donors
Tariffs that exceed costs for other customers served by the SPP or for other non-SPP electricity consumers	Other customers from within a tariff class, from other tariff classes, or from customers whose tariffs are not regulated
Subsidies that lower costs	
Connection cost grants	Government/donors/other customers
Customer contributions in aid of construction	Customers
Discounted purchase price on bulk supply tariff	National utility/government/selling utility's other customers
Waivers of import taxes	Government/donors
Concessional/soft loans	Government/donors
Production tax credit	Government
Tax holidays	Government
Guarantees on SPP loan payments	Government/donors
Guarantees that national utilities will pay for electricity supplied by the SPP	Government/donors
Loan buy-down programs	Governments/donors

Note: SPD = small power distributor; SPP = small power producer.

of the SPP who are charged more than their cost of supply. The first two are external subsidies and the last two are cross-subsidies.

Key Observation

Subsidies available to SPPs and SPDs either increase revenues or decrease costs, and can come from the government, external donors, and customers.

Not all nontariff revenues received by SPPs are subsidies. For example, an SPP may earn additional revenues through the sale of carbon credits, which are not subsidies but payments for the provision of an additional service: a reduction in carbon emissions going into the atmosphere. Another closely related example would be "top-up" payments to feed-in tariffs (FITs), which have been proposed in the Deutsche Bank GET FiT program (Rickerson and others 2012) and is being implemented in Uganda. These are proposed grants for renewable generators that would be separate from any revenues earned from the Clean Development Mechanism (CDM)[3] or other carbon credit programs. The question that arises whenever an SPP receives additional nontariff revenues is: who should benefit from these revenues—the SPP operator, its customers, or both? And should the regulator have any say in that decision, or should it be left solely to the discretion of the SPP operator and those who provide these additional revenues?

Regulating Subsidies: The Key Recommendation

Governments usually mandate or authorize subsidies to meet a social objective such as promoting electrification or encouraging renewable energy. A government's decision to promote these objectives represents government policy making. Most recent regulatory statutes in Africa and elsewhere make it clear that the government's job is to make policy, and the regulator's job is to implement government policy—subject to any legal limits imposed by regulatory and other statutes.

Even if a regulator does not make the initial subsidy decision, its regulatory decisions will often determine whether the subsidy achieves its stated purpose. Ideally, a government should make its policy preferences clear by giving explicit policy guidance to the regulator on how to treat the subsidy. An example of such guidance can be found in the 2006 Rural Electrification Policy of the national government in India. The policy document states that: "If the State Government/State Electricity Regulatory Commission decides to permit a licensee to use assets created with subsidy, it must be ensured that the benefit of [the] capital subsidy is passed on to the consumers" (Government of India and Ministry of Power 2006, section 7.5). It is then the job of the state regulators in India to determine what specific tariff-setting actions are needed to implement this government policy directive.

Key Recommendation

If a subsidy is authorized, mandated, provided, or allowed by the government, the regulator should not take actions that would nullify or reduce the effect of the subsidy. Instead, the regulator should take regulatory actions that help to ensure that the subsidy is delivered to its intended target as efficiently as possible. The regulator, however, should periodically inform the government of the costs and benefits of the subsidy.

We now consider how this general principle could be applied for subsidies intended to reduce connection charges for rural households.

Subsidies for Connection Charges and Costs

It is widely recognized that the biggest single impediment to expanding electrification in Sub-Saharan Africa are connection charges: the payment required from new customers for their initial physical connection to an electricity supplier. (See box 5.1 for a discussion of connection charges versus connection costs.) A recent Africa Electrification Initiative survey found that the minimum connection charges for new on-grid customers served by the national utilities was above $100 for nine Sub-Saharan African countries[4] (see figure 5.1) (Golumbeanu and Barnes 2013). The connection charges of national utilities in Africa are, on average, considerably higher than the connection charges of national utilities in Asia. In fact, in the survey conducted by Golumbeanu and Barnes (2013, 5), they found that "Sub-Saharan Africa had the highest number of countries with connection charges above $100 per customer" for the lowest-available connection option.

Key Definition

A connection charge is the payment required from new customers for their initial physical connection to an electricity supplier.

When a connection charge is high, it acts as a real barrier to electrification because many poor rural households simply do not have the financial capacity to make a large up-front payment. Hence, the paradox is that rural households can generally afford the cost of electricity once they are connected and may see large drops in their monthly energy costs, but they are unable to pay the initial connection charge. Given the inability of households to make the up-front connection payment, it is not uncommon to see villages in Tanzania where only 10–20 percent of village households have signed up to be connected even though the village has been connected to the grid for five or six years (Sawe 2005).

Box 5.1 Connection Costs versus Connection Charges

Connection *costs* are the costs a traditional utility or an SPP incurs to connect new customers. Connection *charges* are the charges or fees that the new customer pays to the utility or SPP to be connected. Customer connection charges are often much lower than utility connection costs because the supplying utility may not charge a new customer the full cost of the connection when it receives a grant or subsidy that covers some portion of connection costs. Or the traditional utility may choose to recover the cost of connecting a new customer in the tariffs charged to all customers, new and existing. When this happens, the cost of connecting a new customer is cross-subsidized by existing customers. But this option is generally not available to a new SPP that proposes to build and operate an isolated mini-grid because all of its customers will be new customers.

Connection costs can differ among African national utilities, for many reasons. The most obvious reason is that construction and equipment costs vary across countries. Also, utilities may simply use different definitions of what constitutes connection costs. For example, one utility may define it in basic terms: the service connection costs to a household, such as, the cost of dropping a wire, additional poles if necessary, a meter, and circuit breakers. Another utility may go further upstream and include an allocated share of the distribution transformer costs (that is, the neighborhood costs). Yet another utility may include the costs of any actual or expected expansion in the distribution or subtransmission networks. In this last case, the customer will pay the highest connection charge because it will pay a share of three cost components: the household, neighborhood, and network connection cost.

From a customer's perspective, the total cost of a connection will be the connection charge of the utility or SPP *plus* the additional costs of installing internal wiring within the house. One recent study in Tanzania estimated that the cost of internal wiring in a rural household would range from $175 in a one-room house to $435 in a four-room house. Hence, in 2011 the total cost of connection for a rural household in Tanzania living in a three-room house located within 30 meters of existing TANESCO facilities would be approximately $680 ($300 for the connection and $380 for the internal wiring) (NRECA 2012, 28).

One way to reduce the cost of internal wiring is to use readyboards—prefabricated electricity distribution boards that typically contain one light point and one outlet point. These eliminate the need to install internal wiring because the two usage points are located on the board and the board itself is placed at a central location in the room. Readyboards are much easier to install in traditional rural houses that have mud, stone, or wood walls. A readyboard might cost $50–75 to install versus $175 for wiring of the same room. A newly connected household might initially use a readyboard and then move to internal wiring in one or more rooms at a later time. Readyboards were the norm in the successful South African program to electrify poor households in urban townships.

Sources: Authors' analysis and NRECA 2012.
Note: Some authors may also include in the definition of connection costs the expenses incurred by the household (for example, internal household wiring expenses) to make use of the new connection (Golumbeanu and Barnes 2013, 2).

Figure 5.1 Minimum Average Connection Charges and Rural Electrification Rates

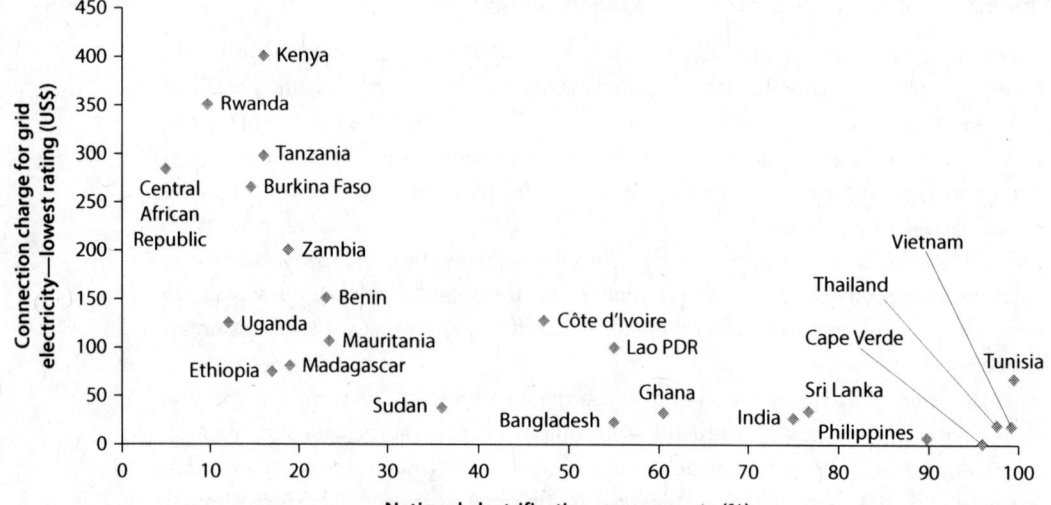

Source: Golumbeanu and Barnes 2013.
Note: The reported numbers are not comparable because they are based on reported data from different years (2005–10). Also, in some countries, they may reflect the cost of a short connection (for example, less than 30 meters), while in other countries the numbers may represent an average of connection charges for dwellings both near and far.

Village electrification (that is, an electricity supply source is available to the village) does not automatically lead to household electrification.[5]

Key Observation

Even after the main grid has arrived in a village, many rural households will not be able to connect because of the high connection charges established by national utilities. Many rural households in Africa cannot afford to make the up-front payment for a connection even though they are able to afford the cost of electricity once they are connected.

Connection Charges in Tanzania: The National Utility versus Mini-Grid Operators

In Tanzania the national utility's connection charges are especially high. In 2011 rural households located within 30 meters of existing distribution facilities that requested a basic single-phase connection consisting of a dropline and a pre-paid meter were charged almost $300 by the Tanzania Electric Supply Company (TANESCO),[6] an amount that is about 43 percent of the annual median rural income in 2007 (NRECA 2012). Moreover, TANESCO will not undertake a connection unless it receives the payment up front and in full.

In contrast, mini-grid operators in Tanzania are offering considerably lower connection charges. For example, the LUMAMA project, a micro-hydropower mini-grid in western Tanzania built with assistance from the Asociación de Cooperación Rural en Africa y América Latina (ACRA), an Italian NGO, does not impose any connection charge on new households—they pay only a T Sh 2,000 ($1.25) fee for processing the application form. LUMAMA also helps with household wiring costs by providing a loan for 50 percent of the household wiring costs, payable over six months through payments that are added to the customer's monthly electric bill (Todeschini 2011).

The Mwenga Hydro project in southern Tanzania charges a connection cost of T Sh 180,000 ($113) for the first 2,600 single-phase connections, and T Sh 385,682 thereafter (Mwenga Hydro Limited 2012). For customers for whom T Sh 180,000 is a barrier, there is an option to make a partial payment of T Sh 100,000 ($63) and then pay off the remainder through a zero-interest loan, with payments spread out over time. Basic household wiring typically costs around 90,000 T Sh ($56) and is not subsidized, but the operator offers so-called readyboards with three prewired electrical outlets at a cost of 60,000 T Sh.

These two examples would seem to imply that mini-grids in Tanzania are more efficient because they are able to connect new customers at lower cost than TANESCO. But this may not be true. It may simply reflect the fact that the mini-grid operators have access to larger grants from donors on a per household basis than the grants that are available to TANESCO. We would need more detailed information on gross connection costs (that is, unsubsidized costs) to reach any firm conclusions about the relative underlying connection costs of TANESCO versus the mini-grid operators.

Two Approaches to Recovering Connection Costs

Among electricity distribution entities, there are two basic approaches to recovering connection costs.

In the first approach, the connection charge is thought of as simply a *service charge* to the new customer, and is kept low to get more households to sign up. The service charge is not intended to recover all capital costs incurred by the utility in connecting the new customer; instead, the connection capital costs are intended to be recovered from all customers (new and existing) over time through tariffs. This seems to be the philosophy of electricity suppliers in many Asian countries (Bangladesh, India, Sri Lanka, Vietnam, Thailand, and the Philippines)—by rolling the capital costs into the regulatory asset base used in setting general retail tariffs, they recover the connection costs from all customers (see figure 5.1).

One can see examples of the same commercial strategy in the mobile-phone sector. In many countries, mobile-phone companies will sell a new smartphone at a low price if a customer is willing to sign a contract to take a minimum amount of monthly service for two or more years. In the United States, a new customer can purchase a 16 GB iPhone 5 for $199 if the customer is willing to sign up for a two-year contract with an obligation to pay for prespecified minimum monthly voice and data usage; however, if the customer wishes to buy an unlocked iPhone 5 without any contract, the purchase price jumps to $649. This is because the mobile company sees no point in offering a large discount if the customer can easily transfer his or her mobile-phone usage to another company or another country, as the $450 subsidy would then never be recovered.

Electricity service is, however, different from mobile-phone service. Most new electricity customers do not sign up for a multiyear contract with a specified minimum, monthly usage. Also, though the customer is not "locked in" by a contract, there is very little risk that he or she will find another supplier unless the country suddenly adopts retail competition. So once the customer signs up, the electricity supplier has a *de facto* monopoly for at least several years. If the electricity supplier opts for the service charge approach, the unrecovered capital costs of the new connection are rolled into the supplier's regulatory rate base for tariff-setting purposes. *This means that the capital costs associated with new connections would be recovered as one element of the retail tariffs charged to all customers (new and old).* To the extent that the tariffs of existing customers are calculated to include the capital costs of connecting new customers, there is a cross-subsidy from existing customers to new customers.

Under the second approach, the cost of connecting a new rural household is recovered in full from the individual customer in a separate connection charge that is paid for by that customer in a single up-front payment or paid over time in separate charges added to the new customer's monthly bill. Most African utilities have opted for this second approach, the full-cost recovery approach.

The second approach is typically justified on two grounds. First, it is pointed out that the retail tariffs for national utilities in most African countries do not recover their overall capital and operating costs. So under current conditions, there is a high risk that any connection costs rolled into overall capital costs used to determine retail tariffs would never be recovered. Instead, the costs for connecting new customers would simply widen the gap between incurred costs and collected revenues. Second, some of the connection costs (for example, droplines and meters) are not common or shared costs—they are incurred to supply one particular customer. Therefore, it is argued that these should be recovered only from those customers whose requests for connection produced these costs.

These are the reasons most often voiced by many African utilities as to why they favor full, up-front recovery of all connection costs. But there is another (often unspoken) reason as to why this is preferred—it is because the utilities' retail tariffs may fall far short of recovering the expected operating costs in rural areas. If a utility doubts that the government will make up the revenue shortfall in serving poor rural households, it will have an economic incentive to drag its feet in signing up new rural customers. Thus, high connection charges (such as those shown in figure 5.1) may simply be an indirect way of discouraging new users from signing up. (This is discussed further in the next section.)

Key Observation

Electricity providers generally adopt one of two strategies to recover the costs of connecting new retail customers. The first, *the service charge approach*, offers low initial connection charges to encourage more customers to sign up. Under this approach, the connection charge does not recover the provider's costs of connecting the new customer. Instead, the provider's connection capital costs are recovered from all customers (new and existing) over time through tariffs. The second strategy, *the full-cost recovery approach*, establishes connection charges that are designed to recover the provider's full cost of connecting new customers, either in a one-time, up-front charge or over time as separate charges added to the customer's monthly bill or prepayment card.

High Connection Charges and the Disincentive to Connect Rural Customers

Many reasons are given for the relatively high cost of household connections of large African national utilities. The most frequently mentioned are: use of costly European engineering standards (for example, oversized conductors for small rural loads), poor procurement practices, corruption in procurement, and the phenomenon of equipment suppliers charging high prices to state-owned national utilities because the national utility has a history of slow payment or nonpayment.[7] The underlying presumption is that the problem of high connection costs is largely one of design, construction, and procurement, and solving these would lead to lower connection costs and connection charges. This presumption ignores the fact that many state-owned utilities in Africa do not have a strong incentive to solve these engineering and procurement issues if they expect that "success" means that they will lose money selling electricity to newly connected rural customers. *So, expensive construction standards and faulty contracting methods may be symptoms of a more fundamental problem: the fact that many state-owned national utilities have few, if any, financial incentives to supply electricity to rural customers.*

Most state-owned African utilities do not have a financial incentive to connect new rural customers even if their initial capital costs are heavily subsidized. Since national utilities in Africa are often required to charge rural customers the same tariff that they charge their urban customers (that is, a uniform national tariff, as discussed in chapter 9) and most new rural household customers will also be eligible for an even lower lifeline or "social" tariff, national utilities will almost always lose money on every kilowatt-hour (kWh) that they sell to newly connected rural customers. An important but often ignored question is: *why would any rational business enterprise want to actively pursue new customers when they are almost certain to lose money on sales to these customers?*

With this question in mind, one might consider the widespread existence of high connection charges among state-owned utilities in Sub-Saharan Africa as a form of "passive resistance." No utility executives who want to keep their jobs at a state-owned utility are going to actively or openly oppose the government's efforts to expand rural electrification, even if they know that success in rural electrification will weaken their company's finances. Therefore, from their perspective, the hidden benefit of high connection charges is that they provide an indirect way of not complying with the government's mandate to electrify rural households.[8]

Even if connection costs for the utility and connection charges for new customers are significantly lowered by grants, such grants will not achieve sustainable electrification if the supplying utility expects that it will still lose money on almost every kWh that it sells to rural customers once the connection is made. This was clearly recognized in a recent memo of donor staff that described the failure of an electrification program in one African country. The memo stated that "[name of the utility] did not make an effort to roll out connections to poor households under this scheme as it had no incentive to connect them, *since actual connection costs were three times higher, and clearly these costs would not be recouped through the lower tariff revenue earned by serving low-income households.*" (Emphasis added.) In our view, it is unlikely that there will be significant acceleration in rural electrification through grid extensions unless national utilities can see a positive economic incentive to make electricity sales to rural customers after the physical connections are made.

Key Observation

The usually cited reasons for high connection charges in rural Sub-Saharan Africa are costly engineering and construction standards, poor or corrupt procurement practices, and overpriced contracts with equipment suppliers. While these are true, *national utilities may intentionally charge high connection fees to rural customers as a way to avoid compliance with government mandates on rural electrification. If the national utility has an underlying financial disincentive to connect new rural customers, lower-cost construction standards and better procurement will not solve the problem.*

In contrast, the incentives are likely to be quite different for SPPs. Unlike state-owned utilities that have a legal obligation to serve the public interest, privately owned SPPs will enter the retail electricity supply business only if they see a reasonable chance to cover their costs and make a profit. If regulators allow SPPs to charge tariffs that are cost recovering (as has been proposed by the Tanzanian regulator and discussed in chapter 9), the SPPs will have a strong incentive to increase the number of connected customers and the number of kWh that they sell to them. In addition, they will also have strong economic incentives to reduce connection costs and connection charges because they need connected customers to make sales. The difference, then, is that SPPs, if given positive economic incentives, will be seeking new customers rather than discouraging them.

Reducing Connection Charges

There are three basic ways to reduce connection charges. The first is to reduce the underlying connection costs that affect the level of connection charges, by undertaking engineering and procurement actions that can lead to lower-cost electrification. The most frequently discussed techniques include: the use of single-phase rather than three-phase distribution systems, single-wire earth return (SWER) shield wires on top of transmission lines to connect villages near transmission lines (avoiding the need to build expensive substations), and locally acquired materials.[9] The second way is for the national utility and SPPs to receive subsidies or grants from outside sources to lower the capital costs of connecting new customers. The third way is for the national utility or the SPP to establish a mechanism that allows new customers to pay for the connection charge in smaller payments over time rather than in one large up-front payment. In the discussion that follows, we focus on the regulatory decisions required for the second and third options.

Key Observation

The three basic ways to reduce connection charges, particularly for rural customers, are: reducing the underlying capital costs by adjusting engineering standards and improving procurement practices, providing subsidies or grants to the national utility and SPPs to reduce the capital costs of connecting new customers, and allowing customers to pay their connection charges over time in smaller monthly installments.

Grants to Lower Connection Costs in Africa

Outside funding for connection cost grants can come from several different sources: the national government, rural electrification agencies or funds, international donors, and the customers who have applied for a new connection and have paid a connection fee. If outside assistance is supplied by the national government or international donors, the assistance typically comes in the form of grants rather than commercial loans. (See box 5.2, describing the World Bank's Global Partnership on Output-Based Aid [GPOBA] program that has subsidized connection charges for new electricity customers in several developing countries.) The grant can be made in-kind or as a cash payment. For example, in Kenya the rural electrification agency builds new distribution facilities to serve previously unserved communities and then hands over these facilities at zero cost to the Kenya Power and Lighting Company (KPLC), the national distribution utility. In contrast, the Ethiopian Electric Power Corporation (EEPCO, the national utility) receives money for new distribution facilities that it builds on its own. The grants are an example of results-based financing: the full amount of the grant will be disbursed only upon independent verification that the household has been connected, with specifications set forth in the grant agreement.[10]

Box 5.2 GPOBA: Output-Based Aid at the World Bank

GPOBA and energy projects. An important program at the World Bank that has provided grants to subsidize the cost of new electricity connections is the Global Partnership on Output-Based Aid (GPOBA), which defines output-based aid as a subsidy payment linked to the achievement of a predefined output such as the installation of a working household connection. In a 2010 World Bank survey, 30 World Bank energy projects—involving both GPOBA and non-GPOBA projects—were found to contain output-based aid components (Mumssen, Johannes, and Kumar 2010). The most common subsidy was a one-off capital cost grant to bring down the initial cost of connection for poor households, but GPOBA has also provided grants for grid extension, installation of solar home systems, and the creation of mini-grids.

Of the 30 World Bank energy projects that were surveyed, 5 of the projects involved OBA subsidies for new mini-grids. The transaction costs of GPOBA or other donors providing OBA to an individual mini-grid operator are prohibitively high, so they reach existing and new mini-grid operators by channeling grant money through rural electrification agencies (REAs) that have ongoing programs to promote mini-grids.

GPOBA intends to use this approach in a major planned project in Uganda. Along with the Government of Uganda and KfW (the German development bank), GPOBA is expected to commit about $16 million to finance 102,000 poor households to become customers of six privately and cooperatively owned distribution entities. The donor grants, ranging from $125 to $167, will cover the costs of four types of "no-pole" connections and the entire estimated connection costs of the distribution entity. The newly connected customers will pay only for a security deposit and internal wiring, estimated to range from $90 to $116. The poorest households, who may not be able to afford the cost of internal wiring, will have a lower-cost option of receiving a readyboard with a load limiter for an up-front cost of $8.

Outputs. When GPOBA provides connection cost grants, the required output is typically defined as a verified working physical connection to the network. The network could be the main grid, a regional grid, or a new isolated mini-grid. In recent and new projects, the output definition has been expanded to include both access and service. An independent outside auditor verifies that a working physical connection was installed and that the newly connected household actually received electricity over the connections for a specified period of time. Ongoing supply and consumption of electricity are typically verified through billing and collection records. For example, in the case of main-grid connections made by the KPLC (Kenya's main distribution company) in the slum areas of Nairobi, the KPLC receives $125 upon independent verification of each household connection and then an additional $100 six months later upon verification that the connection is still in operation and electricity is still being purchased by the newly connected household.

Targeting. GPOBA must ensure that its grants reach the genuinely poor. Conducting surveys on household income to identify poor households can be costly and time consuming. In the proposed Uganda project, a proxy has been developed. A household is eligible to receive a connection grant if either the household has not connected for at least 18 months after grid connections were available in its locality or the household was identified as poor in a poverty mapping exercise for newly electrified areas.

box continues next page

Box 5.2   GPOBA: Output-Based Aid at the World Bank *(continued)*

Grants and commercial sustainability. The OBA grant is designed to lower connection charges for new customers, but *the grant by itself does not guarantee the commercial viability of the enterprise.* That will largely depend on the retail tariffs that the enterprise is allowed to charge by the regulator, the REA, or some other entity that has ongoing tariff-setting responsibility over the grant recipient. GPOBA, like most grant-giving agencies, has only limited influence over the regulatory environment in which the grant recipient will operate. This point was emphasized in the 2010 World Bank survey of OBA initiatives: "OBA schemes are only as sustainable as the environment in which they operate … in order to provide sustainable service over time, tariffs need to be at appropriate levels and subsidies need to be minimized" (Mumssen, Johannes, and Kumar 2010). If the grant recipient's tariffs connect a large number of poor customers who are eligible to purchase electricity under a "lifeline" or "social tariff" that is not cost recovering *and* there is no other mechanism such as tariff cross-subsidies in place to cover the resulting revenue shortfall, then the grant program will increase the number of connections but may not achieve commercially sustainable electrification.

Tanzania's REA (which has received funding from the World Bank and other donors; see http://www.rea.go.tz) currently offers grants of $500 to SPPs for each new rural customer that is connected by suppliers (isolated or connected mini-grid operators) other than the national utility.[11] These grants are typically disbursed in tranches: 40 percent on signing the grant agreement, 40 percent on delivery of the connection materials to the village, and the remaining 20 percent on verification of the actual connections. Because the REA's goal is to maximize new customer connections, it does not distinguish between different sources of generation in giving these grants. In other words, the REA grants are provided on a per connection basis regardless of whether the electricity supplied comes from a renewable generator, a diesel generator, or a hybrid generating system.

A similar arrangement exists in Mali. The Malian Agency for Household Energy and Rural Electrification, AMADER, has provided grants of about $570 per new connection to the operators of more than 50 new isolated mini-grids. Almost all of these are privately owned and currently use diesel generation as their source of electrical supply (Adama and Agalassou 2008). The overarching mandate for AMADER and most other African electrification agencies is electrification. At present, renewable energy is a secondary consideration, especially if they believe that by subsidizing renewable energy they will have less money to subsidize new connections.[12]

If the grants are going to be effective in reaching genuinely poor households, care must be taken to ensure that there are no legal barriers that prevent such households from signing up for grant-subsidized connections. For example, in Bangladesh, it has been reported that in some locations only the head of the household is allowed to apply for a connection. This is a problem as many

Bangladeshi men work in the Gulf states (and send remittances home to their families), and the wife is not allowed to sign for a connection under the rules.

Connection Cost Grants and Extended Payment Programs: Regulatory Issues

Outside grants to bring down the costs of making connections to new households raise three issues for electricity regulators. The first issue is: will the regulator allow for cross-subsidies for the enterprise receiving the grant? Specifically, will the regulator approve higher tariffs for more-affluent households and business customers to cover the revenue shortfall produced by non-cost-recovering tariffs charged to lifeline customers? The second issue is how such grants should be treated in calculating an SPP's maximum allowed revenues (that is, the overall amount of revenue that the SPP will be allowed to recover through its retail tariffs).[13] For connection equipment financed by the grant, will the regulator allow the enterprise that receives the grant to earn an equity return, charge for depreciation, or do both? The third issue is how to treat administrative and financing costs that an SPP incurs when it allows new customers to pay for connection charges over time (with a loan at a subsidized or market interest rate) rather than in one lump-sum payment.[14] These programs are generally referred to as deferred or extended payment programs, and their administrative and financing costs are usually referred to as subsidy delivery costs.

Regulatory Issue 1: Cross-Subsidies

The politics and finances of cross-subsidies are discussed later in this chapter and in chapter 9. Our view is that tariff cross-subsidies will generally be needed, at least in the early years, to achieve commercial sustainability for most mini-grids. Presumably, any outside provider of connection grants will not want to provide grants to a mini-grid that is prohibited from cross-subsidizing among its current or expected customers because it would be the equivalent to giving a gift to an enterprise that is not likely to survive. We recommend that the grant-giving agency should satisfy itself that:

- First, the regulatory statute or rules give the regulator the authority to allow cross-subsidies in the tariffs charged by distribution entities, whether they are connected to the main grid, a regional grid, or operate an isolated mini-grid.
- Second, the regulatory entity has given a commitment (or at least strong indications) that it will use its legal authority to allow for cross-subsidies in the relevant tariffs.
- Third, it is economically realistic to expect that the overall revenue shortfall created by lifeline or social tariffs can be covered charging other customers tariffs that exceed their cost of supply.

Regulatory Issue 2: Capital Grants and Tariff Levels

How should outside grants be considered in the regulator's determination of an SPP's overall allowed revenue used to set retail tariffs? Outside grants are almost always used to finance capital investments. Once an operator makes a capital investment, it normally affects tariff setting in two ways: the depreciation allowed on the investment and the return or profit allowed on the investment. But the tariff rules should be different when the capital investment is funded with an external grant or subsidy. In this situation, we recommend the following general rule for calculating overall revenue requirements: *the SPP should be allowed to take depreciation, but should not be allowed to earn a profit or return on the equity provided by the grant.*[15]

This regulatory rule is justified on two grounds. The first is that any capital equipment (that is, generators, transformers, distribution poles, and wires) will eventually wear out and have to be replaced. By allowing the SPP to take depreciation on capital investments (whether funded from outside grants or the SPP's own funds), the regulator helps to ensure that the SPP will have money to replace the equipment when it wears out. Second, there is no need to provide the SPP with a profit on the outside grant, since it was given as a gift by an external party with no expectation that a profit would be made on the gift. In commenting on the specific case of Kenyan government grants for electrification, a government energy official observed: "If consumers have already paid for the facilities as taxpayers, why should they pay for the same facilities again as electricity consumers?" Hence, our recommended rule is that the SPP's retail tariffs should be set to allow a "return of" (that is, depreciation) but not a "return on" (that is, profit) externally provided capital. To make this recommendation less abstract, let us consider how outside contributions are treated by the electricity regulators of Tanzania, South Africa, and Peru.

The 2008 Tanzanian Electricity Law directly deals with the issue of outside grants. Section 23(2) of the law states that "costs covered by subsidies or grants provided by the Government or donor agencies shall not be reflected in the costs of business operation" (EWURA 2008). A weakness of the Tanzanian law is that the wording is too general. Specifically, the language of the law does not distinguish between depreciation and a return on capital; instead, it appears to prohibit the regulator from including either element in setting SPP tariffs.

South Africa seems to have taken a similar approach. In section 8.16 of the 2008 Electricity Pricing Policy, the South African government gives the following guidance to the electricity regulator:

> Any assets which are not financed by the distributors, but from sources such as: State grants, customer capital contributions and connection fees, developer networks handed to the utilities and networks transferred to new utilities debt free, shall be excluded from the asset base or the purpose of determining depreciation and return on assets and the same way these costs shall be excluded from COS [cost of service] studies. (Government of South Africa 2008)

But the South African government policy recognizes that there needs to be some provision for accumulating funds that can be used for the replacement of grant-funded equipment that wears out. So in the very next paragraph of the policy document, there is a clarification: "The provision for the replacement of these assets when it becomes due shall form part of the Licensee's revenue requirements ..." (Government of South Africa 2008, section 8.16). This seems to be saying that the regulator must set tariffs so that they include a cost element that allows for the eventual replacement of grant-funded equipment, even if the regulator is not allowed to call it depreciation.

Of the three countries, Peru has taken the clearer and more straightforward approach. A Peruvian government decree issued under the Rural Electrification Law prohibits earning a return on outside capital but explicitly directs the regulator to use an annual depreciation allowance of 16.9 percent for any capital equipment provided to isolated mini-grids that was financed through a government grant (Government of Peru 2007, Article 25). The Peruvian approach is consistent with our recommended rule (see box 5.3).

Box 5.3 Peru: Three Subsidies for Rural Electricity Providers

In 2008 approximately 70 percent of rural households in Peru did not have grid-based electricity. To increase rural electrification, the Peruvian government established three types of subsidies for rural electricity providers. The key features of the three subsidies are:

- *Initial capital cost subsidy ($100 million per year)*
  - Provided to isolated rural mini-grids (small power producers, SPPs) (less than 500 kW of installed capacity) and to small, grid-connected distribution systems (SPDs) outside the geographic concession areas of larger private and public utilities
  - Can be no higher than $1,000 per operator and the recipient must provide at least 10 percent of the initial capital cost
  - Selected based on bids for the lowest required subsidy per consumer based on prespecified maximum retail tariffs
  - Funded by the national budget, international loans, the rural electrification fund, and donor grants
- *Operating cost subsidy ($36 million per year)*
  - Reduces the ongoing generation ($23 million) costs and distribution ($13 million) costs of rural providers
  - Bases the subsidy on the regulator's calculation of the distribution costs that would be incurred by an efficient distribution provider serving specified geographic areas with different customer densities
  - Funded by urban electricity customers

- *Consumption subsidy ($31 million per year)*
  - Ensures that rural customers served by SPPs and SPDs pay tariffs that are similar to comparable customers in urban areas
  - Leads to a 50–60 percent reduction in the tariffs of SPP and SPD customers with monthly consumption of 30 kWh or less (for example, the subsidy reduces the tariff for 30 kWh customers from 17.42 cents to 11.91 cents in one low-density rural mini-grid)
  - Funded by a 3 percent surcharge on all consumers whose monthly consumption is 100 kWh or higher per month

box continues next page

Box 5.3 Peru: Three Subsidies for Rural Electricity Providers *(continued)*

The first two subsidies are producer subsidies designed to lower the capital and operating costs for rural providers. The third subsidy is a direct consumer subsidy that is a cross-subsidy because the funding comes from the 3 percent surcharge paid by higher consuming customers. OSINERGMIN, the national electricity regulator, calculates the amount required to be paid by each utility and the amounts to be received by each rural provider. To ensure the financial integrity of the system, the money is channeled through bank accounts that are not accessible to either the regulator or any other government official. Peru is able to offer these three subsidies because it is a richer country with a per capita income of $5,292 in 2010 (World Bank 2012) and its regulatory system has succeeded in setting retail tariffs at cost-recovering levels.

Source: Revolo Acevedo 2009.

Key Recommendation

If an electricity provider receives a grant from an outside entity to reduce its capital costs, the SPP should be allowed to take depreciation on the equity provided by the grant, but should not be allowed to earn a profit or return on this equity.

Regulatory Issue 3: Extended Payment Programs and Tariffs

The third regulatory issue involves how to account for the expenses incurred by an SPP or utility in delivering a connection subsidy to newly connected households. The most common mechanism for reducing the first-cost burden on consumers is to allow them to make an initial down payment for the connection and then to make installment payments on the remaining balance with little or no interest being charged. But most banks and microfinance institutions are reluctant to make such loans (or do so only at high interest rates) because the loan will be used for consumption rather than production that facilitates new income-generating activities. Therefore, by default, most deferred payment programs are typically operated by the electricity supplier—either the utility, an SPP, or an SPD.

We think that extended payment programs operated by mini-grid operators should be expanded beyond just connection costs. For example, "on-bill financing" could also be used to finance electrical equipment for productive uses (grain mills, and so on), to pay for internal household wiring, or even to make improvements to a potential customer's house, such as adding a metal roof—which is sometimes a minimum requirement to receive electricity. Mini-grid operators in Tanzania have pointed out that giving them explicit authority to provide on-bill financing to customers for metal roofs would lead to a more rapid increase in the number of new connections because some poor households are unable to afford the cost of putting a metal roof on their house. Similarly, if mini-grid operators could also finance the purchase of productive-use machinery for their commercial customers, this, too, would lead to more sales. We think that expanding extended payment programs would lead to a win-win outcome because more rural households could be connected, more businesses could expand their income-producing activities, and the mini-grid operator would increase sales and be able to achieve financial viability sooner.

Some have argued that any loans that help to increase electricity usage in rural areas should be made through rural microcredit institutions or regular banks rather than the electricity provider. We disagree. While microcredit institutions should not be excluded from this activity, a rural electricity provider has two big advantages over a microcredit institution. First, it is relatively easy to add loan repayments onto an existing monthly prepaid or postpaid billing system. Second, if the customer fails to repay the loan, the electricity provider can simply turn off the electricity to that customer—an option that is not available to a microcredit institution or a bank.

The expansion of on-bill financing requires both regulatory changes and the availability of financing. Both are under consideration in Tanzania. The Tanzanian electricity regulator is considering a proposal that would allow SPPs to recover the interest subsidy and administrative costs of any expanded on-bill financing programs as a recoverable cost in tariffs. It has also been proposed to external donors in Tanzania that they consider providing grants or subsidized loan programs to mini-grid operators in addition to the grants that they currently provide for initial connections.

An ambitious connection-fee-financing program is being undertaken by EEPCO in Ethiopia, with support from GPOBA,[16] to connect more than 225,000 new households. While EEPCO is a traditional, vertically integrated utility supplier, the same regulatory issues would exist for SPPs that wish to provide extended connection payment plans for new customers. Under the EEPCO program, each newly connected customer is given the option of making a minimum down payment equal to 20 percent of the estimated cost of providing a new connection.[17] If the customer chooses this option, he or she will then pay for the remaining balance of connection costs, without any interest, in 60 equal monthly payments over a five-year period. The customer is, in effect, paying for the connection through an interest-free loan from EEPCO, which incurs both financing and administrative costs to operate this program. EEPCO must borrow money to on-lend money to its new customers. It also must have sufficient working capital to cover the lag between its payments (to acquire and install the equipment [for example, droplines, meters, and poles] to connect customers) and the reimbursement that will be received over time from newly connected customers. The GPOBA grant reimburses EEPCO for the cost of providing interest-free loans and two free compact fluorescent lights (CFLs) to its newly connected customers.

The basic lesson here is that the financing and installation costs of the lending program are real costs. The regulator should allow the operator, whether it is a large vertically integrated utility in Ethiopia or an SPP in Mali, to recover these costs in its retail tariffs if these costs have not been covered through an outside grant.[18]

Key Recommendation

If an electricity provider offers its customers the ability to repay their connection charge, the cost of internal wiring, the cost of improving their house to meet minimum electricity connection standards, and the cost of purchasing electricity-powered appliances and machinery through monthly on-bill installments, the regulator should allow the provider to recover both the financing and administrative costs that it incurs to provide these loans.

Cross-Subsidies in Tariffs

In most general discussions, cross-subsidies are defined to mean a tariff structure where some customers pay more than their costs of supply and other customers pay less than their costs of supply. In developing countries, the three most common forms of cross-subsidies are industrial customers subsidizing residential customers, high-usage residential customers subsidizing low-usage customers, and urban customers subsidizing rural customers.

Key Definition

A cross-subsidy is a tariff structure in which some customers pay more than their costs of supply to subsidize other customers who pay less than their costs of supply.

Economists often criticize cross-subsidies because they distort prices. They argue that cross-subsidies can lead to inefficient outcomes because customers do not see the true costs of being supplied electricity.[19] Similar statements are often made in official government policy pronouncements and laws. For example, India's National Electricity Policy states that: "Cross-subsidies hide inefficiencies and losses in operations. There is urgent need to correct this imbalance without giving tariff shock to consumers. The existing cross-subsidies for other categories of consumers would need to be reduced progressively and gradually" (Government of India and Ministry of Power 2005, section 5.5.3). (See box 5.4 for an example of conflicting language in the 2008 Tanzanian Electricity Law.)

The Politics of Cross-Subsidies in Africa

While cross-subsidies are discouraged in policy statements and prohibited in statutes, it is not uncommon for the policy statements and laws to be ignored

From the Bottom Up • http://dx.doi.org/10.1596/978-1-4648-0093-1

Box 5.4 Cross-Subsidies in Tanzania: The Regulator's Legal Dilemma

Even if a regulator or policy maker decides that the only viable tariff option for promoting small power producers (SPPs) is to allow cross-subsidies, there still may be legal barriers that will need to be addressed. For example, in Tanzania, the 2008 Electricity Law states that: "no customer class should pay more to a licensee than is justified by the costs that it imposes on such a licensee" (EWURA 2008, section 23 (2) (f)). Though this statutory language would seem to clearly preclude approval of cross-subsidies for isolated mini-grids, it has also been argued that this provision of the law cannot be read in isolation from other provisions of the law. In the very same section of the law (the section that specifies tariff-setting principles), there is also a requirement that: "tariffs should allow licensees to recover a fair return on their investment" (EWURA 2008, section 23 (2) (b)). Clearly, the two criteria are in conflict because SPPs will not be able to achieve financial viability by earning a fair return on their investments unless they are allowed to charge tariffs across customer classes that will produce sufficient revenues to earn such a return.

When two legally mandated tariff-setting principles are in direct conflict, it seems reasonable that the regulator should be guided by the government's principal stated policy objectives. In Tanzania the government has emphasized the overriding importance of achieving rapid rural electrification. This, then, would imply that cross-subsidies in SPP tariffs should be allowed because they will achieve commercial sustainability for SPPs that wish to supply rural customers on isolated mini-grids. Another justification for such cross-subsidies is that they are already allowed in the national utility's tariff structure for main-grid customers. Under the Tanzania Electric Supply Company's (TANESCO's) current tariff structure for its main-grid customers, a 2010 utility-sponsored tariff study clearly showed that its business customers cross-subsidized its household customers (Vernstrom 2010). So it seemed reasonable that operators of isolated mini-grids should be given the same pricing flexibility to make them commercially viable. When faced with this dilemma, the Tanzanian electricity regulator decided to accept cross-subsidies. In its June 2012 proposal for "second-generation" SPP rules, EWURA proposed the following rule: "To facilitate commercial sustainability, an SPP or SPD [small power distributor] may propose tariffs for specific customer categories or for customers within a single category, subject to the Authority's approval, that take account of the ability to pay of these customers" (EWURA 2012). This is one of four proposals made by EWURA to promote the financial viability of private- and community-owned isolated mini-grids in Tanzania. The four proposed regulatory actions are discussed in the section of chapter 9 titled "What Can a Regulator Do to Promote the Commercial Viability of Isolated Mini-Grids?"

in practice. It is worth taking a closer look at why this happens. The most plausible explanation is that cross-subsidies continue to be favored because they serve the political needs of presidents and prime ministers and the operational needs of government electrification officials. When it comes to actual implementation on the ground, politics usually takes precedence over policy.

Presidents and Prime Ministers

Presidents and prime ministers typically support uniform national tariffs and cross-subsidies in both public pronouncements and private conversations. In public speeches, they will state that basic fairness requires that everyone in the country should be treated equally. And "equal treatment" requires that every residential electricity consumer should pay the same tariff as any other consumer in the same tariff category or class, regardless of whether the residential consumer is located in the capital or in an isolated rural village. This then implies that there must be a uniform national tariff that eliminates all geographic differences in tariffs.[20] In addition, most presidents and prime ministers will state that the uniform national tariff should also include a lifeline or social tariff component because a low-price social tariff will help to alleviate rural poverty and promote affordable electricity for the poorest of the poor. The uniform national tariff, if combined with a social tariff for low-consumption customers, leads to two cross-subsidies: urban customers subsidizing rural customers and higher-consumption customers subsidizing low-consumption customers.

In private, off-the-record conversations a president or prime minister may also acknowledge three other political benefits produced by the cross-subsidies that are not mentioned in public speeches. The first is that cross-subsidies do not need to be financed through the government budget because the money that supports the cross-subsidies comes from the tariffs of other electricity consumers rather than from the government budget. The second is that cross-subsidies are largely hidden from public view and therefore get little or no attention in parliamentary debates. The third is that they help to produce votes from poor people, the principal beneficiaries of the cross-subsidies.

Government Electrification Officials

Government officials, who are involved in the day-to-day work of promoting rural electrification, support cross-subsidies for other reasons. From their perspective, cross-subsidies have three major practical (as opposed to political) benefits. First, even if the president commits to providing general subsidies from the government's general budget, the subsidies may not always be delivered as promised. Second, cross-subsidies are much easier to deliver because they simply require adjustments in an existing tariff system. They avoid the need to establish and administer a separate new subsidy delivery system. Third, without cross-subsidies, most isolated mini-grids will not be commercially viable because total revenues will fall short of total costs. And this is likely to be true even if the mini-grid operator receives grants to subsidize initial capital costs.

Why Cross-Subsidies are Needed (at Least Initially)

Our focus is on commercial sustainability—SPPs must be commercially viable or they will not be sustainable. Commercial viability cannot be achieved if costs exceed revenues on an ongoing basis. In chapter 9 we examine the likelihood of costs exceeding revenues for a hypothetical isolated mini-grid (Case 1: an isolated SPP that sells at retail) using typical real-world numbers from Tanzania. We simulate financial outcomes under different subsidy and tariff scenarios using a spreadsheet. Financial viability is measured using two key standard financial parameters: the debt service coverage ratio (DSCR) and the internal rate of return (IRR). We assume that the DSCR must be at least 1.44 and the IRR must be above 15 percent for the project to be viewed as commercially viable by potential lenders and developers. The simulations show that the only scenario that achieves these minimum thresholds is the scenario in which the mini-grid operator receives up-front capital-cost grants for more than 50 percent of its investment costs, is allowed to charge tariffs to all customers that exceed Tanzania's current uniform national tariff, and is allowed to charge higher tariffs to its commercial customers to cross-subsidize the tariff charged to its household customers. These simulation results are consistent with what Tanzanian developers have said in private conversations and public forums.

So the threshold decision for regulators and policy makers is: *should mini-grid operators be allowed to charge tariffs that exceed the uniform national tariff and to cross-subsidize residential customers?* In our view, the answer is yes, for three reasons.

First, the decentralized track represented by mini-grids will be a viable and sustainable option only if mini-grids can achieve commercial viability. It is unrealistic to expect that governments and donors will be able to offer a credible commitment to cover any ongoing shortfall in revenues in addition to the up-front capital cost subsidies that they sometimes offer. If mini-grids are going to be commercially viable, their sustainability cannot be based on ongoing external subsidies.

Second, under the centralized track of extensions in the main grid, most national utilities are routinely allowed to cross-subsidize their residential customers by imposing higher (that is, non-cost-justified) tariffs on their commercial and industrial customers. So if a national utility is allowed to cross-subsidize across customer classes or categories, why should that same tariff strategy be denied to SPPs?

Third, when electricity arrives in rural areas, it is often first used for lighting that had previously been supplied by kerosene lanterns. And if the price of kerosene is subsidized, it could be argued on grounds of economic welfare that subsidizing the price of a substitute (that is, electricity) does not distort consumption choices. We recognize that the ideal would be to remove both subsidies over time. But if this "first-best" solution is not available, then allowing SPP operators to use cross-subsidies in their tariff structure seems like a reasonable second-best solution especially if it is the critical factor in determining whether an isolated mini-grid will be a "go" or "no-go" option for isolated villages.

Key Recommendation

Regulators should allow mini-grid operators to charge tariffs that exceed the uniform national tariff if the operators' costs exceed the uniform national tariff and to cross-subsidize residential customers. The three justifications for this recommendation are: it ensures the financial viability of SPPs; national utilities routinely cross-subsidize their residential customers by charging commercial and industrial customers higher tariffs, so that same opportunity should be available to potential mini-grid operators; and electricity is a good substitute for kerosene, which is itself often subsidized, so offering subsidized electricity would not distort consumption choices.

Revenues Earned from Carbon Credits through the Clean Development Mechanism (CDM) or Other Carbon Credit Programs

SPPs operating on both the main grid and isolated mini-grids have the potential to earn carbon emission reduction credits through the CDM, a carbon trading mechanism established under the Kyoto Protocol. For example, if an isolated mini-grid SPP can replace an existing fossil-fuel generator (such as a diesel generator) with a hybrid generating system, the SPP can make a credible argument that its planned operation will reduce carbon emissions and it should therefore be entitled to receive certified emission reduction (CER) credits. These CER credits can provide SPPs with an additional source of revenue, above and beyond the revenue that would be collected through the tariffs paid by its customers. As discussed earlier, CERs are not a type of subsidy; instead, they are payments for the provision of a service by the SPP: the reduction of global carbon emissions against a calculated "business-as-usual" benchmark.

Key Definition

Certified emission reduction (CER) credits are payments offered by the UN's CDM or other emission-abatement programs to entities that are able to offer a reduction in a specified and audited amount of carbon emissions against an estimated "business-as-usual" benchmark.

While the opportunity for an SPP to monetize the value of avoided carbon by earning CERs exists in theory, in practice, it would be prohibitively expensive and time consuming for one small, isolated SPP to submit and process a stand-alone application for a CER credit and to create and implement the required post-approval monitoring system required by the CDM board that approves such applications.[21] To date, the only SPPs that have succeeded in earning CERs are those that have applied jointly as individual projects within a larger SPP program. At this time, there seem to be two approaches that allow individual SPPs to be grouped together in a joint application. The first approach is a bundled application; the second, is known as a program of activity (POA) application.

Key Observation

Applying for CER credits is prohibitively expensive and time consuming for a single SPP. Therefore, individual SPPs can submit a joint application, using either a bundled application or a program of activity (POA) application.

Type 1: A Bundled Small-Scale Application

Bundled applications are feasible when the locations of the SPPs are known in advance and the SPPs will use a single generating technology whose aggregate size is below the threshold level required to meet the United Nations Framework Convention on Climate Change's (UNFCCC's) small-scale requirements. One of the first bundled SPP applications that succeeded in earning carbon credits was a community-based SPP program in the Northern Areas and Chitral (NAC) region of Pakistan. It was developed as an initiative of the Aga Khan Rural Support Programme (AKRSP)[22] over a four-year period, and was projected to create 90 isolated, run-of-the-river, micro- and mini-hydro facilities (Case 1 in chapter 2: isolated SPP that sells at retail) totaling about 15 MW of installed capacity at an average cost of $1,120 per kW of installed capacity. These small, village-level hydro facilities will help to replace existing or new diesel generators. In October 2009 the NAC program received approval from the CDM executive board after close to three years of effort in developing and obtaining final approval of the application. A significant portion of the application and project validation cost of the program was paid for by the World Bank Community Development Carbon Funds, and this cost was only partially recovered from subsequent carbon revenues realized.

With this approval, it is expected that the NAC project will receive CDM revenues that will almost double the average annual revenues attributable to each participating SPP project. The 90 village organizations will provide about 20 percent of the capital costs, mostly through in-kind contributions of labor. The villages agreed to sign over the rights to any carbon credit revenues to the AKRSP in return for ongoing and future technical assistance on the SPP project and other future rural development projects in their villages. Hence, the decision as to who would receive the CDM revenues was one component of a larger package that involved sharing of several different costs and benefits.

Type 2: A Program of Activity (POA) Application

A POA application is an alternative approach for grouping SPPs together in a joint application in situations in which the SPPs are operating under a common program but the mix of technologies and the locations of the SPPs are not known or finalized at the time of initial submission. This approach has been proposed in Tanzania for SPPs: (a) that will be using different generating technologies, (b) that will be operating as both grid and off-grid SPPs, and (c) whose location and technology are *not* known in advance. In Tanzania the REA, with assistance from the World Bank, is taking the lead in preparing the CDM application. The REA is the logical candidate for performing this role because it is already providing other forms of technical assistance to SPPs. In Tanzania, as in more than 15 other Sub-Saharan African countries, the REA is providing "connection grants" and guidance to develop project business plans. This assistance gives the REA detailed knowledge of SPP operations and technologies that can also be used to prepare a POA application.

Ultimately, an REA's decision to undertake a CDM application on behalf of an SPP depends on its projections of costs and revenues. The costs are principally the up-front costs of the application and of preparing initial documentation, the cost of preparing responses to questions received during the application process, as well as ongoing administrative costs such as those of the annual monitoring and verification required to validate that the SPPs are producing electricity as promised. The revenues are the future stream of carbon credit revenues and will depend critically on the market price of CERs at the time the application is approved, unless forward contracts (with or without delivery guaranteed) are obtained to provide certainty about future revenues. As of December 2013, the spot market price of CERs was about $0.50 per ton and there is a high level of uncertainty about future price levels, given the lack of global agreement about future carbon markets. If the price remains at this low level, the expected revenue levels may be too low and the transaction costs are likely to be too high to justify going ahead with an application.

The CER Calculation

In Tanzania it has been estimated that a hydro-based SPP that proposes to produce electricity to sell to an existing, isolated diesel-fired mini-grid would be eligible for 0.88 CER for every megawatt-hour (MWh) that it generates.[23] The assumption here is that the SPP's production of electricity will allow it to replace electricity that would otherwise be generated from the diesel generator. But if the SPP proposes to connect to the main grid, it will earn only 0.55 CER for every MWh that it generates, because the SPP's production will replace a less-polluting mix of the national utility's hydro and fossil-fuel generation. If one assumes that each CER brings in $12 of additional revenue, it has been estimated that the carbon credit revenues in Tanzania would increase an off-grid SPP's wholesale revenues by 4.3 percent per kWh sold and an on-grid SPP's wholesale revenues by 9.9 percent per kWh sold.[24] Even though an off-grid SPP will earn more CERs per kWh produced, the impact on the off-grid SPP's revenues, measured on a percentage basis, is smaller than for on-grid SPPs because the off-grid FIT (24 cents/kWh) allowed by the Tanzanian regulator is much higher than the on-grid FIT (6 cents/kWh) and the off-grid SPP operator may not be able to dispatch his plant as much as he would want because of technical constraints. For example, the diesel generator may have to satisfy minimum production levels to achieve minimum operating efficiency levels. Therefore, an SPP operating on an isolated mini-grid will probably have fewer hours of generation and sales than a comparable on-grid SPP facility (see chapter 8).

From the Bottom Up • http://dx.doi.org/10.1596/978-1-4648-0093-1

Table 5.2 provides estimates of the revenue impact of CER credits for main-grid-connected SPPs in Tanzania and three other African countries at different projected prices for the CER credits. Ideally, the revenue impact of CER credits should be measured as a percentage increase in the SPP's average tariff revenues (regardless of whether these revenues come from wholesale sales, retail sales, or a combination of the two). But since these numbers are not readily available,

Table 5.2 Potential Increase in Electricity Revenues from CDM Credits for Grid-Connected SPPs in Africa

Country	Measured entity	Unit of measure	Carbon price ($/tCO$_2$e)				
			5	10	15	20	25
South Africa	Emissions factor	tCO$_2$e/MWh	1.0481	1.0481	1.0481	1.0481	1.0481
	Likely potential CDM revenues	U.S. cents/kWh	0.524	0.786	1.048	1.310	1.572
	National average tariff	U.S. cents/kWh	7.35	7.35	7.35	7.35	7.35
	CDM revenues as a percentage of the uniform national tariff	%	7	11	14	18	21
Tanzania	Emissions factor	tCO$_2$e/MWh	0.5	0.5	0.5	0.5	0.5
	Likely potential CDM revenues	U.S. cents/kWh	0.125	0.25	0.375	0.5	0.625
	National average tariff	U.S. cents/kWh	9	9	9	9	9
	CDM revenues as a percentage of the uniform national tariff	%	1.4	3	4	6	7
Kenya	Emissions factor	tCO$_2$e/MWh	0.63	0.63	0.63	0.63	0.63
	Likely potential CDM revenues	U.S. cents/kWh	0.158	0.315	0.4725	0.63	0.7875
	National average tariff	U.S. cents/kWh	17	17	17	17	17
	CDM revenues as a percentage of the uniform national tariff	%	0.9	3.5	5.2	6.9	8.7
Ethiopia	Emissions factor	tCO$_2$e/MWh	0.0034	0.0034	0.0034	0.0034	0.0034
	Likely potential CDM revenues	U.S. cents/kWh	0.001	0.0017	0.0025	0.0034	0.0042
	National average tariff	U.S. cents/kWh	16.62	16.62	16.62	16.62	16.62
	CDM revenues as a percentage of the uniform national tariff	%	0.01	0.01	0.02	0.02	0.03

Sources: South Africa: http://www.erc.uct.ac.za/Information/Climate%20change/Climate_change_info3-Carbon_accounting.pdf; Tanzania: http://www.cd4cdm.org/tanzania.htm; Kenya: http://www.kplc.co.ke/ and http://cdm.unfccc.int/; Ethiopia: http://www.jiko-bmu.de/files/basisinformationen/application/pdf/subsaharan_ldcs_cdm_potentials.pdf.
Note: CDM = Clean Development Mechanism; kWh = kilowatt-hour; MWh = megawatt-hour; SPP = small power producer; tCO$_2$e = tonnes of carbon dioxide equivalent.

we used average retail tariffs for the national utility as a proxy for tariff levels within a country. Though this would be a reasonably good proxy for SPPs that are selling just to retail customers on an isolated mini-grid, it provides a low estimate of the revenue impact for SPPs that would be selling just at wholesale to the national utility on the main grid because one would expect that the wholesale FIT would generally be below the average national retail tariff.

Of the four countries, the expected revenue impact of CDM revenues will be greatest in South Africa for two reasons. First, South African SPPs that use renewable energy would be replacing electricity generation systems that currently use a lot of coal. South Africa's emission factor of 1.04 tonnes of carbon/MWh generated on the main grid is almost double the comparable value for Tanzania. Second, South Africa has relatively low average tariffs. Hence, any new CDM revenues will have a greater revenue impact in South Africa than in some other African country with a higher national average tariff.

In three of the countries—Tanzania, Kenya, and Ethiopia—the revenue impact of carbon credits at current CER prices is likely to be small, perhaps in the 0–5 percent range. And these are estimates of gross benefits. A more accurate estimate of the benefits of seeking CERs through the CDM program would require reducing the expected revenue stream by the cost of the mandatory annual monitoring and verification, after subtracting out the initial costs of CDM validation.

There is another possible source of outside revenues that could be larger than CDM revenues. It has been proposed that outside donors provide direct "top-ups" of the FITs of SPPs that are connected to the main grid (Hanley 2010). The top-ups would be calculated as estimates of the additional revenues required per kWh to make a proposed SPP commercially viable. Unlike CERs, top-up payments would not be justified on the basis of reduced carbon emissions. Instead, they would be justified in terms of increasing national security of supply, reducing vulnerability to fluctuations in fossil fuel prices, and reducing the likelihood of generation capacity shortages within a country. (See appendix G for a discussion of questions that would have to be resolved to implement a donor top-up program.) Uganda is considering a top-up program that would be an overlay to existing FITs. In its first round, the Ugandan program will offer about 2 cents/kWh for mini-hydro projects on top of an existing FIT of 9 cents/kWh. The proposed supplemental payments will increase SPP revenues by about 22 percent/kWh. At the time of this writing, top-up agreements have been signed with three SPPs with the expectation that a total of eight agreements will be signed in the first round, leading to 85 MW of SPP-installed capacity. (See www.getfit-uganda.org.)

CER Credits: Should They Affect Electricity Tariffs?

The central issue for electricity regulators is: how should carbon credit revenues be considered in setting retail or wholesale tariffs for an SPP? This general issue raises several subsidiary questions:

(a) *Who should decide how the revenues from carbon credits earned by an SPP project are allocated?*

(b) If the regulator makes this decision, should the revenues be given to:
- The owner of the SPP facility?
- Any party in addition to the owner that provided equity capital either through a monetary or in-kind contribution?
- The wholesale and retail customers of the SPPs?

Regulators have typically *not* been involved in the initial decision on how to allocate CER revenues—it is usually decided beforehand as one element in a larger agreement. However, a regulator has the *de facto* capability to overturn prior allocation decisions among project participants. For example, if a regulator decides that an SPP's customers, rather than the developer, should receive all carbon credit revenues, the regulator can simply reduce the allowed retail or wholesale tariff by an amount equal to the expected annual CER revenues. By doing so, the regulator would effectively "claw back" the CER revenues that would otherwise go to the developer.

Should a regulator change the allocation of CER revenues, either directly or indirectly, through offsetting adjustments to tariffs? *Our recommendation is that the regulator should not modify a previously agreed-to allocation of CER revenues.* It will usually be the case that the CER revenue allocation will be just one component of a larger package of benefits for which there will be various "*quid pro quos.*" If the regulator attempts to change this, the whole package may unravel.[25]

Sometimes the regulator may be explicitly asked by one or all parties to pass judgment on the allocation formula. *We recommend that the regulator adopt the general principle that the revenues should go to those who supplied equity or who took the lead and assumed the risk in developing the project.*

Key Recommendation

Regulators should not modify a previously agreed-to allocation of CER revenues, and if they are asked to create or pass judgment on an allocation formula, they should adopt the general principle that CER revenues should go to those who supplied equity or who took the lead and assumed the risk in developing the project.

We recommend this principle for three reasons. First, it is expensive both in time and money to develop a successful CER application. No rational developer will want to incur the expense of preparing an application if there is an expectation that the regulator may, after the fact, turn over the CER revenues to the SPP's customers by reducing the allowed revenues to be recovered through tariffs by an amount equal to the expected or realized carbon credit revenues.

Second, it is likely that the SPP's retail customers will already be benefitting from subsidized tariffs through external grants received for the project's capital costs. Therefore, rather than further reducing already subsidized tariffs, we think that it would be preferable for the revenues to go either to the developer of the project or the rural energy agency that has provided connection grants to the project. If it is the latter, the revenues could be used to create the equivalent of an electrification "revolving fund" that could fund capital cost subsidies for new SPPs. But if the regulator decides that CER revenues should go to some entity other than the developer or the REA, then both entities need to know this right from the beginning so that the other entity that will receive the revenues should incur the risk and expense of making the application.

Third, if there is uncertainty as to how the regulator will deal with CER revenues when setting tariffs, the developer's application may not be approved. CER credits will only be awarded if the developer can make a convincing case that the CDM revenues are needed to achieve financial viability. This is usually referred to as the financial additionality requirement. But if it is unclear whether the national electricity regulator will allow the developer to retain the CER revenues, then it will become difficult (and perhaps impossible) for the developer to make the argument that CER revenues will help to ensure the project's financial viability.

Advance Payments to Close the Equity Gap

At present, SPP developers in many African countries face an equity gap. This means that SPP operators are unable to acquire sufficient up-front capital to make initial capital investments to get an SPP mini-grid up and running. The current reality is that SPPs are generally not able to finance the total capital cost of mini-grids from their own funds and outside grants. Therefore, if a government wants to develop an SPP program that goes beyond a few pilot projects propped up by major government and donor contributions, it needs to find some way for SPPs to gain access to loans from local commercial banks on a regular basis. But local commercial banks are generally reluctant to loan money to SPPs, unless the SPPs or their outside investors can provide a significant amount of up-front equity.

Key Observation

SPP developers face an equity gap in many African countries, in that they cannot secure sufficient up-front equity capital. This, in turn, makes it difficult, if not impossible, to obtain loans from commercial banks.

Minimum Equity Requirements to Obtain Commercial Bank Loans
In Tanzania commercial banks have stated that they will not provide loans to an SPP operator unless the SPP is able to provide 30–40 percent of its own equity for capital equipment. (See table 5.3 for an overview of possible sources of financing for Tanzanian SPPs.) While the Tanzanian bank's high equity requirements may be reduced in the future as the banks gain more experience with SPPs,

Table 5.3 Sources of Funding for SPPs in Tanzania

Funding source		%	Comments
Debt financing		70	Long-term debt from local banks enabled by the World Bank line of credit
Equity requirement (to be arranged by the project developers)	In-kind equity	5	Valuation of developer's efforts in getting water rights, land, preparatory works, and so on
	Cash equity	10–15	Typical amounts (based on actual data from several projects)
	Connection grants from REA	5	Advance payment of a portion of the $500 per new connection
	Equity gap	5–10	Advance payments of carbon credit revenues + donor grants

Note: REA = rural electrification agency; SPP = small power producer.

the current situation is that very few potential SPP operators in Tanzania can provide 30–40 percent of the overall capital costs of a mini-grid system from their own resources. The situation in Tanzania is not unique. Similar equity requirements have been reported in other developing countries. For example, one of our authors reports: "In Sri Lanka, reported debt-to-equity ratios have been between 50:50 and 80:20. There is no fixed ratio, but most projects are reported to be in the range of 60:40–70:30. A newcomer (small investor, no previous record with the bank) or a new type of SPP (the first wind plant) would be required by bankers to go on 50:50, whereas a good hydro site done by a strong company with other credit history with the bank would be offered 80:20. It's all a matter to be negotiated between the developer and the bank" (Siyambalapitiya 2012).

Other Sources of Equity and Their Regulatory Treatment

To help close this equity gap, Tanzania's REA announced that it is willing to allow SPPs to treat the REA's $500 connection grants as if they were the SPP's own equity. But there is a difference between the REA grants and what might be termed "normal" equity that is supplied by the SPP operator or an outside investor. In the case of the REA grants, the grant is being given as a gift. In return for the grant, the REA expects that the SPP will connect a specified number of new rural customers, but does not expect to earn any returns. So the regulator should treat the grant like any other grant when setting tariffs: the SPP operator should be allowed to take depreciation on the capital financed by the grant, but would not be allowed to earn an equity return on the grant. This is different from the tariff treatment that would be given to "normal" equity supplied by an SPP operator or an outside investor in the project. In this latter case, neither the SPP operator nor the outside investor is providing the equity as a gift, but instead both are expecting a return on their investment. For this normal equity, the regulator should allow both a return on the equity supplied and depreciation on any capital equipment financed by the equity.

Another possible source of advance payments could be carbon revenues. Tanzania's REA and the World Bank are exploring the possibility of securitizing future carbon revenues. Again, this would be an advance payment that would give the developer more money to make up-front investments but unlike the REA connection grants, the securitized carbon revenues would not be a gift (table 5.3). Once the project is up and running, the SPP developer would receive a reduced amount of carbon revenues to reflect the fact that it received an advance payment of a portion of the carbon revenues. So in projecting the financial viability of the SPP, the regulator should project somewhat lower future carbon revenues to recognize the fact that some of the carbon revenues have been "securitized" (that is, received as a prepayment). At the time of this writing, it has been proposed that a "green generation grant" would provide an advance of 70 percent of the carbon revenues that are expected to be generated in the first eight years of operation.

Key Observation

To close the equity gap, two innovative funding mechanisms have been proposed: allowing SPPs to record REA and donor grants as their own equity, and "securitizing" future carbon revenues through a CER program—that is, receiving a larger sum of CER revenues up front, in return for decreased CER revenues later.

Notes

1. A full discussion of energy subsidies can be found in Reiche and Teplitz (2009).

2. An excellent introduction to the theory and practice of targeted consumer subsidies in the water and power sectors of developing countries can be found in Komives and others (2005).

3. A good introductory description of the CDM mechanism can be found at: https://cdm.unfccc.int/about/index.html.

4. These are the average official reported connection charges for large national utilities. The actual out-of-pocket costs for households may be higher if they have to pay a bribe to get to the front of the queue. For example, a villager pointed out, "They [households and businesses seeking a new connection] have to pay money to the engineers and the linemen ... There are separate bribe rates for setting up a pole, a transformer, a wire and a connection" (Lakshmi and Denyer 2012).

5. In contrast, it has been reported that about 33 percent of village households have signed up to lease rechargeable lanterns and battery boxes from the Omnigrid Micropower Company (OMC) in the Indian state of Uttar Pradesh within 45 days after the system was put into place (see box 2.2). This is probably attributable to the fact that OMC's customers do not have to pay a connection charge or security deposit (see Raj 2012).

6. Households within 31–70 meters would be charged between $871; those located within 71–120 meters, $1,288. The higher charges reflect the cost of installing distribution poles. In January 2013, TANESCO reduced its connection fees by 30–75 percent in response to a directive from the Government of Tanzania (www.tanesco.co.tz).

7. Techniques for lowering these costs were discussed in several sessions at the Africa Electrification Initiative (AEI) Dakar workshop. See, in particular, the presentations at the session titled "Low-Cost Solutions for Electrification" (http://go.worldbank.org /WCEDP90SZ0). A comprehensive manual on low-cost electrification techniques can be found in Karhammar and others (2006).

8. The reluctance of state-owned utilities to sell to rural customers is not limited to Africa. The same is true in India. It has been estimated that state-owned enterprises lose, on average, about 8 U.S. cents per kWh supplied in rural areas (Dixit 2012). Therefore, it should not be surprising that even after rural villages are connected, state-owned utilities in India try to minimize the number of hours of electricity that they supply to rural customers so as to minimize their losses. It has been estimated that about 75 percent of grid-connected rural households in India average at least four hours per day of outages. The Central Government has issued statements and policies to try to overcome this "reluctance to supply." For example, the 2006 Rural Electrification Policy states that it will be "necessary that the distribution licensee follows [a] non-discriminatory approach towards the franchisees in case of power supply shortage" (Government of India and Ministry of Power 2006, section 9.11). But in the absence of credible positive incentives (profits) or negative disincentives (penalties) for compliance, it seems unlikely that these government policy directives will have much effect.

9. Engineering techniques for reducing connection costs are discussed more fully in Golumbeanu and Barnes (2013, appendix B).

10. But not all donors tie disbursement of their grants to independently verified connections of individual households or businesses. For example, the ACP-EU Energy Facility II: 2nd Call for Proposals of the European Commission requires that the project seeking a grant provide evidence that the new power lines (whether from a main-grid extension or a new mini-grid) will reach at least 30,000 beneficiaries. A beneficiary appears to be defined as a person who lives in the village. There is no requirement that the person must actually use electricity in her home. A beneficiary can be a direct beneficiary (that is, she and members of her household receive electricity in their home), or an indirect beneficiary (the household now has access to cold sodas in village shops or a computer in a village school). The fact that indirect beneficiaries are included in the output measure may create incentives to connect a village but not to connect poor, low-consumption households—with the result that the cost per achieved connection paid for by the grant may be very high (see http:// ec.europa.eu/europeaid/where/acp/regional-cooperation/energy).

11. In Tanzania, it has recently been proposed that the level of the REA's connection grants should vary among rural energy providers. It has been pointed out that the capital costs of a shared solar micro-grid that provides low power levels of DC electricity at 24 volts to households in a village (www.devergy.com) will be much lower than the capital costs of a hybrid mini-grid (solar, diesel, and batteries) that provides AC electricity. Moreover, the shared solar DC electricity system provides a lower level of service. For example, the electricity that it provides cannot power the operation of most machines. Hence, in terms of the electrification ladder described in chapter 2, the shared solar micro-grid is providing electricity service at a lower step on the ladder. As a general rule, we think that connection grants should be keyed to the level of electricity that the mini-grid can provide using some variation of the electrification ladder framework rather than being based on an analysis of the capital costs of the provider's facilities. In other

words, the grants should be keyed to the outputs provided rather than to the inputs installed.

12. However, the Malian government is currently discussing a new program with the World Bank that would provide grants through AMADER to operators of the existing diesel systems to convert their systems to hybrid generating systems comprising diesel, solar, and battery components.

13. While our focus here is on outside grants, similar tariff-setting issues arise for customer capital contributions and customer connection fees.

14. This is not just a regulatory issue for SPPs. The same regulatory issue exists for large national utilities that have received outside grants and who use the grants to connect new customers either through grid extensions or new off-grid installations.

15. This assumes that an SPP's tariffs are set based on the SPP's actual costs. This need not be the case. Under Tanzanian law, EWURA has the authority to set tariffs on a generic basis. Section 23(4) of Tanzania's 2008 Electricity Law states that: "[EWURA] may prescribe maximum tariffs of a generic nature of simplified tariff methodologies, applicable to licensees or persons exempted under section 18" (EWURA 2008). Entities that are exempt under section 18 are SPPs with an installed generating capacity of less than 1 megawatt (MW) or SPDs serving an off-grid system with a total maximum demand of 1 MW or less. This section would give EWURA the authority to set SPP retail tariffs at the same level as the retail tariffs charged by the national utility (TANESCO), or by shared characteristics (for example, generating technology) of SPPs. (Further discussion of the theory and practice of retail tariff setting for SPPs is given in chapter 9.)

16. Programs that give customers the option of paying for connection costs in smaller separate payments spread out over time are operating or have been proposed by national utilities in Côte d'Ivoire, Senegal, and Kenya.

17. While EEPCO is not an SPP, the regulatory issues would be the same for an SPP.

18. This policy has been adopted by AMADER, the Malian REA. The annex to the concession agreement awarded by AMADER states that one component of the allowed electricity tariff should be "linked to the pre-financing by the operator of the cost of connection, customer interface (circuit breaker, energy meter, etc.) for interior installations, and electrical equipment such as lighting units" (AMADER, undated, article 7.3). Under Malian law, AMADER is both grantor and regulator for isolated mini-grids.

19. But if the consumption of a substitute product such as kerosene is also being subsidized, it is not clear that a cross-subsidy that lowers the cost of consuming electricity is necessarily inefficient.

20. Since it usually costs more to serve a rural customer, a uniform national tariff will lead to geographic cross-subsidies (that is, urban customers subsidizing rural customers).

21. At the time of this writing, it has been estimated that the cost of putting together a CDM application is about $500,000–750,000.

22. A good description of the project can be found in the project design document (PDD) filed with the CDM executive board. See http://cdm.unfccc.int/Projects/DB/DNV -CUK1204739473.81/view.

23. For an SPP that proposes to create an isolated, greenfield mini-grid, the CER would probably have to be calculated on the assumption that the SPP is mostly providing an alternative to kerosene lanterns.

24. If the value of the CER is $12, then an SPP operating on an existing isolated mini-grid would receive $10.56 per megawatt-hour (MWh) generated or $0.01056 per kilowatt-hour (kWh) generated. In 2010 the SPP received $0.246/kWh sold at wholesale on existing isolated mini-grids. Hence, the opportunity to earn CER revenues would increase its wholesale revenues by about 4.3 percent per kWh produced. The percentage increase would be 9.9 percent for SPPs selling on the main grid because the 2010 wholesale tariff for grid connected is much lower ($0.066 instead of $0.246). The overall increase in SPP revenues will depend on the price of the CER credits and the future feed-in tariff (FIT) prices allowed by the regulator for these two types of sales.

25. We would also recommend applying the same rule if an SPP receives top-up payments from donors to raise the effective price received under FITs. Such payments are designed to improve the commercial viability of renewable generation projects. If the regulator "claws back" such payments by reducing the base FIT, it is likely that top-up payments will not be given by donors in the future.

References

Adama, Sissoko, and Alassane Agalassou. 2008. "Mali's Rural Electrification Fund." Presentation at the Sustainable Development Week, Washington, DC, February. http://siteresources.worldbank.org/INTENERGY2/Resources/presentation8.pdf.

AEI (Africa Electrification Initiative). 2012. *Institutional Approaches to Electrification: The Experience of Rural Energy Agencies/Rural Energy Funds in Sub-Saharan Africa*. Washington, DC: World Bank. http://siteresources.worldbank.org /EXTAFRREGTOPENERGY/Resources/717305-1327690230600/8397692 -1327690360341/AEI_Dakar_Workshop_Proceedings_As_of_7-30-12.pdf.

Dixit, Shantanu. 2012. "Powering 1.2 Billion People: Case of India's Access Efforts." Presentation at World Bank Energy Days 2012 Conference, Washington, DC, February 23.

EWURA (Energy and Water Utilities Regulatory Authority). 2008. *Standardized Tariff Methodology for the Sale of Electricity to the Main Grid in Tanzania under Standardized Small Power Purchase Agreements*. Dar es Salaam, Tanzania. http://www.ewura.go.tz /pdf/public%20notices/SPP%20Tariff%20Methodology.pdf.

———. 2012. *The Electricity (Development of Small Power Projects) Rules*. Proposed for Public Consultation. Dar es Salaam, Tanzania.

Faulhaber, Gerald R., and Stephen B. Levinson. 1981. "Subsidy-Free Prices and Anonymous Equity." *American Economic Review* 71 (5): 1083–91.

Golumbeanu, Raluca, and Douglas Barnes. 2013. *Comparisons of Grid Connection Costs and Electricity Access in Developing Countries*. Africa Electrification Initiative, World Bank, Washington, DC.

Government of India, and Ministry of Power. 2005. *National Electricity Policy*. http://218.248.11.68/energy/NationalElectPolicy6.asp?lnk=26.

———. 2006. *Rural Electrification Policy*. http://www.powermin.nic.in/whats_new/pdf /RE%20Policy.pdf.

Government of Peru. 2007. "General Rural Electrification Law." Article 25.

Government of South Africa. 2008. *Electricity Pricing Policy*. http://www.info.gov.za/view /DownloadFileAction?id=94204.

Hanley, Christina. 2010. "Feed-in Tariff Readiness." Presentation at the Renewable Energy Policy Workshop, World Resources Institute, Washington, DC, November 22. http://powerpoints.wri.org/repw_hanley_fit_readiness_panel.pdf.

Karhammar, Ralph, Arun Sanghvi, Eric Fernstrom, Moncef Aissa, Jabesh Arthur, John Tulloch, Ian Davies, Sten Bergman, and Subodh Mathur. 2006. "Sub-Saharan Africa: Introducing Low-Cost Methods in Electricity Distribution Networks." ESMAP Technical Paper, World Bank, Washington, DC.

Komives, Kristin, Vivien Foster, Jonathan Halpern, and Quentin Wodon. 2005. *Water, Electricity, and the Poor: Who Benefits from Utility Subsidies?* Directions in Development Series. Washington, DC: World Bank. http://siteresources.worldbank.org/INTWSS/Resources/Figures.pdf.

Lakshmi, Rama, and Simon Denyer. 2012. "Lack of Power Symbolizes India's Inequalities." *Washington Post*, August 6. http://www.washingtonpost.com/world/asia_pacific/lack-of-power-symbolizes-indias-inequalities/2012/08/06/ecdbef64-df20-11e1-a19c-fcfa365396c8_print.html.

Mumssen, Yogita, Lars Johannes, and Geeta Kumar. 2010. *Output-Based Aid: Lessons Learned and Best Practices.* Directions in Development: Finance. Washington, DC: World Bank. https://openknowledge.worldbank.org/bitstream/handle/10986/2423/536440PUB0outp101Official0Use0Only1.pdf?sequence=1.

Mwenga Hydro Limited. 2012. *Application for Tariff Approval by Mwenga Hydro Ltd (MHL) Submitted to EWURA.* Dar es Salaam, Tanzania.

NRECA (National Rural Electric Cooperative Association). 2012. *Affordability Analysis and Options for a Program to Make the Cost of Rural Household Grid Connections Affordable.* Unpublished draft report, Arlington, VA, June.

Raj, Anil. 2012. "The Micropower Opportunity: Paving the Way for Rural Electrification." PowerPoint Presentation, World Bank, November 15.

Reiche, Kilian, and Witold Teplitz. 2009. "Energy Subsidies: Why, When and How? A Think Piece." GTZ, Eschborn, Germany. http://www.medemip.eu/Calc/FM/MED-EMIP/OtherDownloads/Other_Energy_Topics/201008_Energy-Subsidieswhy_when_and_how.pdf.

Revolo Acevedo, Miguel. 2009. "Mechanism of Subsidies Applied in Peru." Presentation at the AEI Practitioners Workshop, Maputo, Mozambique, June 9. http://siteresources.worldbank.org/EXTAFRREGTOPENERGY/Resources/717305-1264695610003/6743444-1268073611861/11.3Mechanism_subsidies_applied_in_Peru.pdf.

Rickerson, Wilson, Christina Hanley, Chad Laurent, and Chris Greacen. 2012. "Implementing a Global Fund for Feed-in Tariffs in Developing Countries: A Case Study of Tanzania." *Renewable Energy* 49 (Special Issue: Selected Papers from World Renewable Energy Congress—XI): 29–32.

Sawe, Estomih N. 2005. "Rural Energy and Stoves Development in Tanzania." Conference Presentation at the Workshop on Rural Energy, Stoves, and Indoor Air Quality in China, Beijing, January 14.

Siyambalapitiya, Tilak. 2012. Personal e-mail communication. August 22.

Todeschini, Luca. 2011. Personal communication. February.

U.S. DOE (U.S. Department of Energy). 2007. "The Potential Benefits of Distributed Generation and Rate-Related Issues That May Impede Their Expansion." A Study Pursuant to Section 1817 of the Energy Policy Act of 2005. February. http://www.ferc.gov/legal/fed-sta/exp-study.pdf.

Vernstrom, Robert. 2010. *Long-Run Marginal Cost of Service Tariff Study*. Final Report to Tanzania Electric Supply Company, Menlo Park, CA, May. http://www.ewura.go.tz /pdf/Notices/Tariffs%20COSS%20Final_Annex%20to%20the%20TA.pdf.

World Bank. 2012. *World Data Bank*. Washington, DC. October 31. http://databank .worldbank.org/ddp/home.do.

CHAPTER 6

Regulatory Decisions for Grid-Connected Small Power Producers

No contract is ever complete.

—Dr. Edward Kahn, 2005

No performance guarantees, no penalties, build as you please, operate as you please.

—One buyer's criticism of power-purchase agreements of small power producers

Yes, the price is fine, but will I ever actually be paid?

—Developer of a small power producer in Tanzania

Abstract

In chapter 6 we examine the terms and conditions of standardized power-purchase agreements (SPPAs) for small power producers (SPPs) and address deemed energy clauses and performance requirements for SPPs. These are issues of particular concern to SPPs in Sub-Saharan African countries with weak transmission systems, insufficient generating capacity, or both. We also consider the contentious issue of backup tariffs for main-grid-connected SPPs. Backup tariffs can have a major effect on the commercial viability of main-grid-connected SPPs.

In most discussions of grid-connected SPPs, one topic typically gets the most attention: feed-in tariffs (FITs). A FIT refers to the price that an SPP or a larger renewable generator will receive for the wholesale power that it sells to the national utility, the national system operator, or other obligated purchaser of its power. FITs, which are discussed in detail in the next chapter, are obviously important because they directly determine an SPP's revenues. If an SPP's revenues do not cover its costs, the project will simply not be viable. But a FIT, even if set at cost-recovering levels, does not, by itself, guarantee an SPP's commercial viability; it must be one element of a larger regulatory and policy support package. The full package must include the following elements:

- Guaranteed interconnection to the grid with prespecified rules for assigning responsibility for the costs of the interconnection (chapter 8)
- Standardized interconnection and operation procedures (chapter 8)
- Guaranteed purchase of power whenever it is produced by the SPP (usually referred to as a "must-take" or "priority dispatch" requirement) (chapters 6 and 8)[1]
- Physical capability of the purchasing entity to receive (that is, "evacuate") the power (chapter 8)
- A fixed, prespecified pricing formula for the purchase of the SPP's power with a clearly defined adjustment mechanism for the life of the contract (chapter 7)
- A regulatory mechanism through which the utility buyer can recover the costs of wholesale purchases from an SPP; this is usually done through an automatic pass-through of these costs to the buyer's retail customers (chapter 9)
- A standardized power-purchase agreement (SPPA) with a duration at least as long as the prespecified FIT pricing formula (chapter 6)
- Guaranteed sale of backup power by the utility to the SPP when needed because of planned or unplanned outages (chapter 6)

These last two elements are discussed in more detail in this chapter.

Comparing the Purchase Agreements of SPPs and Independent Producers

In most countries with successful SPP programs, the power-purchase agreement (PPA) used by SPPs is a standardized, short, nonnegotiable document. This so-called SPPA (for standardized PPA) is published and available for anyone to study. The SPPA is exactly the same for all SPPs except for a few blank pages to fill in with project-specific information such as the company names of the seller and the buyer, the name and technical details of the power-generating facility, and an electrical diagram showing how the generator is connected to the main grid or to an existing mini-grid. Thailand's SPPA for very small power producers (VSPPs) is only five pages followed by two appendices; Sri Lanka's SPPA is 20 pages with 4 appendices; and Tanzania's is 21 pages with 3 appendices.[2]

By contrast, a typical PPA for a larger generating plant, using either renewable energy or fossil fuels, is project specific, and may have a number of negotiable clauses. A PPA may run to 100 pages or more, as the buyer and the seller and their lawyers enter into lengthy negotiations for several months or years to manage their risks by adding new clauses and conditions. If the buyer undertakes a competitive solicitation, he may issue a draft PPA along with the request for proposal, and this will limit the scope of negotiations. In many countries, PPAs for major independent power producers (IPPs) are usually not published, and carry clauses, known as nondisclosure agreements or NDAs, that explicitly prohibit the signatories from disclosing the contents of the PPA and related documents. But this is not the norm in all countries. For example, in at least several states in the United States, the electricity regulators require that PPAs for major IPP purchases be publicly available documents.

Rationale for Standardizing the Power-Purchase Agreements of SPPs

SPP PPAs need to be standardized for two reasons. The first is that an individual SPP will have little or no negotiating power with a national utility. In most instances, the national utility is the only available buyer (that is, the monopoly buyer), so it is in a position to completely dictate the terms and conditions of the PPA. Also, if the national utility does not wish to purchase electricity from the SPP, it can dictate terms that ensure that the SPP project will not be commercially viable. Therefore, without the intervention of the regulator, the bargaining power of the two parties will be totally lopsided in favor of the national utility. As one SPP developer observed: "without standardized PPAs, we would live in a world of never-ending negotiations." In addition, if a regulator accepts modifications to an SPPA, this in effect allows utilities to use their superior bargaining power to force SPPs to make concessions that are then formally filed as "joint requests" supported by both parties. As another SPP developer observed: "I would be forced to come to the regulator with a smile on my face . . . but it would actually be a grimace."[3]

The second reason is to avoid overwhelming the limited administrative capacity of the national regulator. If an SPP program is successful, it could easily lead to the development of 50 or more SPPs. In Thailand, for example, over 960 PPAs have been signed for small renewable energy projects (EPPO 2012). Neither the regulator nor a government ministry will have the human resources to review a separate PPA for each SPP. More important, the review of dozens or hundreds of different PPAs would not be a good use of the regulator's time. In most developing countries, a regulator's limited time and resources are best spent focusing on improving the national utility, because its performance will have a greater impact on the country's economy. Hence, from a regulator's perspective, the principal advantage of an SPPA document is that it economizes on the regulator's time and resources; he has to conduct only a single major review of one SPPA rather than many separate reviews of different PPAs. It also facilitates financing by local banks that have limited experience with SPPs. The downside of this approach is that one SPPA may not be equally suitable for all potential SPPs. The decision to use an SPPA means that the regulator or government has decided that the societal benefit of getting investments from many SPPs outweighs the cost of losing a few SPPs whose business or technical needs are not accommodated by the SPPA.

Key Recommendation

PPAs should be standardized for all SPPs to eliminate the imbalance in bargaining power that SPPs face if they have to individually negotiate a PPA, to avoid overburdening the national electricity regulator with individual PPA reviews, and to ease the work of local banks that are considering loans to SPPs.

The Length of Power-Purchase Agreements

The SPPAs of SPPs are typically shorter than the specific PPAs used by IPPs, for several reasons:

- Some issues that would normally have to be dealt with in a regular PPA are addressed in SPP regulations or rules issued by the regulator or some other government agency. This allows the SPPA to make reference to these government or regulatory documents rather than repeating the same text in the PPA.
- SPPs that sell onto the main grid are usually "must-take" plants (that is, if the SPP is generating electricity, the utility must accept it). This simplifies the commercial and operational relationship between the buyer and the seller.
- In the case of SPPs, both the SPP and the buyer are typically domestic entities. Hence, there is no need to provide for international dispute resolution in case of disagreements.
- SPP financing is usually provided by domestic banks, which will often have ongoing business relationships with the seller and the buyer and, therefore, are generally willing to accept a lower level of specificity in the SPPA than international banks would of a PPA.

Obtaining Bank Financing on the Basis of Standardized Power-Purchase Agreements

Banks generally prefer to finance projects based on tightly defined contracts between their borrower (the SPP) and the buyer (utility). From a legal point of view, banks tend to view the SPPA as a somewhat weak agreement. Nevertheless, SPPAs have a track record of being financeable, bankable contracts in countries such as Sri Lanka and Thailand. Bankers generally appreciate the fact that the document is standardized and fully disclosed up front when they review loan applications. If there were a history of buyers and sellers amending the SPPA, this would generally harm the spirit of the agreement, and prompt the banks to be more cautious on accepting the agreement as bankable. To date, it is estimated that more than 100 SPPs in Sri Lanka and approximately 250 VSPPs and 60 SPPs in Thailand have received bank loans on the basis of the SPPA. In both countries (as well as in Tanzania), SPPs have benefited from government credit support programs that allowed banks to lengthen the duration of the loans and to lower the interest rate that is charged. Even with these support programs, however, the risk of nonpayment of the loan by the borrower remains solely with the bank and not with the government. Also, in both Asian countries, the SPP and VSPP programs have had the advantage of the buying entities being commercially solvent, which is not the case in many Sub-Saharan African countries.

In Sri Lanka it is generally recognized that in the initial years, the SPPA and the project profitability indices were not the only criteria on which bankers decided to lend to SPPs; the borrower's track record with the bank for their other non-SPP businesses was an overriding factor. The common wisdom was that for a good client of a bank, even a badly designed SPP with low profitability could be financed, and the strengths and weaknesses of the PPA were of secondary importance. There has now been a history of more than 15 years of adherence to the terms and conditions of the SPPA in Sri Lanka. So despite the frequent debates on FIT values (discussed more fully in the next chapter), local banks in these countries clearly now have confidence in the SPPA as evidenced by the fact that they routinely make loans based on its terms and conditions.

In Tanzania the situation is different. Several potential SPP developers complained to the national electricity regulator at a June 2012 public meeting that the current PPA for grid-connected SPPs was not bankable on a project finance basis by the standards of non-Tanzanian financial institutions.[4] Among the weaknesses that they pointed to in the current SPPA were: currency risk, since the payments would be in Tanzanian shillings but debt payments would be in hard currencies like U.S. dollars or euros; lack of indexing for the FIT price floor; and no required payments if the buying utility was unable to receive energy from the SPP because of problems on the buyer's transmission grid (see the later discussion of "deemed energy" clauses). In a private conversation, a representative of a local Tanzanian bank stated that "80 percent of the problem was the insolvency of the buying entity Tanzania Electric Supply Company (TANESCO)" and "20 percent of the problem was in the other terms of the PPA."

A similar situation exists in Uganda, but the government has indicated that it is willing to guarantee payments to SPPs by the government-owned buying entity (the Uganda Electricity Transmission Company Limited, a state-owned enterprise) by purchasing a partial risk guarantee (PRG) for these payments from the World Bank. If implemented, this would greatly reduce the risk to SPPs that they will not be paid and to banks that the SPPs to whom they provide loans will be unable to make interest and principal payments on those loans.[5] In March 2013 initial discussions began for setting up a similar PRG for SPPs in Tanzania. But until this is put in place it remains to be seen whether SPP development in Tanzania will be limited to existing companies that are in a position to use their balance sheets to back up their loan applications.

Key Observation

In Sri Lanka and Thailand, SPPs have obtained loans from local banks on the basis of SPPAs. SPP developers in Africa have stated that for an SPPA to be "bankable" (that is, to be able to get loans from commercial banks), it must contain provisions that deal with currency risks, fully adjust tariffs for input cost inflation, and cover the risk of lost revenues if the buying utility is unable to receive SPP energy because of problems on its grid.

Optimal Duration of a Power-Purchase Agreement for a Grid-Connected SPP
A PPA for a grid-connected SPP should have a duration long enough to repay the project debt. This would be required by any bank that is considering making a loan to an SPP. While the term of project loans vary from project to project (with a typical minimum of 7–10 years), PPAs and licenses in Tanzania have been set

for 15 years, and SPPs have requested that this be extended to 20 years. The PPA should also have a duration at least as long as the mandated availability of a FIT. When it is shorter, it creates an anomaly: the SPP is offered a specified price, but the national utility has no legal obligation to purchase at that price once the PPA expires.

This is the case in Thailand. There, SPPAs have a duration of one year but FITs are for longer periods (10 years for solar and wind power and 7 years for all other renewable energy sources). So far this has not caused problems because developers have trusted the purchasing utilities to renew these contracts, and the utilities have done so. But this is a major inconsistency; in our view PPA durations should be commensurate with FIT durations, if not longer.

Key Recommendation

The PPAs of grid-connected SPPs should have a duration long enough to repay the project debt, and at least as long as the mandated availability of a FIT. This will make it possible for SPPs to obtain project financing, make their scheduled loan payments, and ensure that the PPA tracks the FIT.

SPPs versus IPPs Connected to the Grid: The Buyer's Purchase Obligations

As noted above, an SPP is a "must-take" or "priority dispatch" facility: if the SPP facility is generating electricity, the utility must accept it. In the power industry, such power plants are also described as nondispatchable, because the system operator cannot dispatch the output of the power plant as it sees fit to balance overall available electricity generation and purchases with overall load. To meet overall customer demand on the system, the system operator must accept the electricity produced by all must-take power plants, and then call up other dispatchable power plants in the order of their operating costs (this process is referred to as merit-order dispatch) to provide any additional needed supplies. If the SPP power plant is must-take, then the buyer must take the electricity produced at all times.[6]

Hence, investing in and running an SPP has a unique advantage not available to most businesses: by signing the SPPA, the buyer unconditionally agrees to purchase the entire production of the facility, at a predefined price or formula, for the entire period of the agreement, with no restrictions imposed. Utilities complain that even if the electricity produced by SPPs is not required by the grid, or if it is not competitive with the prices of other generators at any given time, the power still has to be purchased, because of the must-take clause in the SPPA. In Sri Lanka this situation occurs once every few years, when larger storage reservoirs spill water because of heavy rainfall. When several hydropower reservoirs spill and the output of other large power plants are adequate to meet customer demand, there is no need for additional energy supplied by the SPPs. Buying electricity from SPPs in such situations will cause more water to spill out of the reservoirs, yet the buyer has to pay SPPs for their electricity even though the same would have been available from its own hydro plants at zero cost.

The PPAs of larger IPPs are written differently. A typical PPA with a large IPP will state that the buyer determines when and to what level of output the IPP power plant will be operated every hour. A dispatch schedule is provided to the IPP in advance by the grid system operator, who reserves the right to order a shutdown, require the power plant to remain on standby, and start up to produce electricity again when needed (subject to limits on the number of separate start-ups during a specified time period). Through this mechanism, the buyer attempts to run its own power plants and the plants of IPPs at different levels of output, so that the total cost of meeting the overall customer demand at any given moment will be the lowest possible. Under this arrangement, an IPP typically receives a capacity payment that ensures that it can recover costs and earn a profit even if it is not dispatched. In the event that the buying utility has a surplus of electricity from its own power plants, the IPPs are not dispatched, but under typical "take or pay" provisions, the buyer makes a capacity payment to the IPP regardless of whether the IPP is called on to produce electricity.

Key Observation

Typical SPPAs for grid-connected SPPs stipulate that the buyer agrees to purchase the entire output of electricity, for the entire period of the agreement, with no restrictions imposed. In contrast, the PPAs of larger IPPs specify that the IPP will be dispatchable, which means that the buyer has the right to determine (within limits) when and to what level of output the IPP will operate at every hour. In return, the IPP is guaranteed "capacity payments" if its plant is available, even though it may not be called on to generate electricity. In contrast, an SPP earns money only if it generates and sells electricity. Unlike IPPs, SPPs generally do not receive capacity payments.

Should a Buyer Be Required to Accept a Grid-Connected SPP's Entire Output at All Times?

We recommend that the purchasing utility have a "must-take" obligation to purchase the SPP's output for two reasons:

- SPPs typically generate from a renewable energy source, which cannot be stored (for example, run-of-river hydro, wind, solar); others use waste heat from an industrial process. The electricity produced by cogeneration SPPs is strongly linked to an underlying industrial process; therefore, the SPP has no control over the primary energy resource. If the energy source is not used, it would most likely go to waste. SPPs using biomass are the only exception because the resource can be stored and used when required.

From the Bottom Up • http://dx.doi.org/10.1596/978-1-4648-0093-1

- SPP tariffs are generally based on only the electricity produced; if there is no electricity production for whatever reason, no payments are made to the SPP. In contrast, a PPA for a larger IPP almost always carries capacity payments for remaining on standby (that is, being ready to generate).

Key Recommendation

PPAs for grid-connected SPPs should have a "must-take" clause that obligates the utility to purchase all of the SPPs' electrical power output, because SPPs typically have no control over the amount of electricity they produce at any given time, and they usually receive payments based solely on electricity produced, rather than receiving capacity payments just for being ready to generate electricity.

Should the Power-Purchase Agreement Include a "Deemed Energy" Clause?

Deemed energy refers to a situation in which a main-grid-connected SPP seller is able to produce electricity, but the buyer is unable to receive it. A "deemed energy" clause in the SPPA obligates the buyer to provide compensation for electricity that the SPP was capable of producing but the buyer was unable to receive.

Physical Conditions That Prevent a Utility from Receiving an SPP's Energy

In Africa and elsewhere, large (usually national) utilities may not be able to receive an SPP's electricity for different reasons. The three most common reasons (given below) and the relative importance of each may vary from country to country and between different regions within a single country.

Insufficient Overall Generation Capacity

Many African countries do not have enough installed generation capacity to meet overall customer demands. And even if they do, they still may not be able to generate sufficient electricity at all times of the year (for example, when there is insufficient water at hydroelectric dams to generate electricity, as has happened in Tanzania and Kenya). When either of these conditions exists, the national utility will be unable to meet the total customer demand and, if nothing is done, there is a risk of an uncontrolled nationwide blackout. To avoid this, most utilities engage in rolling blackouts—the utility intentionally sheds load in one or more geographic areas to bring demand and supply back into balance. To spread out the inconvenience to customers, the utility may rotate the blackouts among several different geographic areas and switch off one or more outgoing distribution lines from high-voltage substations. Any SPPs connected to these distribution lines (typically at voltages ranging from 11 kilovolts [kV] to 38 kV) will be automatically shut down and will not be able to sell electricity to the buying utility, even though they were ready to do so.

Insufficient Capacity or Damage to the Local Distribution Network

The distribution system comprises the equipment that carries electricity from the transmission lines to end-use consumers, and includes substations as well as distribution lines that generally operate at 33 kV or less. In some cases, the customer loads have grown over the years to the point that they overload the substation's transformers, causing voltage sag or even prompting the whole substation to trip off line. While the addition of distributed generation in the form of an SPP helps alleviate this condition (especially if the SPP's generation is not intermittent), the addition may not be sufficient to resolve the problem. High loads or faults can cause breakers on these lines to trip, or produce voltage or frequency deviations that will force SPPs' relays to trip. In addition, SPPs are often connected to the national grid via relatively long overhead distribution lines which can be damaged by falling trees, theft of valuable materials, animals, and high winds, resulting in the SPP not being able to sell its electricity to the buying utility.

Weaknesses on the Receiving Utility's Transmission Grid

Many countries in Africa experience transmission line outages that may force the buying utility to fully or partially curtail its purchases from one or more SPPs. While some risks of damage can be minimized by route selection, the use of higher-quality materials, and regular maintenance of the lines' right-of-way, outages may still occur especially in bad weather. Repairs may take a few hours to several days, and an SPP will not be able to sell electricity to the buying utility if it is connected to a section of the transmission system that has gone down and needs to be repaired.

Conflicting Views of the Small Power Developer, the National Utility Manager, and the Regulator on Deemed Energy Clauses

As soon as one starts to discuss deemed energy clauses, it quickly becomes apparent that SPP operators, utility buyers, and electricity regulators have widely divergent views on the need for and workability of such clauses. The following are typical comments of the key stakeholders.

From a grid-connected SPP operator

This is the ultimate indignity. I have to shut down because of a disturbance or lack of capacity on the national utility's system. Even though the utility caused the problem, I am the one who gets hurt. And I get hurt in two ways. First, I lose revenues because I am not able to make sales. Second, I have to pay the utility's high demand and energy charges under its backup tariff because I need its electricity to restart my generator each time the shutdown ends. So the bottom line is that it was totally the utility's fault but I'm the one who suffers. Let's be honest with each other. If there is no deemed energy clause in the PPA, then the buying utility's obligation to take the energy that I produce is nothing more than a joke. This is totally unfair.

From a buying utility

This SPP developer has no right to talk about fairness. The PPA gives him a very sweet deal. He gets payments from me but he does not accept any obligations for himself. His SPP plant can come online whenever he is ready, regardless of my needs. And then he gives me no commitments on the timing and amount of the energy that he will produce. We have seen SPPs going out of service for months, and we do not get any compensation. And when his plant is available, I have to be ready to take whatever energy he produces. Sure, it is true that larger IPPs get capacity payments regardless of whether or not I take their energy. But IPPs, unlike SPPs, have assumed obligations and commitments that SPPs are unwilling or unable to offer. So if he wants a deemed energy clause in the PPA, he should be willing to offer an availability commitment just like the IPPs. Finally, let's not forget that these SPP developers made their investments with open eyes. I never hid the fact that there are weaknesses on my transmission and distribution grids or that there are times when I need to initiate rolling blackouts so the entire country does not get blacked out. None of this was a secret. And he chose to go ahead with his investment. So he is the last person in the world who should be raising any issues of fairness.

From a national electricity regulator

Look, the reality is that I have very limited regulatory resources at my disposal. I can't be asked to adjudicate a dispute as to who was at fault every time some SPP says that he was ready to make a sale but the buying utility couldn't receive that power for one reason or another. If I have to get involved every time there is a disagreement over who is at fault, I won't have any time left for the more important tasks of setting tariffs and getting better performance from the national utility and other big players in the sector.

Three Proposed Solutions

In discussions of deemed energy clauses and SPPs, three options are usually proposed:

- *Option 1: Compensation if the buyer was at fault.* Under this option, the SPP would be paid a full or a partial kilowatt-hour (kWh) charge whenever it is determined that the seller could not make the sale because the buyer was at fault. The key words here are *it is determined*. Who makes that determination and how often will it have to be made? It is likely that many such disputes will be brought to the regulator by SPP developers. In response, we would expect buyers to argue that the interruption was the SPP's fault or the result of an "act of God." Even if the regulator is able to sift through the evidence and make a timely decision on the cause of the interruption, there would still be a need for accurate records as to how many hours the buyer's system was down as well as plausible estimates of the amount of energy that the SPP could have produced during the downtime. Accurate records of this kind are generally unavailable or too difficult to analyze in a timely manner. This option could quickly overwhelm the regulator with the need to adjudicate hundreds of disputes and greatly increase transaction costs for all parties— the buyer, the seller, and the regulator—which is fundamentally inconsistent with the overall strategy of the SPP regulatory system, which is to minimize transaction costs.

- *Option 2: "No-fault" compensation.* Under this option, the SPP would be compensated with a payment if *any problem* on the buyer's system made it impossible for the buyer to receive delivery of the SPP's electricity, without any attempt to determine whether the buyer was at fault. The rationale for this approach is that the alternative of deciding whether the buyer was responsible for the fault (option 1) would be too time consuming and costly. Instead, payments would be required after a certain prespecified period of time. At first glance, this no-fault approach seems to be the one taken in the proposed SPPA for Kenya:

Any such stoppage shall, where the stoppage exceeds seven (7) days, thereafter entitle the Seller to payment by the Buyer for Deemed Generated Energy for the period in excess of seven (7) days as hereinafter provided. (ECA and Ramboll Management Consulting 2012, paragraph 6.14)

But in fact, the proposal for Kenya has two caveats that would block automatic no-fault payments by the buyer. First, the buyer would not need to make any payment to the SPP if it undertakes planned maintenance in accordance with prudent operating practice. Second, the buyer would have "no obligation to pay for Deemed Generated Energy if the failure or inability of the Buyer to receive delivery of electrical energy from the [SPP] Plan is caused by Force Majeure" (paragraph 6.16.2). Hence, the proposed deemed generated energy clause in Kenya is not a purely no-fault clause. It comes with two caveats, and the *force majeure* exception is likely to lead to many disputes as to whether the fault was really beyond the buyer's control. Thus, like option 1 it could be difficult and expensive to administer.

- *Option 3: No deemed generated energy clause.* This is the approach taken in Thailand, Sri Lanka, and Tanzania, where the SPPAs do not include a deemed generated energy clause. In the two Asian countries, deemed energy has been a nonissue because both countries have generally adequate and stable transmission grids that can reliably receive delivery of SPP power. But this is not the case in Tanzania, where one new SPP reported 40 line trips in a five-month period mostly caused by a local substation that was overtaxed by growing municipal loads. What is unclear is whether the SPP operator knew of this risk before making his investment; if he did, it could be argued that he made the investments with open eyes.

All things being equal, we favor this last option because we believe that it would be very costly and time consuming to implement either of the first two options. If there is no deemed energy clause in the SPPA, however, we think that the buying utility must have an obligation to provide historical data on the frequency and duration of interruptions at the substation to which the SPP wishes to connect, and this should be provided to the SPP developer early in the development process before he spends significant amounts of money. The buying utility should also indicate actions that could be taken to reduce interruptions.

Key Recommendation

In African countries with weak transmission systems, PPAs with SPPs should not include a deemed generated energy clause because they will be difficult to administer and will increase regulatory transaction costs. But the purchasing utility should be required to provide historical data on the frequency and duration of interruptions at the substation to which the SPP wishes to connect. This will prevent unnecessary cost and time burdens for all parties involved, and give the SPP additional information on which to make its investment decision.

Performance Requirements for Small and Independent Power Producers in Typical Power-Purchase Agreements

The SPPAs of SPPs are most often favorable to SPPs because they do not include penalties for nonperformance once the SPP becomes operational (this is the case in Thailand, Sri Lanka, Tanzania, and many other countries). Once the power plant is commissioned, it can be operated when and at what power output level the developer pleases. If there are design flaws in the plant, the SPP does not have to provide its declared output. If there are any shortcomings in maintenance, the plant may shut down for long periods. In all these situations, the SPP does not pay any damages or penalties to the buyer, who has agreed, in effect, to take as much or as little electricity as the SPP can produce. In summary, an SPP can "walk away" from a power plant project at any time during the approval process, after signing the SPPA, or even after the plant has begun operations.

In contrast, in a conventional PPA for IPPs there are specific clauses that require "liquidated damages"[7] to be paid to the buyer if there are delays in construction. When the IPP plant is tested on the commissioning date, if the plant does not produce the declared output, then the IPP has to pay damages to the buyer. During power plant operation, there are specific clauses to ensure the power plant is available to produce electricity when requested by the buyer; if it is not available, penalty clauses will be triggered. Some PPAs for IPPs also include additional clauses to help ensure that the IPP produces electricity when it is needed by the buying utility. Such clauses might require: "maintenance of operational records in addition to the common guidelines for operation, or monthly reports" or might "specify to the minute how quickly status questions should be answered, identify a required amount of 'spinning' reserves, specify when the generators can go offline, specify staffing levels at the project control room,[8] or require an independent review of calibration results and maintenance procedures" (Ferrey and Cabraal 2005, 252). Such clauses are not needed in an SPPA because individual SPPs are not guaranteeing their performance. The strongest performance incentive for an SPP is economic rather than legal—the SPP does not get paid if it does not produce.

Key Recommendation

PPAs for SPPs need not have performance requirements. In the absence of capacity payments, SPPs already have a strong incentive to produce power up to their full potential because they will be paid based on the electricity they produce.

Protections for Buyer and Seller in Case of a Change of Law

Some changes in the legal environment may impact the SPP or the buyer. Typical examples are adjustments to the laws and regulations that affect income (changes to taxation policies and tax exemptions) and approval requirements (such as new permits to be obtained). Usually, if the buyer (such as a state-owned utility), the government, or the regulator initiates a change of law, a "change in law" clause would generally protect the buyer and the SPP seller from the impacts of such new laws. For example, in Sri Lanka, in the new electricity law that was approved in 2009—13 years after the SPP program began in 1996—specific transition clauses were included in the law to ensure existing SPPs and new SPPs would not be adversely affected. But it is common for other agencies and ministries of government to introduce new laws without adequate consultations, which do not ensure that the interests of SPPs and the buyer are safeguarded.

The SPPA for SPPs usually does not provide any special protection against change of law. Designers of SPP programs may have decided that such clauses would cause unnecessary complications, and if negotiations are allowed, the standardized character of the SPPA may be lost. For example, after signing an SPPA and committing funds to a project on the basis of certain assumptions on tax exemptions or rebates on customs duties on imported equipment, it is possible that the finance ministry may announce a general withdrawal of tax exemptions granted to investment projects, affecting both existing and new SPPs. Similarly, when a standardized FIT is based on estimates of the technology-specific costs of producing electricity, and SPPAs are signed based on these estimates, the government may announce new concessions (such as a tax holiday or an exception on a customs duty), and the SPP developer would reap windfall profits, leaving the buyer to take the blame for paying exorbitant prices. In contrast, a PPA for a conventional power plant will typically include clauses that trigger negotiations if a change of law occurs. If the buyer is a state-owned utility, the PPA usually states that all changes to taxation or customs duties will simply be passed on to the buyer by raising an additional invoice. If the regulator approves the PPA, then this implies the regulator will also allow these higher costs to be passed on to the retail customers of the buying utility.

Key Observation

SPPAs for SPPs usually do not provide any special protection against changes of law.

Should the National Utility Be Given Positive Incentives to Buy from SPPs?

An example of a positive incentive can be found in the VSPP program in Thailand. For projects over 1 megawatt (MW), the buying utility is obligated to pay for only 98 percent of the power it receives; the other 2 percent is "free" electricity from the utility's perspective. This arrangement provides some incentive to the buying utility to interconnect VSPP generators, helping to overcome utility resistance related to the transaction costs of interconnecting and maintaining billing arrangements with small generators. Thai SPP generators generally felt that 2 percent was a reasonable price to pay to help ensure that utilities were also on board.

In general, we believe that a regulatory regime comprising positive incentives ("carrots") as well as negative incentives such as fines ("sticks") is more likely to be effective in soliciting meaningful and cooperative participation by all parties involved. Incentives as in the Thai example above are effective when they create situations in which all parties win from SPPs coming online and generating electricity. This is especially important for utilities, who may feel they have little to gain from an SPP program. If utility leadership is not friendly to SPPs, as both the buyer and network operator the utility has many tools at its disposal to stall or thwart SPPs from running smoothly.

Key Recommendation

Regulators should consider giving utilities the "carrot" of a positive incentive to purchase electricity from SPPs. This is likely to be more effective than a regulatory regime comprising entirely "sticks."

Should the National Utility's Purchases from SPPs Be Exempt from Public Procurement Regulations?

As an anticorruption measure, many government entities including state-owned utilities are required to go through a competitive public procurement process for the purchase of many goods and services. The principal rationale for this legal requirement is to reduce the likelihood of corruption. And for larger power purchases, it clearly has the potential for producing better prices for the buyer.

But if this requirement were to be applied to SPP purchases, it could be time consuming, very costly for small systems to participate, and introduce considerable uncertainty into the SPP project development process.

In our view, public procurement regulations *should not apply* to electricity purchased by utilities under an SPP program for three reasons. First, and most important, the price of the electricity is already fixed under the SPP regulations, so the need to ensure that the utility is not paying too high a price no longer exists. In other words, the buying utility has little or no influence on the price that it pays for SPP power. Second, in programs in which the total SPP MW are not capped, electricity generated by a particular SPP is nonexclusive—it does not preclude another SPP generator from generating and selling to the buying utility. Third, it would significantly increase transaction costs for SPP developers, constraining the number of projects developed.

Key Recommendation

Public procurement regulations should not apply to electricity that utilities purchase from SPPs. They are not necessary because the price of electricity is already fixed under PPAs, FITs, or other SPP regulations; electricity produced by one SPP does not prevent another SPP from selling its electricity to the utility; and the regulations would significantly increase transaction costs for SPP developers.

How Can Regulators Help Ensure That SPPs Are Actually Paid for the Electricity They Sell?

In some countries, nonpayment by utilities has been a substantial and serious issue. This reflects the fact that many state-owned utilities in Africa are commercially insolvent. These utilities accept electricity generated by SPPs, but fail to pay invoices submitted by SPPs (or pay only after very long delays), which has a deadly effect on the SPP industry: existing SPPs have inadequate revenue to pay their loans and are threatened with bankruptcy. If a country's utility develops a reputation for failing to pay, SPP projects in the pipeline may be endangered as project developers and partners such as equity financers become nervous and abandon projects midstream.

In addressing this problem, one option is to consider the approaches that larger IPPs take to address the threat of nonpayment. One tactic is simply to threaten to shut down if not paid. Because these are large projects, typically with installed capacities of hundreds of MWs, IPPs inherently hold more bargaining power than small SPPs in this regard. If one or more IPPs threaten to shut down in response to nonpayment, the threat may carry the credible and substantial consequence of nationwide blackouts and brownouts. SPPs, as relatively small players, are less credible if they try to make this threat: the loss of one or even several SPPs will have a relatively minor impact on overall power availability in the country.

From the Bottom Up • http://dx.doi.org/10.1596/978-1-4648-0093-1

Another tactic used by developers of IPPs is to require that the recipient government sign a guarantee obligating the government to make payments to the IPP if the off-taking utility fails to do so. Often the guarantee is backed up by a PRG issued by the World Bank, International Finance Corporation (IFC), or other international development agency, which covers private lenders against risk of default by the government in honoring its guarantee. PRGs have been issued for a number of proposed IPP projects in Kenya and will also be a component in the SPP program for Uganda and the Deutsche Bank's GET FiT Program: Global Energy Transfer Feed-in Tariff Program for Developing Countries (2010).

PRGs are a form of insurance. Insurance is usually purchased to protect against an event that has some probability of happening or not happening. But if the buying utility is already insolvent, nonpayment is almost a certainty. It is also not probable that the SPP could make a case for being at the front of the payment queue relative to other suppliers of goods and services. In such circumstances, it is not clear that an international financial organization would be willing to offer a PRG. The problem highlights the crucial point that a financially unhealthy utility off-taker strongly undermines opportunities for any SPPs, whose business plans are based on significant sales to a national utility.

Tariffs for Backup Power Purchased by the SPP

A backup or standby tariff compensates the national utility for providing electricity to an SPP when the SPP is not generating electricity, or not generating enough electricity to meet its loads.

The SPP may need to buy backup power for one of several reasons:

- The SPP's generator is too small to meet its own or its retail customers' demand.
- The SPP's generator may need an external source of power to start or restart after it was shut down because of a planned or unplanned outage.
- The SPP's own retail customers and/or the SPP facility load consumes power while the SPP is not generating for whatever reason.

How Should Backup Power Tariffs Be Designed?

While no one doubts the need for backup power, there is less consensus about how to set backup power tariffs, especially the demand charge. Most utilities levy some kind of demand charge for all types of larger customers,[9] the justification being that the utility had to invest in transmission and distribution capacity to deliver this power, and it needs to pay for this investment whether electricity is used intermittently or continuously. A demand charge can be based on a customer's measured peak demand (peak kW or kilovolt-amperes [kVA] per month), or a contracted peak demand with surcharges if actual power demand exceeds the contracted amount. In most countries, smaller customers do not pay a demand charge because it would cost too much money to meter their peak demand and to educate them about the rationale for such charges.

When considering how backup tariffs should apply to SPPs, important questions include the following:

- Under what circumstances should SPPs be charged a demand charge?
- If they are charged a demand charge, do they deserve lower demand charges than regular large customers? If so, how much lower should the demand charge be?
- Does the backup service have to meet certain standards before an SPP is charged for the service?

Different answers to these questions have been considered, or adopted, by regulators in countries that have operating SPPs. The principal approaches are described below.

Option 1: No Special Backup Tariffs for SPPs

One approach used by some countries (including Thailand, Kenya, and Sri Lanka) is to charge SPPs the same demand charge that would be charged any other commercial or industrial customer.[10] For example, a rice mill in Thailand has a steam-fired turbine using rice husk as a fuel. When the power plant is down for maintenance, the rice mill needs to draw a peak demand of 1 MW of electricity from the grid for a four-hour period to keep its milling equipment running. This would trigger a peak demand charge[11] of:

$$1,000 \text{ kW} \times \$6.53 \text{ per kW} = \$6,530 \text{ for the month}$$

and an energy charge of:

$$4 \text{ hours} \times 1,000 \text{ kW} \times \$0.0874/\text{kWh} = \$349.60.$$

The backup tariff also contains a "ratchet" clause: once you, as a customer, hit a certain level of peak demand, you have to continue paying for that demand even though the peak was a one-time event. In the case of the rice mill, its peak electricity demand during this one four-hour period triggers a minimum demand charge of 70 percent of the maximum demand charge (70% × \$6,530 = \$4,571) for each of the next 12 months (see table 6.1).

From a utility perspective, the ability to treat SPPs as if they were like any other large customer has the advantage of simplicity: fewer types of customer classes and a simpler tariff schedule. But from an SPP's perspective, it is being forced to pay demand charges month after month, even though it only needed the power for just a few hours in one month. Its one-time peak consumption has, in effect, "ratcheted" the SPP into paying demand charges for a full 12 months, which can lead to a big increase in the SPP project's operating costs, making it commercially unviable. The SPP would argue that the ratchet is not fair. The supplying utility would respond more or less as follows: "Look, I am sorry if you are no longer commercially viable. But I, too, have a business. You seem to be

Table 6.1 Examples of Backup Power Charges

Utility (country)	Backup tariff
Tenaga Nasional Berhad (TNB) (Malaysia)	For cogeneration facilities, a top-up charge or a backup charge is applicable. Monthly maximum demand charge: $8.50/kW for top-up customers (in which the facility's power consumption exceeds self-production). $4.50/kW for backup customers (electricity consumed from the utility powers the entire facility). Energy charge: about $0.10 per kWh. (TNB 2013)
Provincial Electricity Authority (Thailand)	The SPP is considered yet another customer and placed in the appropriate category: small general service (<30 kW), medium general service (30 kW < × < 1 MW), or large general service (>1 MW) (Government of Thailand, Provincial Electricity Authority 2000). Rates vary depending on the voltage level of service and whether or not the customer is on a time-of-use (TOU) meter. To take a representative sample for a non-TOU MV customer (22–33 kV): Demand charge: $6.53/kW. Energy charge: $0.0874/kWh. An SPP that uses backup power infrequently would fall under this important ratchet clause: minimum charge is 70 percent of the maximum demand charge during the previous 12-month period ending with the current month.
TANESCO (Tanzania)	Through 2012, the SPP was treated the same as just another commercial customer, with the charge varying by voltage level of interconnection. The charge for customers connected at 11 kV is: Demand charge: 14,520 T Sh/kVA (about $9.20/kVA). Energy charge: 118/kWh (about $0.075/kWh). The demand charge is subject to the "highest maximum demand rule" and the billing demand is "the higher of kVA Maximum Demand during the meter reading period and 75% of the highest kVA Maximum Demand for the preceding three months" (TANESCO 2012). As of December 2013, the regulator EWURA is considering a rule change in which backup power customers with load factors below 15 percent are charged only for energy.
KPLC (Kenya)	The SPP is treated the same as other customers. Tariffs depend on the voltage of connection. To take a representative example, customers connected at 33 kV pay: Demand charge: K Sh 200/kVA ($2.20/kVA). Energy charge: 4.49/kWh (cents 4.95/kWh) (KPLC 2008).
CEB (Sri Lanka)	The SPP is considered yet another customer in the commercial category. Tariffs for commercial customers are higher than the tariffs for industrial customers. Demand charge: $6 per kVA/month. Energy charge: approximately $0.10/kWh (CEB 2012).
Long Island Power Authority (United States)	"Rate 680: Commercial, Supplemental Back-Up Service" is available for customers with a load factor of 10 percent or lower. Rates vary with voltage level of interconnection. For primary voltage (11 kV): Demand charge: $2.10/kW (compared to demand charges of $8.95–10.12/kW/month for regular commercial customers) plus a monthly service charge of $99. Energy charges: based on time of use, and varying from $0.02/kWh (off peak) to $0.34/kWh (on peak). (Long Island Power Authority 2012)

Source: Authors' compilation from cited sources.
Note: The table does not include monthly service charges, fuel surcharges, taxes, or other fees. CEB = Ceylon Electricity Board; KPLC = Kenya Power and Lighting Company; K Sh = Kenya shilling; kV = kilovolt; kVA = kilovolt-amps; kW = kilowatt; kWh = kilowatt-hours; MV = medium voltage; MW = megawatt; SPP = small power producer; TANESCO = Tanzania Electric Supply Company; TOU = time of use; T Sh = Tanzania shilling.

ignoring the fact that I have to make capital investments if I am going to be able to supply your demands. The equipment has to be in place regardless of whether your peak demand lasts for 10 minutes or 30 days. And the capital costs of that equipment continue long after your peak demand goes away."

Option 2: Reduced Demand Charges for Backup Customers

A second option gives SPPs a lower demand charge than regular industrial and commercial customers. The rationale for the discount is that SPPs have a very different pattern of demand: they use backup electricity only rarely, whereas most industrial customers draw peak or near-peak amounts consistently. In the case of a backup customer, the utility's substation and distribution infrastructure are available for other users most of the time. And by injecting power close to loads, SPPs, unlike other industrial and commercial customers who do not have their own generators, can help the supplying utility to reduce or delay the need for distribution infrastructure upgrades. The low load factor and the ability to supply power locally justifies charging SPPs less per peak kW than other large customers.

Key Definition

The *load factor* is defined as the ratio of average load to peak load in a specified time period, which is often the billing period (one month).

The amount of the discount is based on estimations of the extent to which the assets of the utility that are required to serve an infrequent backup power event are shared among other customers the rest of the time. In Sri Lanka this takes the form of a simple rule—the demand charge for backup supplies shall be 50 percent of the demand charge for a regular customer, provided that the SPP's monthly power import load factor[12] does not exceed 15 percent. For the Long Island Power Authority in the United States, a backup power customer at secondary voltage pays a demand charge of $2.10/kW/month (compared to a demand charge of $8.95–10.84 for "regular" commercial customers), provided that its load factor is no higher than 10 percent (Long Island Power Authority 2012).

By fixing a maximum load factor, utilities prevent regular customers from taking advantage of the lower charges for maximum demand (that is, by registering as backup customers and then not operating their generating facilities). A typical industrial or commercial customer would have a monthly load factor of between 25 and 75 percent and therefore would not be able to get under the 10–15 percent load factor ceiling. Some utilities also create a default transfer or trigger mechanism. Say, for example, that a backup customer records a monthly load factor of more than 15 percent for two consecutive months; he would automatically be placed on the regular customer tariff. A separate assessment would be required if the customer asks to be returned to the backup tariff.

This approach helps make backup power more affordable for SPPs. But as the example in box 6.1 indicates, there are circumstances, especially in countries with weak power grids, under which discounts to backup customers—such as the 50 percent demand charge for 15 percent or lower capacity factor—are arguably still not fair to SPPs. The SPP's need for backup supplies is primarily triggered by weaknesses on the supplying utility's transmission grid, rather than any actions or conditions on the SPP's system. Options 3 and 4 on the pages that follow address this situation.

Box 6.1 Unplanned SPP Outages Caused by Weak Distribution Networks

A sugar factory in Tanzania has a cogeneration small power producer (SPP), delivering steam and electricity to the factory for sugar production and selling surplus power to the grid. The power plant has an output of 12 megawatts (MW), and the sugar factory uses about 8 MW for its own needs during normal operations. The balance of 4 MW is fed to the grid under a standardized power-purchase agreement (SPPA) and a standardized tariff. At least once a week, power quality problems on the buying utility's network cause the SPP's protection relays to trip, taking the SPP offline. To get the SPP's cogeneration unit back online, the SPP must purchase several hundred kilowatts (kW) of electricity from the utility for around 10 minutes.

From the perspective of a grid-connected SPP operator, this is an indignity. The SPP had to shut down because of a disturbance or power quality issue on the utility's system, and then had to pay the utility high demand and energy charges for electricity needed to restart the SPP generator. One SPP operator in an East African country complained of 40 such relay trips in five months, all caused, he said, by faults and utility power disturbances mostly related to a local substation that was overtaxed by a growing municipal load and the utility's inability to find money to pay for an expansion in the capacity of the substation.

Another project in the same country, a 4 MW hydropower project, faces an even worse situation: a 55 kilometer (km) transmission line has been built to connect to the buying utility, and along the way the SPP has electrified 16 villages who are now its retail customers. When the hydropower generator trips off because of low frequency or voltage on the buying utility's transmission grid, the SPP remains off until grid voltage and frequency are restored within specifications, and then requires about 15 minutes to synchronize. During the time that power is flowing from the buying utility but the hydropower SPP is not yet reconnected, the utility's backup power will be supplying the villages' entire load, incurring a peak load of several hundred kW. When the hydropower project connects a nearby tea factory owned by a sister company, peak demand drawn from the utility when the hydropower plants trip offline will rise to over 1 MW and the SPP will have to pay high backup charges even though the problem was created by operational weaknesses on the buying utility's system. A single 15-minute, 1 MW backup power event triggers a monthly backup charge of T Sh 14.52 million, or about $9,000 under Tanzania Electric Supply Company's (TANESCO's) T3 (high-voltage maximum demand) electricity tariffs (TANESCO 2012).

Option 3: No Demand Charge, but a Higher Energy Charge

For countries with weak utility grids—that is, in which disturbances on the buying utility's lines are the primary cause of backup power events for SPPs—we believe it is fair to establish a special tariff for backup power that includes only an energy charge (per kWh) and no demand charge. This tariff would only be available to SPPs that show a power import load factor of 15 percent or less. A fair tariff would be equal to or somewhat higher than the tariff that the SPP receives per kWh for sales to the utility. For example, the tariff might be set at the utility's blended retail tariff (total utility revenues from electricity sales divided by total kWh sold). Such a tariff would incentivize the SPP to minimize use of backup power, but would also not impose cripplingly expensive charges for events that are not the fault of the SPP, as is currently the case in Tanzania.

In considering backup power tariffs for SPPs, regulators and utilities should consider the big picture of multiple SPPs connected and generating. A utility representative might point out, "If an SPP draws backup power from us, this contributes to the peak demand that we must pay to IPPs that supply our wholesale power." This is an incomplete perspective. When multiple SPPs are online in a country, the chance that they would all simultaneously trip offline and require backup power is negligible. Instead of considering that the loss of a single SPP incurs demand costs, utilities should recognize that countrywide the aggregate SPP net contribution, at every moment, reduces the utility's need to draw power from large IPPs—even if one or more SPPs trip offline and require backup power. The take-home message is that having more SPPs in the system *reduces* peak power purchase requirements from IPPs and generally reduces other capacity costs incurred by utilities (transmission line investments, and so on).

Our recommendation to not charge a demand charge in this case emerged after considerable discussion and analysis. One alternative (discussed next as option 4) would be to try to analyze each backup power event to ascertain who was at fault (the SPP or the utility). In practice, this option requires extensive custom data logging and analysis and/or meter programming, which is costly to implement and runs a higher risk of mistakes and disagreements. Conversely, charging only an energy (kWh) charge is easily implemented with standard unidirectional kWh meters.

While energy-only charges may not fully capture the cost to the utility of serving the backup power event, utilities should recognize that they are also not compensating SPPs for their contribution to reduction of peak power constraints in the hours of the year in which they are operating. For most SPPs online today (hydropower, biomass), capacity factors are high and contributions to reducing peak load burdens faced by utilities are significant and consistent.

As long as SPPs are incentivized to minimize duration and energy imported in backup power events through, as we recommend, higher-than-normal energy charges, the overall benefit of such a backup power tariff arrangement will be win-win for SPPs and utilities alike.

Option 4: Backup Charges Only Occur When Backup Power Has Met Power Quality Standards for 30 Minutes or More

Another approach to setting fair backup power tariffs in countries with weak grids would be to assess a backup charge only when power delivery is measured to be of good electrical quality for a certain minimum time period (for example, 30 minutes). The examples in box 6.1 are of situations in which the majority of disconnections are triggered by low utility power quality. When utility power ultimately returns, it still may be of low quality. Even when frequency and voltage return to standard levels, resynchronization requires at least 15 minutes. Either of these circumstances leads to high recorded peak power flows from the utility to the SPP—for an event caused by the national utility, not the SPP.

To address this problem, the Tanzanian electricity regulator considered—but ultimately rejected—an option that peak power charges accrue only after 30 minutes of electricity flow that meets approved national standards for voltage and frequency. This option was rejected because it has metering requirements that were difficult to implement in practice, and out of concern that the system could be manipulated by deliberate actions (such as operating a high-power-draw device, such as a welder, intermittently) on the part of the SPP to disturb power quality periodically to reset the 30-minute clock and to avoid paying demand charges.

Key Recommendation

When determining a backup tariff for SPPs, utilities and utility regulators should consider granting a special, lower backup power tariff to backup customers whose import load factor is less than 15 percent. In countries with strong grids, a 50 percent discount (or more) compared with regular demand charges is reasonable for SPPs and has precedent in other countries. Countries in which SPPs trip offline because of low power quality on the national grid should consider implementing a backup power tariff with no demand charge, but with a charge for energy (kWh) that is higher than for regular large customers.

Should the SPP Have the Option of Not Entering a Backup Capacity Contract?

We think the SPP should have the right not to take backup power. Instead, the SPP might (a) agree to the conditions laid down by the utility, and accept a special backup tariff, if offered; (b) sign a regular contract just like any other industrial or commercial customer, or (c) not take any contract for backup purchases from the grid. In the third case, the SPP would need to own a quick-start generator to provide its own backup and starting power when the grid is not operational.

Notes

1. This is true for an SPP that is connected to the national grid. But a "must-take" requirement is often not feasible for SPPs that are connected to an isolated grid (see chapter 8).

2. Links to English-language versions of the three PPAs can be found on the Africa Electrification Initiative website at: http://ppp.worldbank.org/public-private -partnership/sector/energy/energy-power-agreements/power-purchase-agreements.

3. In Tanzania one SPP developer recently proposed that the "standardized PPA" be available as a safety net, and that the SPP and national utility should be free to negotiate a PPA that meets their particular needs and that the regulator should routinely accept this particularized PPA unless there is reason to believe that it will hurt third parties (for example, captive customers of the buyer) who are not direct parties to the sale. But if the two parties are unable to reach agreement on a particularized PPA, the SPP developer would always have the legal right to demand that the buyer accept the SPPA developed by the regulator. Hence, the SPPA would always be available to the SPP developer as a backup option.

4. Comments made by two SPP developers at a June 18, 2012, consultative meeting called by the national electricity regulator on proposed second-generation SPP rules changes.

5. Any PPA, whether short or long, inevitably creates an allocation of risk between the buyer and the seller. A detailed analysis (prepared by Ben Gerritsen of Castalia Strategic Advisors) of how different risks would be allocated in a proposed SPPA for a grid-connected, mini-hydro SPP in Rwanda is provided in appendix E.

6. But in the case of SPPs serving an isolated mini-grid, this "must-take" condition is not feasible. The buyer on an isolated mini-grid will be able to accept the full electrical supply of the SPP only if it does not exceed the moment-to-moment electrical demand on the system.

7. "Liquidated damages" are costs to a party to agreement, owing to nonperformance by the other party. For example, if a power plant does not demonstrate its capability to produce the rated output, the seller would be required to pay liquidated damages to the buyer. The penalty or a formula to calculate the penalty would be included in the agreement. Such clauses are not generally included in SPPAs for SPPs.

8. Information required may include status within 15 minutes of a request, a daily plan for the next day, weekly updates to the annual outage and maintenance plans, explanation of cause of outage of circuit breakers, and annual monthly generation estimates with prompt updates.

9. Larger customers can be defined by peak load (for example, exceeding 30 kilovolt-amperes [kVA] in Thailand or 25 kVA in Jamaica) (Government of Thailand, Provincial Electricity Authority 2000), by kWh used per month (for example, exceeding 15,000 kWh per month in Kenya for low-voltage customers) (KPLC 2008), and/ or *de facto* by voltage connection (demand charges apply to all customers connected at voltages greater than 415 volts in Kenya) (KPLC 2008).

10. When a utility calculates the cost of supply to each customer category, usually the regular maximum demand of all the customers in that category will be assessed, and the costs of investing and maintaining generation, transmission, and distribution assets will be apportioned among the regular customers. Utilities use different methods to do this. Some consider the contribution to peak demand by each customer category,

then calculate the cost of supply (for capacity) for a typical customer in the category, and then define the demand charge accordingly. Some other utilities may use the marginal capacity cost to fix the demand charge at each voltage level in the network.

11. A customer pays the peak demand charge even if the customer's peak demand does not coincide with system peak.

12. Load factor is defined as the ratio of average load to peak load. In a given billing period, total kWh = average kW × billing duration [hours], so for a given billing duration, load factor = total energy consumed [kWh]/peak power consumed [kW] × billing duration [hours]. For example, a customer drawing 4,000 kWh over a 30-day billing with a peak load of 40 kW has a load factor of 4,000/(30 × 24 × 40) = 13.89%.

References

CEB (Ceylon Electricity Board). 2012. *How Your Bill Is Calculated.* http://www.ceb.lk/sub/knowledge/billcalculation.html.

Deutsche Bank. 2010. *GET FiT Program: Global Energy Transfer Feed-in Tariff Program for Developing Countries.* http://www.ipfa.org/news/12564/db-climate-change-advisors.

ECA (Economic Consulting Associates), and Ramboll Management Consulting. 2012. *Standardized Non-negotiable Power Purchase Agreement for Renewable Generators of Less Than 10 MW.* Prepared for Kenyan Ministry of Energy, London.

EPPO (Energy Policy and Planning Office). 2012. *Status of Purchase of Electricity from VSPP March 2012.* http://www.eppo.go.th/power/data/index.html.

Ferrey, Steven, and Anil Cabraal. 2005. *Renewable Power in Developing Countries: Winning the War on Global Warming.* Tulsa, OK: PennWell Books.

Government of Thailand, and Provincial Electricity Authority. 2000. *Electricity Rates.* http://www.pea.co.th/th/eng/downloadable/electricityrates.pdf.

KPLC (Kenya Power and Lighting Company). 2008. *Rates and Tariffs.* http://www.kplc.co.ke/index.php?id=45.

Long Island Power Authority. 2012. *Common Commercial Electric Rates.* http://www.lipower.org/pdfs/account/rates_comm.pdf.

TANESCO (Tanzania Electric Supply Company). 2012. *Electricity Charges.* http://www.tanesco.co.tz/index.php?option=com_content&view=article&id=63&Itemid=205.

TNB (Tenaga Nasional Berhad). 2013. *Pricing & Tariff.* (accessed June 9, 2013), http://www.tnb.com.my/business/for-industrial/pricing-tariff.html.

Grid-Connected SPPs: Creating Workable Feed-In Tariffs

Certainty is paramount.

—IFC OFFICIAL, WORLD BANK WORKSHOP, JANUARY 30, 2012

We need a system of regulation that allows environmentalists to make a living as capitalists and which does not bankrupt the national utility in the process.

—AFRICAN GOVERNMENT OFFICIAL

Abstract

Chapter 7 presents the two principal methods currently used to set feed-in tariffs (FITs) for small power producers (SPPs) connected to the grid. Those methods are (a) the "avoided-cost method," which is based on an estimate of the value of electricity generation to the utility or society, and (b) the "seller's cost" method, which is based on an estimate of the levelized cost of generation using a given technology. The chapter discusses capacity payments, periodic adjustment mechanisms, and calculation of FITs. It identifies implementation problems with both avoided cost and technology-specific, cost-reflective FITs, addresses donor-funded FIT "top-up" programs, and recommends a two-phase approach for implementing FITs in African countries that is keyed to "walking up the renewable energy supply curve."

What Are Feed-In Tariffs?

A feed-in tariff is commonly defined as a tariff-support mechanism for renewable energy generators or cogenerators in which the generator is *guaranteed a certain rate of payment*, usually over a long period, *for every kilowatt-hour (kWh) generated and fed into the grid* (Ren21 2012, 56).[1] In most instances, FITs for small power producers are set administratively rather than competitively as the outcome of a structured bidding process. FITs and accompanying legal and technical support elements have been implemented in 92 states, provinces, or countries (Ren21 2012). A 2008 study concluded that "well-adapted feed-in tariff regimes are generally the most efficient and effective support schemes for promoting renewable electricity" (European Commission 2008, 3). That conclusion was echoed in studies by the Deutsche Bank Group (2009) and the International Energy Agency (2008).

Key Definition

A feed-in tariff (FIT) is a tariff-support mechanism for renewable energy generators or cogenerators in which the generator is guaranteed a certain rate of payment, usually over a long period, for every kWh generated and fed into the grid.

As a mechanism for supporting renewable energy,[2] FITs are successful for several key reasons: (a) they are simple; (b) they guarantee a revenue stream that makes it easier for the developer to convince lenders and donors that a project is commercially viable; (c) they impose few restrictions, allowing many different types of projects and entities to participate; (d) they eliminate the need for project-specific negotiations; (e) they incentivize actual electricity generation, not just installed capacity built not to be used but to receive up-front investment grants. If the project does not generate electricity, it does not receive FIT payments. But FITs are useful only if a developer can acquire the initial funds to invest in the project to get it up and running.[3]

In this chapter, we discuss:

- The two principal methods for calculating FITs
- The major design issues in implementing FITs
- The major questions that need to be answered to create a donor-financed "top-up" program
- A recommended two-stage policy and regulatory strategy for promoting grid-based SPPs in developing countries in Africa and elsewhere

Readers interested in the details of these FIT strategy decisions, including implementation issues and economic consequences, are encouraged to see the description and analysis of Sri Lanka, Tanzania, and South Africa's experience with FITs in appendix F.

The Two Principal Methods for Setting FITs in Developing Countries

There are two main methods for setting FITs in developing countries.[4] The first is to base the FIT payments on the value of the corresponding generation to the utility or society. This approach requires estimating the costs that the utility or society will avoid by purchasing power generated from a renewable source.

From the Bottom Up · http://dx.doi.org/10.1596/978-1-4648-0093-1

This first method is known as the avoided-cost method. The second method is to base the FIT payments on the estimated cost of generation for each designated renewable energy technology, assuming that the developer has made a least-cost investment and will operate efficiently. This second method, which has become the more common, is known as the standardized, cost-reflective, technology-specific method. The key distinction between the two methods is that the first is based on an estimate of the buyer's or society's projected avoided cost and the second on estimates of an efficient seller's costs. Under either method, once the benchmark number is set by the regulator or a ministry, the SPP will have a strong incentive to produce at a lower price in order to increase its profits.[5] We now consider each of the two methods in detail.[6]

Key Observation

Feed-in tariffs (FITs) are calculated based on estimates of the value to the utility or society of the electricity generated, computed in terms of the costs avoided by purchasing from an eligible generator, or on the levelized cost of electricity generation for each eligible renewable or cogeneration technology.

Method 1: Avoided Cost (Value to the Purchasing Utility or Society)

Under the first method, FITs are based on estimates of the value of the renewable energy to the purchasing utility or to the country as a whole. In Tanzania, for example, FITs are set based on a calculation of the national utility's financial avoided cost. Avoided costs have been defined in Tanzania as the projected financial costs that the national utility would incur to supply power purchased from SPPs using other conventional sources, either by generating the electricity itself or purchasing it from an independent power producer (IPP).

A typology of different approaches for measuring avoided costs is presented and discussed in box 7.1.

Key Definition

The three categories of avoided costs are financial, economic, and social. Financial avoided costs are based on how much it would cost the utility to generate the electricity provided by the renewable generator. Economic avoided costs are based on how much it would cost the national economy to replace the electricity generated by the renewable generator. Economic avoided costs do not include subsidies or taxes because these are internal transfers within the national economy. Social avoided costs are calculated as the economic avoided costs plus the environmental and health costs that would be incurred locally and globally if the electricity had to be generated from a source other than the SPP.

Box 7.1 Three Varieties of Avoided Cost: Financial, Economic, and Social

Avoided costs are computed in one of three ways: financial, economic, or social. All three types of calculations include some or all of the following components:

- Avoided variable cost of generation
- Avoided cost of generation capacity
- Avoided operation and maintenance costs of generation
- Avoided cost of transmission
- Avoided local environmental damage cost of generation (SOx, NOx, and particulates)
- Avoided global environmental damage cost of generation (carbon emissions)

In calculating a feed-in tariff (FIT) based on avoided cost, the most common approach is to focus on financial avoided costs. Financial avoided costs are based on the prices that the purchasing utility actually pays for fuel, labor, and other products used to generate the electricity that will be displaced by the renewable generator. These prices are the numbers that appear on actual invoices and will include taxes and subsidies. It is very common for a government to subsidize the fuel costs of a national utility as an indirect way to lower the final retail electricity prices charged by the utility. A calculation of financial avoided costs would use this lower subsidized price.

As shown in table B7.1.1, both financial and economic avoided costs include the same four components. The chief difference between the two is that calculations of economic avoided costs use estimates of costs to the overall economy rather than the costs paid by the purchasing utility. For example, if the government subsidizes the cost of natural gas used by the national utility in generating or buying electricity from natural-gas-fired generating plants, the calculation of economic avoided cost would include that subsidy as part of the true cost to the economy of providing natural gas. The true cost to the economy might be estimated by using the "border price" for natural gas—the price that this country would receive at its border if its natural gas were purchased by buyers from other countries. A secondary difference is that financial cost includes taxes, whereas economic cost does not (as a tax is a transfer of money within an economy).

Social avoided cost adds several other elements to the calculation. These elements include local and global environmental damage and social and health impacts caused by the production of electricity.[a] These damages are usually referred to as "externalities" because they are

Table B7.1.1 Components of Avoided Costs

Avoided cost	Financial	Economic	Social
Variable cost of generation	✔	✔	✔
Cost of generation capacity	✔	✔	✔
Cost of generation operation and maintenance	✔	✔	✔
Cost of transmission	✔	✔	✔
Cost of local environmental and social damage			✔
Cost of global environmental and social damage			✔

box continues next page

Box 7.1 Three Varieties of Avoided Cost: Financial, Economic, and Social *(continued)*

costs that producers of electricity impose on others. In most countries, to date electricity producers have not been obliged to pay compensation for the environmental damages and social impacts they create.

The generally accepted rule is that it is beneficial to a country to pay a renewable generator a FIT price that is less than or equal to the social avoided cost. That is, less than or equal to the cost to society of generating a kilowatt-hour (kWh) from the most expensive power plant dispatched at that moment. While the principle is easy to state, it is rarely implemented for two reasons. First, the damage costs may not be easy to estimate, and the estimate may not be persuasive or convincing in the eyes of all parties affected. Second, even if the true costs of electricity production are calculated in a sound and convincing way, it is often politically difficult to get someone to pay for them. Neither reason is a very good excuse for setting these costs equal to zero. But that is often what is done.

Source: Based in part on Meier 2010.
a. Displacement of populations from inundation and loss of livelihood because of impacts to fisheries are examples of social impacts of hydroelectric dam projects. Increased acid rain from sulfur dioxide and increases in respiratory illness from particulates in coal smoke are examples of social impacts from coal projects.

Key Observation

Under economic theory, it is accepted that economic welfare is maximized if the feed-in tariff (FIT) for renewable generators and cogenerators is set at the social avoided cost. However, countries rarely take this tack because the damage costs are often difficult to calculate and it is often politically difficult to incorporate these costs into electricity tariffs.

If an avoided-cost approach is used, all generally eligible renewable energy generators get the same FIT. In other words, the avoided-cost approach results in a "standard offer" or a "single rate" FIT that is not differentiated by generating technology. This does not imply that all SPP technologies will be equally viable under the avoided-cost approach. In countries with tariffs set at the purchasing utility's avoided costs, it is usually the case that only a few projects will be commercially viable—usually larger-scale biomass cogeneration and particularly good small hydropower sites. Other important renewable energy technologies will not be commercially viable if the FIT is set at the utility's avoided costs. For example, wind power and solar power will generally be nonviable because their costs will almost always be higher than the utility's avoided-cost-based FIT, except in special situations in which a utility's avoided costs are dominated by the price of liquid fuels.

Figure 7.1 provides an estimate of the renewable energy technologies that will and will not be commercially viable under the FIT established by Tanzania's electricity regulator for main-grid-connected SPPs using either renewable energy or cogeneration.[7] The average FIT tariff of 96 Tanzanian shillings (T Sh) for SPPs connected to the main grid was equivalent to US$0.072 when first announced. Since the first avoided-cost numbers were published in 2009, Tanzania's energy

Figure 7.1 Estimated 15-Year Levelized Costs of Various Forms of Renewable Energy Compared with the Published Main Grid FIT in 2012 (per kWh of Generic Plants in 2008)

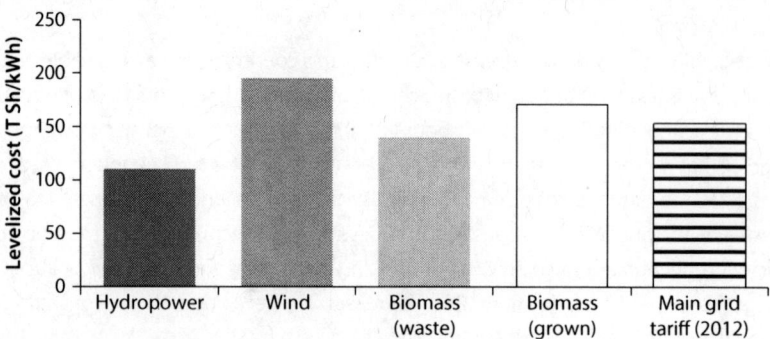

Source: Siyambalapitiya 2007.
Note: At 2012 tariffs, hydropower and biomass from agricultural residues are expected to be commercially viable. FIT = feed-in tariff; kWh = kilowatt-hour; T Sh = Tanzanian shillings.

regulator has received numerous complaints from renewable energy developers that the FITs are too low.

Key Observation

The avoided-cost approach to setting feed-in tariffs (FITs) is technology neutral: all eligible renewable energy generators receive the same FIT.

As discussed in box 7.1, the correct way to calculate avoided costs would be to include all social costs rather than only the utility's financial costs. Doing so would capture the economic value of avoided pollution and greenhouse gas emissions offset by renewable energy generation. Although it can be difficult to place a precise value on crop damage from acid rain and on human sickness and mortality caused by air pollution, many studies that have undertaken to do so have found that they are significant (Sundqvist 2000). However, the Energy and Water Utilities Regulatory Authority (EWURA), the Tanzanian electricity regulator, decided to use Tanzania Electric Supply Company's (TANESCO's) financial costs as the basis for computing the FIT because there was no mechanism available for reimbursing TANESCO if it were required to pay SPPs a higher price based on avoided social costs.

Method 2: Standardized, Cost-Reflective, Technology-Specific Calculation
The second method of calculating FITs is, unlike the first, technology-specific. Standardized, cost-reflective, technology-specific FITs are now used in numerous European countries, states of the United States, and provinces of Canada, plus

Kenya, Malaysia, Thailand, and Turkey (Ren21 2011). Under the cost-reflective, technology-specific method, tariffs are set at a level such that a well-run renewable energy generation business will earn a reasonable profit using the specified renewable technology. The target profit level is set by policy makers, regulators, or program administrators. For example, in 2009, National Energy Regulator of South Africa (NERSA) proposed an after-tax return on equity of 17 percent in setting new values for FITs. (However, NERSA FITs were never implemented; see appendix F.)

This method requires the estimation of several parameters—among them capital structure, the capacity factor,[8] the cost of different pieces of capital equipment, and interest rates for loans. Because the FIT has to be standardized, the costs are based not on the actual or projected costs of each enterprise that will be receiving the FIT payments but on estimates for a hypothetical well-run enterprise. Not surprisingly, the assumptions made in the course of setting these different parameters often trigger major debates (box 7.2).

Because this second method is technology-specific, it involves multiple FITs that reflect the levelized cost (that is, the discounted lifetime cost) of electricity produced from that technology.[9] Because different technologies have different levelized costs, some, such as solar electricity, if included in a FIT tariff system, receive higher tariffs than others, such as biomass or small hydropower. In addition to technology type, tariffs are often differentiated based on project size, the quality of the resource, or other project-specific variables.

This approach has the advantage of being designed to assure project investors of a reasonable and relatively certain rate of return, thus creating conditions for strong market growth in a variety of technologies. Germany, Spain, Thailand, and others that have adopted this policy have seen quick and substantial expansion of renewable energy capacity. The FITs may move up or down over time as the technology or renewable fuel costs change or because the regulator gains access to better information. If power-purchase agreements (PPAs) are signed at an earlier, higher price, good regulatory practice requires that these contracts should be "grandfathered" (that is, not subject to after-the-fact modification by the regulator). Similarly, if an SPP signed a PPA at an earlier, lower price, it should not be eligible to receive a higher FIT offered to later PPAs.

Key Observation

The standardized, cost-reflective approach to setting feed-in tariffs (FITs) is not technology neutral. Instead, based on a technology-specific formula and a variety of technical and financial parameters, FITs are calculated to allow a theoretical, well-run renewable generator to earn a reasonable profit. Disputes often arise over the assumptions used in setting the various parameters. Box 7.2 highlights the key disputed parameters.

Box 7.2 Disputed Parameters in Computing Cost-Reflective, Technology-Specific FITs

Generally, a regulator or some other party presumed to be neutral will calculate the feed-in tariff (FIT) for each renewable energy technology that is eligible for a FIT. The key parameters in any cost-reflective, technology-specific FIT calculation are:

Cost parameters: (a) the initial investment cost, (b) annual maintenance costs, (c) annual fuel costs and related transport costs (in the case of biomass and waste)

Technical parameter: (d) the annual capacity factor

Financial parameters: (e) debt-equity ratio, (f) interest rate and loan tenure, (g) allowable annual return on equity (or internal rate of return, IRR), (h) the escalation factor for maintenance costs and fuel costs, (i) discount rate (if the FIT is levelized for the full contract period or for only a portion of that period).

Of these nine parameters, the five that are most likely to be disputed by developers are: (a) initial investment cost, (b) annual maintenance costs, (c) annual fuel costs, (d) the capacity factor, and (g) allowable return on equity. If there is no direct or indirect political pressure from developers, the regulator will typically use conservative (that is, lower) figures for the first three parameters. This will have the effect of lowering the FIT. In response, the developers will argue that the assumed numbers are unrealistically low and do not reflect real-world conditions.

In the case of mini-hydro projects in Africa, some of the initial capital cost estimates were based on projected costs per kilowatt of installed capacity of $1,700–2,000. These estimates were based largely on installed capacity costs observed in a number of Asian projects. But the early experience in Tanzania and Uganda has shown costs in the range of $2,500–3,500 for recent projects. African developers have argued that regulators and their consultants should recognize these higher African costs and not blindly transfer Asian capital costs to Africa.

With regard to the capacity factor, the regulator is likely to use a higher value, and developers will usually argue that the selected figure is unrealistically high. The higher the assumed capacity factor, the lower will be the calculated unit costs, because a higher capacity factor will result in the project's fixed costs being divided by more units of production. This, in turn, will lead to a lower FIT. For resource-limited technologies with no storage (such as run-of-the-river hydro facilities), the capacity factor will be site-specific. However, the achieved capacity factor is not a totally external decision. It will depend very much on the megawatt size of the generation facility that the developer chooses to build. (See box F.1 for a discussion of incentives in Sri Lanka that may cause an initial developer to "oversize" a mini-hydro facility.) For biomass-based systems, the capacity factor will be affected by the plant's availability, which will, in turn, depend on fuel availability, storage capacity, and the frequency and quality of maintenance work.

The IRR is usually the most-disputed parameter. It is typically calculated relative to a baseline of the most secure investment in a given country, such as long-term government securities or treasury bonds. The starting point in the calculation will be the pretax yield (interest rate) on the government's long-term bonds. The key question is: How much of an increment (premium) above this reference value should a small power producer (SPP) developer be offered because of the risk factors involved in developing and operating the small power project?

box continues next page

Box 7.2 Disputed Parameters in Computing Cost-Reflective, Technology-Specific FITs *(continued)*

The typical project risks are (a) technology and resource risk, (b) development risk, and (c) contractual and payment risk. If a small power project is managed in such a manner that the contractual and payment risk is minimized—that is, if the purchasing utility has a history of paying SPPs on time and of adhering to contractual conditions, if the government avoids making after-the-fact changes to SPP regulations, and if disputes and arbitration cases are infrequent—then arguably the remaining risks are all under the direct control of the SPP developer. But often there will be disagreements about the facts, especially in the early years of an SPP program. The utility that is required to buy SPP power will argue that the IRR offered to SPPs need not be as high as the IRR that would be demanded by investors on other high-risk commercial businesses because SPPs are given a purchase guarantee for a long, fixed period (typically 15 years) and a reasonably predictable price (indeed, fully predictable if the FIT is cost-reflective). SPPs will argue that the guaranteed offtake at a price that covers their costs is of little value if the purchasing utility is commercially insolvent (as is the case with many government-owned national utilities in Africa) or if it is frequently not able to receive the SPP's power because of instability or physical constraints on the distribution or transmission grid. To justify a higher return, SPPs will point out that SPPs in other countries have been given higher IRRs or that larger independent power producers (IPPs) in their country have been granted higher returns, either explicitly or implicitly.

Local Currency versus Hard Currency

Regardless of whether the FIT is calculated as an avoided cost or technology-specific tariff, an issue of key importance is whether it is paid in local currency or in hard currency (U.S. dollars or euros). Developers generally favor tariffs paid in the same currency as their bank loans. For projects with foreign financing, tariffs paid in hard currency or indexed to hard currency are very desirable because they remove the risk of local-currency devaluation. Utilities strongly prefer SPP tariffs denominated in local currency. "Our revenues," they say, "are in local currency, and we cannot bear the risk of payments that must be in dollars." Some countries have opted for FITs payable in hard currency (Uganda, Kenya), others (Sri Lanka, Tanzania, Thailand) in local currency with no indexation for currency valuation.[10]

Key Observation

SPP developers generally favor FITs paid in the same currency as their bank loans. Developers with projects that use foreign financing strongly prefer FITs paid in hard currencies (U.S. dollars or euros) to reduce the risk of devaluation of the local currency against the currency required to make loan repayments. Utilities, on the other hand, strongly prefer FITs paid in local currency, because their revenues are received in local currency.

Major FIT Implementation Questions and Issues

Some implementation issues arise regardless of whether the FIT is based on the buying utility's avoided costs or on the SPP's technology costs. Other issues that arise differ between these two types of FITs. We first consider issues pertaining to all FITs, and then look separately at implementation issues for avoided-cost and technology-specific FITs.

Should FITs Include Consideration of Capacity Value?

A renewable generator can help a purchasing utility avoid both energy and capacity costs. Energy costs are avoided whenever purchases from the renewable generator allow the purchasing utility to reduce the amount of electricity generated from its own generators or purchased from other, more expensive supply sources. Capacity costs are what a utility pays to assure that there is enough power produced during "peak" hours. Capacity cost includes fixed costs such as debt payments and equity return, depreciation, and fixed labor costs of operating a power plant. If grid-connected renewable energy generators, in aggregate, predictably generate electricity during hours of peak electricity usage, then utilities are able to reduce the capacity they must procure from IPPs or by building their own power plants.

Power from many renewable generators is not dispatchable because their ability to generate is contingent upon availability of intermittent renewable energy resources (wind blowing, sunlight shining). Hence, on an individual plant basis, it is hard to justify including a full capacity credit in the avoided-cost FIT. But it is also clear that a portfolio of renewable energy projects has a certain capacity value. So it is not unreasonable for a regulator to consider the capacity benefits of such a portfolio when determining the tariff.

An alternative approach is to give renewable generators a premium for every kilowatt-hour that they generate during the buying utility's peak period. Hence, the premium is based on actual performance rather than a before-the-fact projection of when individual SPPs or a group of SPPs using the same technology might be available. This FIT tariff system incentivizes renewable energy generators to generate during peak times by adding a peak tariff premium paid at times of the day when the system typically experiences peak load. In Thailand renewable energy generators[11] receive a premium of about $0.06 per kWh during the peak period (weekdays from 9 am to 10 pm). The higher peak-period tariff reflects the benefits to Thailand of having to build and dispatch fewer peaking power plants.[12]

This brings up an interesting characteristic of solar power in Thailand (shared to a large extent by wind power). Thailand's peak load is driven by air conditioning on hot sunny days. Solar photovoltaic (PV) is classically a "nonfirm" generation source (with a capacity factor in Thailand of approximately 15 percent), but it is very reliable in terms of meeting peak loads, which always occur on bright days when PVs are producing at or near their maximum capacity.

Key Recommendation

Regulators should consider giving renewable generators a premium tariff rate for every kilowatt-hour they generate during the buying utility's peak periods. This makes the premium performance based and reflects the capacity value of having a portfolio of renewable generators connected to the grid.

Should FIT Prices Be Adjusted over Time?

A workable FIT must satisfy two basic requirements. First, it must allow an efficient seller of power produced by the targeted renewable or hybrid technology to be commercially viable. Second, it must not bankrupt the buyer in the process of helping the seller. When a regulator or other government entity sets an initial FIT price, it is (hopefully) based on the best available estimate of the seller's expected production costs or the buyer's avoided costs. But costs inevitably change over time. For example, they may go down as technologies improve over time or when component costs fall (as has happened with solar PV cells over the last several years) or world oil prices decline. Or on the contrary, costs may go up because of inflation, changes in exchange rates, or increases in world oil prices. The only thing that can be known with certainty is that an SPP's costs will change over time, and it is difficult to predict what these changes will be. This implies that FITs generally cannot be set in stone for all existing and future projects. Instead, they should include some type of adjustment mechanism to accommodate changes.[13]

If it is decided that FITs may need to be adjusted, a number of implementation questions have to be addressed:

- For existing projects:
  - What should be adjusted? Should the adjustments be to the overall FIT, or just to some of its underlying components (for example, the operation and maintenance [O&M] component, which is an ongoing expense, not a fixed up-front expense)?
  - If it is deemed necessary to have ceilings and floors, should provisions be made to adjust them over time?
  - How often should adjustments be made?
- For new projects:
  - Should there be planned periodic adjustments to the starting value of FITs for new projects?
  - Should the periodic adjustments be based on revised cost studies or on preset price trajectories keyed to one or more observable parameters?

To make these implementation issues more concrete, we consider how adjustment mechanisms could be designed for FITs calculated by the avoided-cost method and by the technology-specific method. We will do this by drawing on the actual experience of Sri Lanka and Tanzania.

Adjustments to Avoided-Cost FITs for Existing Projects

In the case of avoided-cost FITs, tariff adjustments are caught between two competing goals. On the one hand, if the regulator adheres strictly to the goal of setting the FIT equal to current estimates of avoided cost, tariffs should be regularly adjusted (annually or even more frequently) because the utility's avoided cost of electricity will change with prevailing fossil fuel prices, cost changes brought about by recent investments, or cost increases resulting from new contracts with IPPs to provide long-term or emergency power. But on the other hand, rapid and unpredictable changes to FITs will tend to discourage new investment because it creates major risks for the SPP developer and its financers. It also hurts the country's economy because the volatility in SPP tariffs will create a barrier to more stable energy costs that renewable energy portfolio additions can provide. In the cases of Sri Lanka (for PPAs signed between 1996 and 1999) and Tanzania, the initial use of annual adjustment mechanisms was motivated by the goal of closely tracking changes in avoided costs from year to year, an arrangement that was deemed consistent with the avoided-cost approach, even though it led to more volatile FIT prices.

Both Sri Lanka and Tanzania initially chose to adjust existing FITs annually in accordance with updated annual estimates of the purchasing utility's avoided cost. In the absence of floors and ceilings, this meant that every SPP with an avoided-cost FIT would get the same price in any given year regardless of when the PPA was originally signed. Both countries soon discovered that such an adjustment mechanism could lead to big swings in FIT values that had nothing to do with changes in an SPP's costs.

Both countries decided that the unconstrained adjustments to FITs based on annual calculations of avoided cost would produce too much volatility and too much risk for both buyers and sellers. Therefore, they decided to adopt measures to reduce that volatility. In Tanzania, the first measure (adopted even before the first FITs were implemented) was to limit the high and low prices that FITs could reach through the use of a ceiling and a floor. Sri Lanka adopted a floor price but no ceiling price. To further limit FIT tariff volatility, the tariff-setting formula was changed from an annual calculation of avoided cost to one based on an annually calculated three-year rolling average. The benefit of using a rolling average is that it lowers volatility for both seller and buyer by lowering the risk that avoided cost will be significantly higher or lower than in prior years.

Ceilings and floors. When there is a floor, downward adjustments in the calculated avoided-cost FIT can go only so low. When there is a ceiling, upward adjustments can go only so high. It is usually the case that the floor and ceiling prices are not symmetric (that is, not the same percentage above and below the initial FIT). Typically, the floor price will be set at or slightly below the initial FIT price, whereas the ceiling price often will be considerably higher than the initial price.

Tanzania and Sri Lanka differ in the ceiling prices that accompany their avoided-cost FIT. In Tanzania, FIT values are allowed to rise and fall with changes in the three-year rolling average of calculations of avoided costs, but at the outset they are capped at 150 percent of the tariff in force when the PPA was signed. However, the ceiling itself is not a fixed cap in Tanzania. Instead, it has its own separate adjustment mechanism, increasing with changes in the consumer price index (CPI). In contrast, avoided-cost FITs in Sri Lanka were never subject to a cap. As the Sri Lankan economy grew, and the country had to import more fuel oil to generate electricity, the FIT paid to SPPs was effectively tied to world oil prices. By 2010, the FITs for PPAs signed in 1996 were 2.1 times higher than their original values, leading to significant windfall profits for those SPPs who signed these early contracts (Siyambalapitiya 2012).

With avoided-cost FITs, price floors are especially important to bankers and project financiers because they lower risk by ensuring that the SPP will have enough revenue to pay off loans and provide a return on investment. But utility buyers who are required to buy from SPPs argue that this added financial security for the SPP and its financiers comes at a cost to them. They point out that price floors may force them to pay FIT prices that are higher than their current avoided costs. In other words, they may end up paying too much for the power produced by the SPP.

Let us take a closer look at how price floors have been implemented in Sri Lanka and Tanzania. Sri Lanka has created a price floor equal to 90 percent of the tariff established in the year when the project's PPA was signed. Tanzania has set a price floor at 100 percent of the tariff in the year when the project's PPA was signed. The Tanzanian floor helps to eliminate the downside risk for developers that avoided costs will drop well below what they were at the PPA signing, but a big risk remains: the SPP's costs may rise because of inflation or local-currency devaluation. This, in turn, could compromise the SPP's ability to meet payments on foreign debt since the avoided-cost calculations, which are based on the buyer's costs, are not designed to capture changes in the seller's costs. SPP developers in Tanzania have argued that the floor should be further adjusted for domestic inflation and depreciation of the Tanzanian shilling relative to the dollar or euro. Without such adjustments, they contend that the floor will be too low and fail to provide adequate protection.

In an avoided-cost FIT system, price floors can also lead to different prices for SPP projects that sign PPAs in different years. For example, if an SPP in Tanzania signs a PPA in 2012, its price floor will be 26 percent higher than if the same SPP had signed its PPA in 2011. If the buying utility's avoided costs drop markedly in some future year from the 2012 value, the SPP that signed in 2012 would be protected from a drop in the FIT price based on post-2012 avoided-cost calculations.

A hybrid approach to adjusting avoided-cost FITs. A hybrid approach would be to base the initial value of the FIT on the avoided-cost calculation but tie any subsequent annual adjustments to one or more inflation indices that approximate changes in the SPP's own variable costs rather than changes in the avoided costs of the buying utility. Some countries that use technology-specific FITs use this adjustment mechanism (see next section). Our recommendation in the final section of this chapter is that it should also be applied to FITs that are initially keyed to the buyer's avoided cost. This hybrid approach has several advantages:

- It ensures that SPP purchases are affordable to the buying utility.
- It eliminates the need for makeshift arrangements (ceilings, floors, and multi-year averaging) to deal with large fluctuations in annual avoided costs.
- It facilitates financing by providing SPP developers, and those who finance them, with significant revenue stability while also protecting against unreimbursed cost increases.

Adjustments to Cost-Reflective, Technology-Specific FITs for Existing Projects

Some countries that adopt cost-reflective FITs include automatic tariff-adjustment mechanisms to account for changes in O&M costs. The rationale for limiting the coverage of the adjustment clause to O&M costs is that these are the costs that are most likely to change during the life of the project and are largely beyond the control of the project operator. In contrast, no provision is made for adjusting the project's capital-related costs (debt repayments, depreciation on capital, and return on equity) because these costs are likely to remain fixed over the life of the project if an SPP project has been domestically financed.

Uganda. In Uganda, the regulator proposed automatic adjustments to O&M costs of SPPs using the following formula:

$$FIT_y = [FIT_{y-1} * (1-w)] + \left[FIT_{y-1} * \frac{CPI_y}{CPI_{y-1}} * w \right] \qquad (7.1)$$

Where
FIT_y is the applicable FIT in year y,
FIT_{y-1} is the applicable FIT in the previous year,
CPI is the core producer price index for the United States as published by the Bureau of Labor Statistics, and
w is the share of O&M in the initial FIT in year y (ERA 2011, 6).

The Ugandan approach provides a clear and understandable mechanism to protect the economic viability of a project against the risks of domestic inflation. A project developer's nightmare is that its costs will increase significantly while its tariffs (and therefore its revenues) remain fixed. This would be a particular concern for renewable energy generators that have substantial fuel and O&M costs. While the partial-adjustment mechanism has been used mostly in countries using cost-reflective, technology-specific FITs (such as Uganda and Sri Lanka), we believe that a similar mechanism could also be used in countries that have opted for avoided-cost FITs.[14] (See the discussion at the end of this chapter on a recommended two-phase strategy for FITs.)

One weakness of Uganda's adjustment mechanism is that it uses the U.S. CPI to measure the likely inflation of SPP O&M costs. This implicitly assumes that domestic inflation in Uganda tracks domestic inflation in the United States, which has not been the case. Over the last five years, Uganda's inflation fluctuated between 4 and 19 percent, whereas U.S. inflation never rose above 3.8 percent. Another problem in using the U.S. CPI is that it measures the inflation rate in the goods and services purchased by typical American households. But inflation in prices of U.S. consumer goods and services may not be a good proxy for inflation in the prices of industrial and commercial goods and services purchased by an SPP, whether the SPP is located in Uganda or in the United States. Presumably, the Ugandan regulator recognized these weaknesses but still chose the U.S. CPI because the U.S. index is readily available on the Internet and is considered to be more reliable than comparable statistics from Uganda.

Sri Lanka. As in Uganda, the adjustment formula for cost-reflective SPPs in Sri Lanka allows for automatic adjustments in O&M after the SPP goes into commercial operation. The costs that can be automatically adjusted are called "escalable" costs.[15] For example, the Sri Lankan government has specified that the total costs of a new hydro project whose PPA was signed in 2011 should be broken down as follows between fixed and escalable costs for project developers who chose the tiered cost-reflective tariff option:

12.64 Sri Lankan rupees (SL Rs) per kWh (fixed)

1.61 SL Rs per kWh (escalable)

Total 14.25 SL Rs (US$0.128)

Unlike Uganda, Sri Lanka uses a domestic measure of inflation: the Sri Lanka CPI. In addition, the country applies a second adjustment index to O&M costs, namely a measure of the strength of the Sri Lankan currency relative to the U.S. dollar. The annual adjustment factor applied to O&M costs is a simple average of the two indices. The rationale for using both indices is that some of an SPP's O&M expenses (for example, the replacement of a transformer) may have to be purchased from abroad. Hence, the prices of these purchases will be more directly affected by fluctuations in the value of the Sri Lankan rupee relative to other currencies than by the rate of domestic inflation in Sri Lanka. It should also be noted that the proportion of escalable costs to total costs is about 11 percent. Hence, even a 10 percent increase in O&M cost (from 1.61 to 1.77 SL Rs) will lead to only a 1 percent increase in the project's FIT.

In Sri Lanka, biomass projects are eligible for an additional adjustment factor. Such projects get an automatic adjustment factor for their fuel costs in addition to their O&M expenses. The breakdown of fixed and escalable costs for a biomass project (using sustainably grown biomass, not waste biomass) that signed a PPA in 2011 was:

7.58 SL Rs/kWh (fixed)

1.29 SL Rs/kWh (escalable O&M costs)

9.10 SL Rs/kWh (escalable fuel costs)

Total 17.97 SL Rs/kWh (US$0.162)

From the Bottom Up • http://dx.doi.org/10.1596/978-1-4648-0093-1

For biomass projects, the proportion of escalable costs to total costs is 42 percent. This means that a larger proportion of the total costs of a biomass project will be eligible for automatic adjustments that will affect the allowed FIT price.

In Sri Lanka, most SPP developers have been able to finance their loans in Sri Lankan rupees with loans from Sri Lankan banks. Since the principal payments, and to a large extent, the interest payments, are fixed in Sri Lankan rupees over the life of the loan, there is no need to include an adjustment mechanism for loan repayments in the FIT formula. However, this policy would have to be reconsidered if SPP developers were to obtain loans from outside Sri Lanka, when, presumably, the loan payments would be fixed in a non-Sri Lankan currency. If the Sri Lankan rupee depreciates relative to the currency in which the loan was received, the developer would need more Sri Lankan rupees to make the same required loan payments. This, then, would quickly raise the question of whether the SPP developer or the Ceylon Electricity Board (CEB), the Sri Lankan national utility, and its customers should bear the risk of currency fluctuations.

In Sri Lanka, the FIT pricing formula for SPP projects that are eligible for automatic tariff adjustments is included in an appendix to the PPA. Since the numbers that go into the adjustment formula are publicly available, the annual adjustments are normally calculated by CEB, subject to review by the SPPs. The regulator does not get involved in these annual adjustment calculations unless there is a dispute.

Key Recommendation

FIT prices should be adjusted over time in existing SPP projects. In the case of FITs based on avoided cost, regulators should tie subsequent annual adjustments to one or more national inflation indices that approximate changes in the SPP's own variable costs rather than changes in the avoided costs of the buying utility. In the case of cost-reflective, technology-specific FITs, regulators should implement automatic tariff adjustments to certain variable costs to the SPP based on national inflation and currency valuations.

Setting FIT Prices for New Projects under Cost-Reflective, Technology-Specific FITs

In the previous section, we described adjustment mechanisms for the FIT prices of SPP that are operational. In this section, we examine options for setting the initial FIT prices of new SPP projects. For new projects, the initial prices for technology-specific FITs usually track changes in expected costs in one of three ways: periodic adjustments, capacity-dependent adjustments, and tariff degression factors.

Periodic Adjustments of Technology-Specific FITs

Most countries with technology-specific FITs revise FITs for new projects periodically. An annual review is common, but sometimes adjustments are made every two or three years. In the Czech Republic the Energy Regulatory Office sets the FIT levels every year, with the adjusted tariff applying to new projects. These tariffs are guaranteed not to decrease by more than 5 percent compared with the level paid to plants that began operation the previous year. This rule helps ensure stability and investment security.

Capacity-Dependent Adjustment of Technology-Specific FITs

A variation on periodic adjustments is to have evaluation and adjustments triggered after a certain target capacity of generation using a specific technology has been installed. In Portugal, for example, tariffs for a renewable energy technology are revised downward when a certain capacity of power plants is reached (PV, 150 MW; biomass, 150 MW; biogas, 50 MW) (Government of Portugal 2005, article 2, annex ii, part 18c–e). The adjustments apply to the initial tariffs of new projects, not to the current tariffs of existing projects.

Tariff Degression of Technology-Specific FITs

A tariff degression is an automatic adjustment sometimes used in conjunction with periodic or capacity-dependent reviews. To understand the degression mechanism, recall that the FIT level typically depends on the year in which an SPP is commissioned. Following a preset degression schedule, each year the tariff for new SPPs is decreased by a certain percentage. The degression schedule affects the initial FITs for new projects coming online in later years. It does not affect those for plants that are already operational. Thus, under a tariff degression mechanism, the later a plant is commissioned, the lower its initial tariff will be. The degression mechanism lowers risks of overcompensation, while at the same time providing incentives for technology improvements and cost reductions because manufacturers and developers know that the FITs for projects commissioned in future years will be lower. Ideally, the rate of degression should be set at an empirically derived cost-reduction rate for each technology. Germany, France (for wind energy), and Italy (for PV) use tariff degression (Klein and others 2008).

In the German renewable energy law (EEG), degressions for solar electricity are tied to the recent deployment capacities in the previous 12 months. For example, the benchmark degression for solar electricity in Germany is 1 percent per month. The target for solar is 2,500 to 3,000 MW per year. If the target is exceeded and 4,500 MW is installed during the previous 12-month period, then the degression increases to 1.8 percent per month. If 6,500 MW per year is installed, the degression climbs to 2.5 percent per month. Conversely, if less than 1,500 MW is installed, the degression falls to 0 percent (Bundestag 2012).

Key Recommendation

Initial FITs should be adjusted over time for new SPP projects subject to cost-reflective, technology-specific FITs. Regulators have several choices. They can make periodic adjustments through a regular review process; they can evaluate and adjust FITs after a certain target quantity of renewable energy or a specific generating technology has been installed; or they can implement a tariff degression schedule based on a prespecified cost-reduction rate for different technologies.

Should There Be a Cap on Eligibility for a FIT?

Where FITs substantially exceed avoided costs, some countries have taken the step of limiting total installed capacity for each technology in order to limit the overall financial impact of the FIT on ratepayers or taxpayers. This can include lower quantity caps on more expensive renewable energy technologies, such as solar PV, and higher quantity caps on lower-cost technologies, such as mini-hydro and biomass power. For example, Uganda limited cumulative PV capacity in 2011 to 2 MW and 7.5 MW in 2014. Similarly, mini-hydro was limited to 60 MW in 2011 and 270 MW in 2014. This strategy gained ground after the experience of Spain in 2008, where applications to produce 3,000 MW of solar electricity far exceeded expectations. The overwhelming response to the incentive program created problems for the buying utility's budget because the solar PVs were eligible for a relatively high FIT (Gonzalez and Johnson 2009).

A similar unmanageable rush occurred in Thailand as well. Thailand had signed PPAs for more than 1,700 MW of solar electricity before the government curtailed the program. Because solar electricity was eligible to receive a FIT of $0.33 per kWh, the Thai government became concerned about the effect on ratepayers. It responded by announcing that new applications were not accepted "until further notice," and by reducing the solar FIT by two baht ($0.06) per kWh for projects in the pipeline that had not yet signed PPAs (Tongsopit and Greacen 2012). The Czech Republic, Italy, and Australia have suffered similar "solar gold rushes" and have had to lower the FIT for solar PV or impose overall capacity or payment limits (or combinations thereof) (Ren21 2011).

Solar electricity is peculiar in that it is difficult, if not impossible, to pick a stable tariff that will bring about just the right amount of investment. Below some threshold, solar electricity simply is not financially viable. Above that threshold, it becomes widely viable in many locations all at once because sunlight resources are likely to be very similar across broad areas of the country. Other renewables do not share these characteristics. For example, economic sites for small hydro installations are limited, and rice husk and bagasse generators are limited by mill production. There is a finite amount of low-hanging fruit in these two technologies and when it is gone, it's gone, whereas potential solar sites with cheap, unshaded land are generally very plentiful. The effect is even more pronounced with the feedback effect of economies of scale (it is cheaper per MW to build 100 MW than 1 MW) and further complicated by world prices for solar PV systems, which have fluctuated but in the last several years have dropped rapidly.

While quantity caps on installed capacity provide a strong safeguard against high policy costs, they can have a chilling effect on investment. Investors are unlikely to know how quickly the caps will be reached, and considering the long development gestation of many SPP projects in Africa, many investors will have legitimate concerns about whether their particular project will be sufficiently developed to qualify for the FIT before the cap is subscribed. From a developer's perspective, the "all-or-nothing" characteristic of a quantity cap introduces high risks. Investors and project developers ask, "How can I risk putting in several years' worth of work if there is a significant chance that the cap will be reached and my project will be worthless?"

Absolute quantity caps can also lead to a project development pipeline filled with a number of unviable projects as renewable energy developers rush to enter speculative bids to secure their place in the queue. This squeezes out better projects that may take time to emerge, and also feeds speculation in letters of intent (LOIs) and PPAs that can turn public opinion against an SPP policy. For the reasons described above, hard quantity caps can constrain the amount of renewable energy development that can occur, reducing prospects for market growth, creating disincentives for serious developers, and reducing the chances that renewable energy targets will be met on schedule.

We think that some mechanism for adjusting FITs, depending on capacity, is important to provide comfort to the buying utility and the public that they will not be overwhelmed with the financial burden of large quantities of higher-cost SPP projects. However, policy makers should weigh options carefully. One option that avoids the "all-or-nothing" cliff of a quantity cap is an arrangement in which achieving a specified target triggers a revision in the policy or an adjustment of technology-specific FITs (described in the earlier section "Capacity-Dependent Adjustment of Technology-Specific FITs"). From the developer's perspective, this changes the "cliff" into a "manageable slope." Developers have strong incentives to get projects done in order to receive higher tariffs. And if certain overall quantity thresholds are reached, the FIT does not disappear, it is simply smaller. We think that some mechanism for adjusting or curtailing FITs depending on capacity (either a cap on physical capacity, total overall FIT payments, or technology-specific premium payments, or else a capacity-dependent tariff adjustment) has merit because it provides comfort to the buying utility that it will not be overwhelmed with the financial burden of higher-cost SPP projects.[16]

Key Recommendation

Regulators should weigh options carefully when considering an overall quantity cap on installed capacity or total overall FIT payments. A less heavy-handed policy tool is building in adjustments to adjust feed-in tariffs or future projects based on recent installed capacity.

Who Calculates FITs?

One question that has arisen in both Tanzania and Sri Lanka is who calculates FITs? Initially in Tanzania (in 2008 and 2009), a working group convened by the Ministry of Energy and Minerals performed the annual SPP tariff calculations based on TANESCO's estimated avoided costs. The working group included members of utilities, government, and industry. In practice, however, it proved difficult to convene the working group and reach a consensus in a timely fashion. In addition, it was recognized that individual members of the working group would inevitably make recommendations based on their particular commercial interests—SPP representatives would want higher values and the TANESCO representative would want lower values. The regulator therefore concluded that it would be best to limit the working group to an advisory rather than a decision-making role. The calculations are now performed by EWURA, the country's multisectoral regulator. EWURA's procedure is to request current information from TANESCO. If TANESCO fails to provide that information in a timely manner, EWURA makes the calculations with the best information available to it. EWURA then announces the new values for public comment, makes adjustments, and ultimately approves the tariff levels.

Sri Lanka has faced somewhat similar challenges. Until the Public Utilities Commission of Sri Lanka (PUCSL), the new electricity regulator, became operational in 2010, FITs were calculated by CEB (for avoided-cost-based FITs that applied to PPAs signed between 1996 and 2007) or by the Ministry of Power and Energy (for cost-reflective, technology-specific FITs that applied to PPAs signed after 2007). CEB continues to make annual calculations of avoided costs, and these numbers are still used to make annual adjustments to FITs in PPAs signed with SPPs prior to 2007. Calculations for the post-2007, cost-reflective, technology-specific FITs have been handled by PUCSL since 2010. Both arrangements have generated controversy. For avoided-cost FITs, renewable generators have complained that CEB has a conflict of interest because it has a clear incentive to arrive at low estimates of avoided costs so as to pay a lower FIT to the SPPs. There have also been complaints about PUCSL's calculations. Some have complained that the recent FIT values announced by PUCSL were unjustifiably high and out of sync with levels that PUCSL proposed in its public consultation. As of 2011, the Sri Lankan FITs for SPPs using mini-hydro, wind, and biomass were among the highest in the group of 10 countries examined in a recent World Bank study (Elizondo Azuela and Barroso 2011).

Who Should Pay the Extra Costs of FITs?

Like their counterparts everywhere in the world, African regulators and policy makers are under pressure from developers of currently higher cost wind and solar projects to adopt cost-reflective, technology-specific FITs. And like their counterparts around the globe, they face the challenge of finding the money to fund a FIT that will be high enough to enable renewable generators small and large to recover their operating and capital costs. But that challenge is more acute in Sub-Saharan Africa than elsewhere. Unlike most utilities in developed countries and many developing countries, almost all state-owned utilities in Sub-Saharan Africa are, at present, commercially insolvent. Simply put, this means that the typical African state-owned utility is not collecting enough revenue to cover its costs. In the most complete survey performed to date on the financial condition of Sub-Saharan African utilities, the Africa Infrastructure Country Diagnostic (AICD) concluded that only 10 of 21 national utilities in Sub-Saharan Africa were allowed to charge tariffs that covered their operating costs, and only 6 of 21 national utilities could charge tariffs that covered operating *and* capital costs (Eberhard and others 2008, 29). So the dilemma for most African national utilities is this: How can they be expected to pay for the currently higher costs of most renewable generating technologies if they are not even covering their current operating and capital costs?

Key Observation

Finding the money to pay for the extra costs of technology-specific, cost-reflective FITs is particularly challenging in Africa, where almost all state-owned utilities are already commercially insolvent and therefore incapable of paying higher FITs.

To make the problem more concrete, consider the comments that one might hear in private, off-the-record conversations with three key stakeholders: a renewable energy developer, an official at a national utility obligated to buy power from the renewable generator, and an official at the country's ministry of energy that is responsible for setting FIT policy. To make their views more realistic, we have assumed that all three individuals are discussing the current situation in Tanzania.

A renewable energy developer

Look, I am happy that renewable energy has suddenly become very fashionable. The bilateral donors and the international financial organizations are falling over themselves to be "green." But when all is said and done, all these grand plans and strategies will not amount to anything concrete unless our projects are allowed to be commercially sustainable. And this is not going to happen unless the government or regulator allows me to charge tariffs that will cover my costs. Let's be honest with each other—an average FIT in Tanzania of $0.072 per kWh based on someone's estimates of TANESCO's avoided cost won't be workable except for a handful of mini-hydro plants with especially good sites and maybe a few biomass projects. I need to receive a FIT that covers all my costs in the same way that TANESCO needs retail tariffs that enable it to cover its costs.

Also, it is clearly absurd to offer SPPs an average price of $0.072 per kWh based on some theoretical long-term, least-cost plan when less than a year after this calculation was made TANESCO was cutting off power for 12 hours a day throughout the country and seeking permission to buy emergency power from a diesel generator at a price that it keeps secret but which is probably at least $0.40 per kWh. I hope that the government and regulator realize that there is a fundamental and logical disconnect between the low prices that TANESCO offers to pay me and the very high economic costs that the country is experiencing on a daily basis because TANESCO does not have sufficient power supplies.

And please don't tell me that SPPs are seeking special treatment because they are the only generators seeking or receiving subsidies. This conveniently ignores the fact that government has been giving subsidized gas for the Songo-Songo natural gas plant for eight years and has also paid all the capacity charges for another IPP, the IPTL plant, because its capacity charges were so high that they could not be included in retail tariffs. My point is very simple: If fossil fuel plants are routinely subsidized, why is it wrong for renewable generators to receive similar subsidies?

A national utility official

I have lost count of the number of times that I have been visited by renewable energy developers. And they all have the same complaint, namely, that my company should be paying them a higher FIT because they claim that they can provide large societal benefits to Tanzania. All of them talk about their contribution to reducing global warming, how they can help to prevent blackouts by providing an additional source of supply, and how they can help Tanzania avoid the need to pay high oil prices if there is another crisis in the Middle East. I give all of them the same answer: You are talking to the wrong person. You should be talking to the minister of energy and minerals or the minister of environment.

At the end of the day, I need to earn revenues that allow me to cover my costs. In 2010 when we applied for a general tariff increase with the national electricity regulator, we produced a detailed study that showed that our average cost per kWh was 159.9 T Sh and our average revenue per kWh was 118.8 T Sh. That means that our revenues fell short of our costs by close to 25 percent. And this shortfall would be even larger if we were forced to cover depreciation or a return on the government's equity. So the sad reality is that we are already losing money on almost every kWh that we sell.

And yet these renewable energy developers have the gall to ask me to do something that will clearly worsen this gap by demanding that I pay them more than it would cost me to buy electricity from nonrenewable sources. Moreover, their projects are generally intermittent and the SPPA contract is "must take." This means that if they produce electricity, we have to buy it. But they are not penalized for not generating. Is this fair? Is this rational? Look, if the President or Minister of Energy orders me to buy high-priced electricity from these renewable generators because they have decided that it is in our national interest to do so, that's fine, and I will obey orders, but the government will need to provide me with a reliable source of funding to pay for these higher costs so our financial situation does not get even worse. It is not fair to ask my company to bear the burden of paying higher prices to obtain national or societal benefits for Tanzania when the reality is that right now we are not even covering our out-of-pocket costs.

A Ministry of Energy official

In the past, donors used to tell us that they wanted to help us create a "world free of poverty." But if they are really serious about this goal, they should provide the funds that will allow us to extend reliable and low-cost electricity to the thousands of families in this country who do not have grid-based electricity. That's where the poverty is. At present, these families are forced to pay for expensive and poor-quality lighting from kerosene lamps and expensive electricity from car batteries. For these families, the most important thing is to get access to reliable and reasonably priced grid-based electricity.

In recent years, the donors have changed their tune. They say that they fully support us on the need to "scale up access," but now the electricity must be supplied from "green" electrons. But the reality is that most families in this country, whether they do or do not have electricity, don't care whether that electricity is produced with green, yellow, or black electrons.

My concern as a ministry official is that the more money that we are forced to spend on higher-priced renewable energy, the less we will have to construct the distribution lines and transformers needed to deliver that electricity. Every time a donor insists that its money must be used for some renewable energy project or program, I can't help but think that some of this money could have been used for the more important need of providing households with connections to the grid. The donors tell us that we can do both. But if their grants and loans come with the condition that we must buy higher-priced renewable energy, I am convinced that there will be less money to pay for our highest priority, which is basic electrification.

Donor Top-Ups of FITs

One obvious solution to the problem debated by these three individuals would be to obtain outside funding to pay the extra costs of buying power from renewable suppliers. And, in fact, for several years now, there have been proposals presented at numerous international conferences to create programs that would allow high- and middle-income donors to make premium (or "top-up") payments in developing countries that would raise the FITs paid to renewable generators above the utility's avoided cost of procuring electricity from conventional sources (Rickerson and others 2012).

The top-up payments would be grants. They would allow African countries to pursue forms of renewable generation that would otherwise be too expensive. Within Africa, Uganda is the country that has taken the lead in trying to develop such a program (*Daily Mirror*, December 21, 2012). It has been proposed that European donors provide top-up payments totaling about $25 million for up to

15 small grid-connected renewable generators that could provide up to 125 MW of installed capacity (Mutambi 2012). We think that the possibility of top-up payments by donor countries combined with some guarantee by the World Bank or some other organization that the payments will be made has considerable potential for accelerating the development of renewable generators in Africa. In appendix G, we discuss some of the implementation issues that would need to be resolved to put such a program into place.

Key Recommendation

To fund the higher costs associated with technology-specific, cost-reflective feed-in tariffs (FITs), regulators should seek outside funding in the form of grants to cover the extra costs, thereby "topping up" the FITs and making them financially viable. Appendix G describes some of the key implementation questions that regulators will need to address to create viable top-up programs.

Walking Up the Renewable Energy Supply Curve: A Recommended Strategy

No single SPP policy will be equally workable for all African countries. Every country has unique circumstances that will require individualized policies and strategies. That said, it is worthwhile to consider a general two-phase approach that starts with FITs that are approximately the same as (or below) the buying utility's avoided costs and moves toward a second phase in which some of the FITs are allowed to exceed the buying utility's avoided costs when funds for the incremental costs of these higher tariffs become available. The essence of this two-phase strategy is that it allows a country "to walk its way up the renewable energy supply curve."

Two important features of the two-phase strategy should be highlighted. First, in Phase I, FITs for SPPs may be either a single FIT set equal to the buying utility's avoided cost (Tanzania) or—for technologies whose levelized costs are at or below the buying utility's avoided cost—separate, cost-reflective, technology-specific FITs (Uganda).[17] Both of these approaches avoid the need to obtain external funding for top-ups. A second important feature that both approaches should share is a common adjustment mechanism applied to initial FIT values once an SPP is in operation. Even when the initial Phase I FIT is set equal to the buying utility's avoided cost, any later adjustments to that initial tariff should be based on general indexed changes in the SPP's operating costs (for example, O&M costs and fuel costs in the case of biomass projects). This same adjustment mechanism will also be applied to SPP projects whose initial FIT prices are based on technology-specific, cost-based calculations.

The advantage of using this hybrid approach—allowing initial FIT values to be set equal to avoided cost or technology-specific tariffs with later adjustments tied to domestic inflation or currency-depreciation indices that affect SPP costs—is that it avoids the need for floors and ceilings and for three-year averaging of avoided-cost values to dampen upward and downward gyrations in annual FIT levels, as has been necessary in both Sri Lanka and Tanzania.

Key Recommendation

When designing a FIT policy, regulators should consider implementing a two-phased approach that starts by setting FITs approximately equal to (or below) the buying utility's avoided costs and moves toward a second phase in which some of the FITs are allowed to exceed those avoided costs when funds for the incremental costs of higher tariffs become available. This approach allows the country to "walk up the renewable energy supply curve."

Benefits of a Phased Strategy

This two-phase approach produces several concrete benefits. The first comes from having relatively stable tariffs (in both phases), rather than tariffs that go up and down with the buying utility's avoided costs (which are usually driven by volatility in the price of fossil fuels). Relatively stable tariffs are a benefit for both SPP investors and the national economy. For the SPP developer, fixed tariffs reduce risk by providing more revenue certainty. This, in turn, lowers the cost and difficulty of securing project finance. For the national economy, stable tariffs provide a hedge against fuel-price volatility.

A second benefit of starting with stable tariffs that are at or below the utility's avoided costs is that these tariffs do not require subsidies from taxpayers or rate-payers. This, in turn, substantially lowers the political challenge of obtaining government and utility support for the SPP program. The SPP program can move ahead without getting bogged down in debates about where the money for incremental FIT costs will come from because the incremental FIT costs will be zero.

A third benefit of a phased strategy is that the pool of potential low-cost renewable energy projects can be tapped as soon as possible. The program is not distracted and delayed by debates over whether high-cost renewable generation projects are in the country's best interests. In countries like Tanzania and Uganda, low-cost renewable projects typically include biomass power from agro-industrial residues (sugarcane, rice husk, sisal, and others) and some small hydropower. The owners of existing agro-industrial plants are often ready to take on the challenge of building and interconnecting SPP generators even with "low" tariffs that would be commercially impossible for solar and wind genera-tors. It often makes commercial sense for a sugar factory to invest in bagasse-fired cogeneration even at relatively low FITs because (a) it has an abundant supply of free fuel (a factory by-product); (b) substantial quantities of steam and electricity are needed in factory operation; (c) the economies of scale are substantial (a steam turbine/boiler that can export electricity to a utility is only marginally more costly than one that is sufficient only to meet the factory's own demand).

A fourth benefit is that success in operating an SPP program under these Phase I conditions sets the stage for Phase II in at least three ways:

- *The slow initial pace allows graceful scale-up of utility and government ability to respond to SPP applications.* Under stable tariffs that are set at or below the buying utility's avoided-cost levels, it is unlikely there will be a deluge of renewable energy projects seeking SPP status so that they can sell to the national utility. This has an important practical advantage in the early years of a program: an initial slow pace helps both the utility and government put into place new administrative procedures to handle SPP application processing and interconnection requests to reduce or eliminate administrative bottle-necks. This early experience will serve the utility and government well when they later need to scale-up to accommodate a larger pipeline of applications if/when the country is able to move to Phase II.
- *Financial institutions learn about risk.* An early small pipeline of projects can help rural electrification agencies and domestic banks and other financial institutions to learn the process of providing financing to SPP projects. The confidence and experience gained from these early projects can, in turn, be invaluable in lowering risks and reducing red tape when the pipeline expands.
- *A constituency for political support can be built.* Pioneering SPP generators in Phase I can start building a constituency to provide political support for ramping up the SPP program. Similarly, successful Phase I experience can be leveraged to build domestic and international support for the transition to Phase II.

Timing, communications, and perceptions are important throughout a two-phase process. One particular pitfall to avoid is the situation in which developers of low-cost generation (biomass or small hydro) make a strategic choice not to invest in projects during Phase I because of the expectation that Phase II tariffs will be higher. The fact that FIT prices for wind and solar have gone down over time in many countries, both developed and developing, has tended to make this situation less likely.

Phase I Specifics

Regulators can move forward without having to figure out where funds for incremental costs of renewable energy might come from by initially setting the FIT for grid-connected SPPs that use a renewable or cogeneration generating technology at or below the purchasing utility's financial avoided cost.

At this stage, the FIT will be technology neutral if it is calculated based on avoided cost. In other words, all eligible SPPs receive the same FIT. If, however, the FIT is calculated on a technology-specific basis, only those technologies for which the levelized costs are below the buying utility's avoided cost will be eligible for the FIT.

From the Bottom Up • http://dx.doi.org/10.1596/978-1-4648-0093-1

The FIT should be locked in on the date that the SPP and the purchasing utility sign the PPA (except for prespecified inflation adjustments to SPP costs that are affected by inflation or currency depreciation).

Avoided-cost calculations should be updated every year, using averaged avoided costs over a three-year period to smooth out abrupt year-to-year changes caused by fossil fuel price volatility or other factors. These calculations should produce a single FIT for the upcoming three-year period if computed on an avoided-cost basis or, if computed on a technology-specific basis, determine which technologies will be eligible to receive technology-specific FITs. For the purpose of determining tariffs, avoided-cost calculations should include generation and transmission costs whenever the electricity from distributed SPPs helps reduce the transmission burden by being consumed locally.

A decision will need to be made about whether the tariff will be paid in hard currency (as is done in Ecuador, Honduras, Kenya, and Nicaragua) or local currency (as is done in Sri Lanka, Tanzania, and Thailand) and whether to build in adjustments based on a CPI or producer price index to account for inflation.

The government entity that makes the final decision on the FIT method should be advised by a committee of electricity sector stakeholders. That committee should have advisory rather than decision-making authority. The final decisions must rest with an independent government entity such as the regulator, because it is inevitable that individual committee members representing different stakeholder groups will espouse their particular commercial interests.

The FIT should be one element of a larger policy and regulatory support package designed to minimize transaction costs and processing time for buyers and sellers. Other elements of the package include the following:

- Guaranteed interconnection to the grid with prespecified rules for assigning responsibility for the costs of interconnection
- Standardized interconnection and operation procedures
- Guaranteed purchase of power produced by any SPP connected to the main grid (usually referred to as a "must-take" requirement)
- A prespecified FIT formula with the same duration as the standardized PPA to eliminate any uncertainty about FIT values during the life of the PPA
- A fixed, prespecified pricing formula with a clearly defined automatic price adjustment mechanism for the life of the contract
- An automatic retail-tariff-adjustment mechanism in which the utility buyer is compensated for the incremental costs of a FIT, if any, to allow for the automatic pass-through of the costs of SPP purchases to retail customers
- The national utility should be required to create an SPP cell that provides "one-stop shopping" for all SPPs that wish to connect to its grid.
- The government should create a line of credit that assists domestic banks to make loans on better terms (for example, lower interest rates, longer durations, and lower collateral requirements) to SPPs.

- If the national utility is not allowed to charge cost-recovering tariffs for its retail customers, the government should establish a guarantee mechanism to ensure that SPPs will be paid for their output.
- The government or an electrification agency should provide capital cost grants to SPPs to encourage them to provide retail electricity service to households that do not have access to grid-supplied electricity. The SPPs should be allowed to define these grants as equity when applying for loans.
- The regulator or some other government entity should put into place a regularly updated, publicly accessible database that keeps track of all SPPs that have applied to the program, those that have signed PPAs, those that have begun commercial operations, the amount of power they generate and supply to the grid, project GPS coordinates (latitude and longitude), names of project developers, and the date on which each commences commercial operations. In addition, the regulator or some other government entity should require that the national utility and other large distribution entities periodically provide information on expansion plans. For example, in Tanzania the regulator has proposed that TANESCO and other large distribution entities should annually: "issue a document indicating the names of the villages and districts to which the [distribution entity] intends to expand its distribution system to serve new customers in the coming 12 months, 24 months, and 36 months." (EWURA 2013, section 58).

If the Phase I FIT measures described above are implemented well, they stand a good chance of creating a smooth-running FIT program in which the private sector, government, and utilities can work together effectively to establish a pipeline of SPP generators.

Phase II

In Phase II, the administrative machinery developed and fine-tuned in Phase I creates a foundation for an expansion of the technologies covered by the SPP FIT program. *The change in Phase II that increases both volume and coverage will be to set technology-specific tariffs that provide for purchase prices (FITs) that are higher than the utility's avoided costs in the case of more expensive technologies (wind, solar, and so on).* What makes this possible is a top-up or "FIT premium payment" secured from a bilateral or international donor, or from some other source.

The guiding tariff-setting principle in Phase II will be to set tariffs at a level at which an SPP developer can earn a reasonable return on the operating and capital costs of an efficiently built and operated project using any technology covered by the program. That said, there is considerable room for discretion on the part of policy makers. In many African countries, for example, grid-connected solar PV will likely be seen as a costly extravagance. Hence, policy makers may choose not to include it on the initial list of technologies included in the Phase II FIT. Alternatively, policy makers may elect to leave the door open for PV generators, but set the FIT at a "safe" (low) level that will allow PV installations to come in selectively as the price of the technology drops. Or policy makers may provide a FIT for PV but combine it with tariff degression levels tied to annual solar capacity additions to avoid the risk of a gold rush.

In setting technology-specific tariffs, policy makers will want to continue to gather data on technology costs, financing costs, and availability of renewable energy resources. They will also want to encourage feedback from developers on the tariff levels at which they believe they can operate a viable business, recognizing that developers will always have an incentive to recommend high FITs. In some cases, such as biomass and small hydro, even small increases in FIT prices over avoided costs may greatly increase the number of commercially viable projects.

The key challenge in ramping up to Phase II is to determine a reliable and timely mechanism to provide funds for premium payments above the buying utility's avoided cost. Most high- and middle-income countries with technology-specific FITs have developed mechanisms that pass the incremental costs of FITs on to customers through a small per-kWh surcharge on electricity consumption. Some countries use taxpayer funds. But the low-income countries of Sub-Saharan Africa probably will not be able to adopt either of those approaches for many years, because, in most of Sub-Saharan Africa the government-owned national utility is not allowed to charge retail tariffs that even cover its current operating costs. Moreover, many government officials do not view renewable energy as a high priority. Given the limited funding that is available in their countries, it should not be surprising that government officials prefer to use scarce government monies to fund schools, roads, health clinics, or malaria control, all of which have more immediate and more obvious benefits.

Phase II Specifics

As Phase II builds on Phase I, all of the Phase I specifics apply. Phase II, however, introduces a FIT tariff premium, or top-up. In Phase II, FITs must be periodically recalculated (typically once every several years based on experiences in other countries). As with Phase I FITs, tariffs are relatively stable (with minor adjustments for inflation in operational costs). But recalculation of initial FIT values for new projects will be needed to keep pace with changes in technology and resource costs so that new projects are made viable but not unnecessarily or unreasonably expensive. It is also worthwhile to consider including a technology-specific "tariff degression" in order to track and encourage technological cost reductions, as discussed earlier in this chapter.

Notes

1. In the 1980s, FIT usually had a more general meaning, referring simply to the tariff or payment that an SPP or other generator received for producing electricity that was fed back into the grid. But over the last several decades, the term has taken on the more specific meaning of a *guaranteed, long-term* payment for renewable generators and cogenerators. For example, a 2010 report funded by the U.S. Department of Energy takes the position that the minimum elements of a FIT include "(a) guaranteed access to the grid; (b) stable, long-term purchase agreements (typically about 15–20 years);

and (c) payment levels based on the costs of [renewable energy] generation" (Couture and others 2010, vi).

2. Feed-in tariffs are not the only possible support mechanism. Other major support mechanisms include renewable portfolio standards (RPSs), competitive auctions, and fiscal incentives (capital subsidy or grants, investment or production tax credits, reductions in sales, value-added and other taxes, and energy production payments). RPSs require that some or all entities that sell electricity at retail acquire a specified percentage of their supplies from specified renewable energy sources. Auctions mandate competitive acquisition for all or certain specified renewable technologies. Tax incentives usually involve accelerated depreciation and other tax and investment incentives. We focus on FITs because they are the most widely used support mechanism in Africa, they are relatively easy to implement, and they have proven to be very effective in catalyzing the deployment of SPPs.

3. The issues involved in acquiring the initial capital to finance an SPP project are discussed in chapter 5.

4. FITs can also be tied to spot market prices. But so far spot markets have not been produced by any of the power sector reforms in Sub-Saharan Africa, so it is currently not a relevant option for Sub-Saharan African countries.

5. These two FIT categories, "avoided cost" and "technology-specific," are not necessarily exclusive; there can be crossover systems that combine the two methods. For example, Thailand uses a fixed technology-specific "adder" on top of a utility-avoided cost tariff that adjusts automatically every three months. See Tongsopit and Greacen (2012). Moreover, utilities and governments in some countries are starting to incorporate FIT components that pay a premium when or where electricity is needed most. For example, time-of-use components of FITs reward generation during peak periods (Thailand), while seasonal variations reflect higher value of electricity during the drought season (Tanzania). Some utilities pay more for customer-owned generation located on feeders that allow utilities to defer investment in transmission or distribution upgrades. Though not yet widely practiced, FITs with these price signals have the potential to substantially increase the penetration of renewable energy that utilities can accommodate, while also lowering the investment requirements. One challenge is to balance the value of these types of tariffs with the increased administrative burden of more complicated tariff structures. Fortunately, the recent widespread use of AMR (automatic meter reading) meters substantially eases the logistical challenges in accommodating tariffs that vary in time and space (Lovins and Rocky Mountain Institute 2011).

6. Both methods require calculations by the regulator or another government entity. Hence, the FITs are based on administrative calculations rather than market outcomes. In contrast, South Africa has announced that FITs for SPPs between 1 and 5 MW will be determined by the prices bid in a competitive procurement. See appendix F.

7. Figure 7.1 does not apply to SPPs connected to one of the existing isolated mini-grids operated by TANESCO. If an SPP chooses to operate on one of these existing isolated mini-grids that are currently supplied by diesel generators, it is eligible to receive a FIT that is a weighted average of the avoided costs of the diesel generator and the long-run marginal costs (LRMC) of generation supply on the main grid. In Tanzania, the FIT for SPPs selling at wholesale on isolated mini-grids was US$0.24 in 2009.

8. The capacity factor of a power plant is the ratio of the actual output of a power plant over a period of time (typically one year) and its potential output if it had operated at full rated (nameplate) capacity the entire time.

9. Probably the single best source for current estimates of levelized renewable energy costs by technology can be found in IRENA (2013). The IRENA levelized cost calculations are based on information from 8,000 medium to large renewable projects from around the world. However, some information is provided on the costs of small hydro plants. The IRENA cost numbers do not include estimates of balancing costs imposed on the buying utility by the variable generation patterns of some renewable technologies.

10. For information on Uganda, see ERA (2011); on Kenya, see Kenya Ministry of Energy (2010); on Tanzania, see EWURA (2008).

11. This tariff arrangement applies to generators that are rated below 10 MW.

12. Peaking plants are generating plants that are designed to operate for a limited number of hours when the overall demand on the system reaches a peak. Typically, peaking plants have relatively low capital costs and relatively high energy costs.

13. Tariff adjustment mechanisms are not unique to FITs. They are very common in economic regulation. Usually, they are used when future costs are volatile and difficult to predict and are largely beyond the control of the entity whose tariffs are being controlled by a regulator. See Graves, Hanser, and Basheda (2006).

14. France and Greece have also used inflation-adjustment mechanisms for FITs that were calculated using levelized cost estimates on a technology-by-technology basis. See Gipe (2011).

15. "Escalable" is a word commonly used in Sri Lanka and India to refer to costs that can be adjusted. It is derived from the English word "escalation." When referring to cost adjustment mechanisms in Sri Lanka and India, we will use "escalable" instead of "scalable," the word that is more commonly used in other countries.

16. A good survey of mechanisms for limiting the overall cost of paying FITs that exceed the buying utility's avoided cost can be found in Couture and others (2010).

17. The same approach was recently recommended by consultants to the government of Kenya. Their specific recommendation is that where the calculated value exceeds the long-run marginal cost (LRMC) of generation, as calculated in the country's least-cost power development (LCPD) process, the FIT that is offered for that technology will be the LRMC (currently $0.1186/kWh). The LCPD provides estimates of the entire system's LRMC, a cost concept closely related to avoided cost. The rationale for creating this cap on FIT prices is that "electricity consumers will not be financially penalized through the introduction of small-scale renewables on the grid" (ECA and Ramboll Management Consulting 2012, 18).

References

Bundestag. 2012. Act on Granting Priority to Renewable Energy Sources (Renewable Energy Sources Act EEG). http://www.erneuerbare-energien.de/fileadmin/ee-import/files/english/pdf/application/pdf/eeg_2012_en_bf.pdf.

Couture, Toby, Karlynn Cory, Claire Kreycik, and Emily Williams. 2010. *Policymaker's Guide to Feed-in Tariff Policy Design*. Technical Report, National Renewable Energy Laboratory. http://www.nrel.gov/docs/fy10osti/44849.pdf.

Daily Mirror. 2012. "Signing of Delegated Cooperation Agreement between the Government of Norway and KFW for the GET FiT Program." December 21.

Deutsche Bank Group. 2009. *Global Climate Change Policy Tracker: An Investor's Assessment.* DB Climate Change Advisors. http://www.dbcca.com/dbcca/EN/_media /Global_Climate_Change_Policy_Tracker_Exec_Summary.pdf.

Eberhard, Anton, Vivien Foster, Cecilia Briceño-Garmendia, Fatimata Ouedraogo, Daniel Camos, and Maria Shkaratan. 2008. "Underpowered: The State of the Power Sector in Sub-Saharan Africa." Background Paper, Africa Infrastructure Country Diagnostic, World Bank, Washington, DC. https://openknowledge.worldbank.org/handle /10986/7833.

ECA (Economic Consulting Associates), and Ramboll Management Consulting. 2012. *Technical and Economic Study for Development of Small Scale Grid Connected Renewable Energy in Kenya.* London.

Elizondo Azuela, Gabriela, and Luiz Augusto Barroso. 2011. "Design and Performance of Policy Instruments to Promote the Development of Renewable Energy: Emerging Experience in Selected Developing Countries." Energy and Mining Sector Board Discussion Paper, World Bank, Washington, DC. http://siteresources.worldbank.org /EXTENERGY2/Resources/DiscPaper22.pdf.

ERA (Electricity Regulatory Authority). 2011. *Uganda Renewable Energy Feed-in Tariff (REFIT) Phase 2 Approved Guidelines for 2011–2012.* Government of Uganda. http:// www.era.or.ug/Pdf/Approved_Uganda%20REFIT%20Guidelines%20V4%20(2).pdf.

European Commission. 2008. *The Support of Electricity from Renewable Energy Sources.* Accompanying Document to the Proposal for a Directive of the European Parliament and of the Council on the Promotion of the Use of Energy from Renewable Sources, Commission Staff Working Document, European Commission, Brussels, January 23. http://ec.europa.eu/energy/climate_actions/doc/2008_res_working_document _en.pdf.

EWURA (Energy and Water Utilities Regulatory Authority). 2008. *Standardized Tariff Methodology for the Sale of Electricity to the Main Grid in Tanzania under Standardized Small Power Purchase Agreements.* http://www.ewura.go.tz/pdf/public%20notices /SPP%20Tariff%20Methodology.pdf.

————. 2013. *The Electricity (Development of Small Power Projects) Rules. Proposed for Public Consultation.* June 13.

Gaddis, Isis. 2012. "Only 14% of Tanzanians Have Electricity: What Can Be Done?" *End Poverty* (blog), October 31. http://blogs.worldbank.org/africacan/node/2187.

Gipe, Paul. 2011. "Model Advanced Renewable Tariff Policy." *Wind-Works.org*, February 7. http://www.wind-works.org/FeedLaws/USA/Model/ModelAdvancedRenewable TariffLegislation.html.

Gonzalez, Angel, and Keith Johnson. 2009. "Spain's Solar-Power Collapse Dims Subsidy Model." *Wall Street Journal*, September 8. http://online.wsj.com/article /SB125193815050081615.html.

Gosling, Melanie. 2011. "Government's U-turn on Wind Energy Rates." *Cape Times*, June 20, Sec. E.

Government of Portugal, Ministério das Actividades Económicas e do Trabalho. 2005. *Decreto-Lei N. 33-A/2005.* http://dre.pt/pdf1s/2005/02/033A01/00020009.pdf.

Graves, Frank, Philip Hanser, and Greg Basheda. 2006. *Electric Utility Automatic Adjustment Clauses: Benefits and Design Considerations.* Edison Electric Institute, Washington, DC, November. http://www.eei.org/whatwedo/PublicPolicyAdvocacy/StateRegulation /Documents/adjustment_clauses.pdf.

IEA (International Energy Agency). 2008. *Deploying Renewables: Principles for Effective Policies*. Paris: International Energy Agency. http://www.iea.org/publications /freepublications/publication/name,15746,en.html.

IRENA (International Renewable Energy Agency). 2013. *Renewable Power Generation Costs in 2012: An Overview*. Bonn, Germany: IRENA.

Kenya Ministry of Energy. 2010. *Feed-in-Tariffs Policy for Wind, Biomass, Small Hydros, Geothermal, Biogas and Solar Resource Generated Electricity*. http://www.erc.go.ke/erc /fitpolicy.pdf.

Klein, Arne, Benjamin Pfluger, Anne Held, Mario Ragwitz, Gustav Resch, and Thomas Faber. 2008. *Evaluation of Different Feed-in Tariff Design Options: Best Practice Paper for the International Feed-in Cooperation*. Karlsruhe, Germany: Fraunhofer Institute Systems and Innovation Research.

Lovins, Amory, and Rocky Mountain Institute. 2011. *Reinventing Fire*. White River Junction, VT: Chelsea Green.

Meier, Peter. 2010. *Economic and Financial Analysis of Grid-Connected Renewable Energy Generation*. World Bank and the Government of Vietnam, Ministry of Industry and Trade.

Mutambi, Benon M. 2012. "How to Make Energy Financing (and Opportunities) a Reality in Uganda." Presentation to Energy Business Dialogue Uganda: Creating Commitment and Momentum for Increased Energy Access, Kampala, Uganda, December 13.

REN21. 2011. *Renewables 2011 Global Status Report*. Paris: REN21 Secretariat. http:// www.ren21.net/Portals/0/documents/Resources/GSR2011_FINAL.pdf.

———. 2012. *Renewables 2012 Global Status Report*. Paris: REN21 Secretariat. http:// www.ren21.net/Portals/0/documents/Resources/GSR2012_low%20res_FINAL.pdf.

Rickerson, Wilson, Christina Hanley, Chad Laurent, and Chris Greacen. 2012. "Implementing a Global Fund for Feed-in Tariffs in Developing Countries: A Case Study of Tanzania." *Renewable Energy* 49, Special Issue: Selected Papers from World Renewable Energy Congress—XI (March 20): 29–32.

Siyambalapitiya, Tilak. 2007. *Standardised Small Power Purchase Tariffs for Tanzania*. Tanzania Ministry of Energy and Minerals, Final Report, September.

———. 2012. Personal e-mail communication. August 22.

Sundqvist, Thomas. 2000. "Electricity Externality Studies: Do the Numbers Make Sense?" Licentiate Thesis, Lulea Tekniska Universitet, Lulea, Sweden.

Tongsopit, Sopitsuda, and Chris Greacen. 2012. *Thailand's Renewable Energy Policy: FiTs and Opportunities for International Support*. Palang Thai, May 31. http://www .palangthai.org/docs/ThailandFiTtongsopit&greacen.pdf.

The Technical and Economic Rules Governing Grid-Integration Interconnections and Operations

Abstract

In chapter 8 we provide a primer on basic engineering terms and concepts relevant to grid-interconnected small power producers (SPPs). We also discuss the technical and commercial rules that govern the connection and operation of SPPs on the national grid or an existing isolated mini-grid. We conclude by discussing factors to be considered when interconnecting SPP generators to isolated mini-grids.

Basic Terms and Concepts

The general term *interconnection* refers to all the physical equipment needed to connect a new generator to an existing grid. Key terms used to describe the points of connection involved are presented in figure 8.1.

SPP generators usually produce electricity at a lower voltage than the large generators on a grid. Typical generating voltages for small power producers are in the range of 400 volts (V)–3.3 kilovolts (kV). When integrated into a larger grid, the SPP's power output will almost always be raised to a higher voltage, usually in the range of 11 kV–110 kV. The exact level will depend on the country's electrical standards and the voltage level of the existing network in the area where the SPP is located.

The SPP's last switch or circuit breaker is the *point of interconnection (POI)*. Beyond this point, all technical matters are the responsibility of the utility. The meter that gauges the SPP's sales to the utility that owns or operates the main grid is located at the *point of supply (POS)*. Beyond this metering point, the ownership of the grid and the power received from the SPP rest with the utility.

In most cases, the POI and POS are adjacent to each other. In some cases, however, the buying utility designates a POS at a location that is farther upstream (toward the grid, away from the SPP) from the POI. For example, in a negotiated

Figure 8.1 Terms and Concepts Relevant to the Interconnection of Distributed Generation

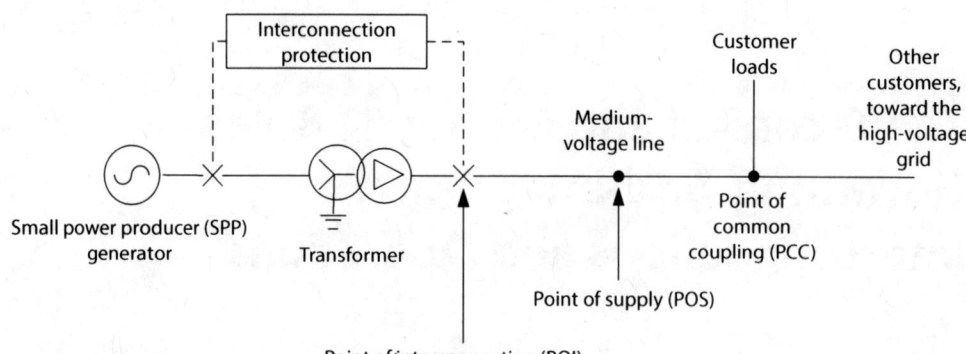

power-purchase agreement (PPA), if the lines to reach the grid are long, the POS may be located where a line reaches the buying utility's main grid. In this case the SPP, rather than the utility, will be responsible for any line losses that are incurred in transmitting electricity on the long line. The SPP will be paid for the energy that arrives at the POS and flows upstream to the utility's grid.

Other customers, lines, or SPPs are connected beyond the *point of common coupling (PCC)*. This point is so defined to ensure that power quality is maintained at both the PCC and beyond. Electricity grids operate at a range of voltages, usually specified in the codes that govern the grid and distribution. The line from the SPP up to the PCC may violate such specifications if there is good reason, and if the equipment can withstand such violations. Beyond the PCC, however, no such violations are allowed because other power plants, customers, and lines will be connected.

Key Definition

The *point of interconnection* (POI) is the point beyond which all technical matters are the responsibility of the utility. The *point of supply* (POS) is the metering point at which the SPP sells power to the utility that owns or operates the main grid, beyond which the ownership of the grid and the power received from the SPP are with the utility. The *point of common coupling* (PCC) is the point on the grid beyond which other lines, customers, or other SPPs are connected.

Standardizing the Process for SPPs to Interconnect to a National or Regional Grid

Should the process by which SPPs connect to a large grid be standardized? The answer to this question is an unequivocal yes. Allowing SPPs to connect to a regional or national grid is a new activity for most traditional utilities in

Africa and elsewhere. At best, traditional utilities will be receptive to connecting SPPs but will have limited experience with the processes and engineering standards required to ensure efficient and technically reliable interconnections. At worst, they may be opposed to purchasing from SPPs and may be tempted to create an application process and to specify technical parameters that will make it difficult for SPPs to connect to their grids. Therefore, just as the regulator requires a standardized PPA (see chapter 6), it should also impose standardized interconnection guidelines that spell out both the application process and the mandated technical standards. Any connection and operations manual for SPPs developed by the purchasing utility should also be consistent with technical guidelines issued by the regulator. Sometime in 2014, it is expected that Tanzania Electric Supply Company (TANESCO) will release a connection and operations manual specifically designed for SPPs.

In most instances, SPPs will be relatively small in capacity, compared to the larger, utility-scale generators serving a national or regional grid. This means that the overall approach to interconnecting an SPP can generally be simpler than that for a large generator. Regulators do not have to reinvent the wheel in creating their guidelines. Standardized guidelines for connecting SPPs to larger grids now exist or have been proposed in Tanzania, Kenya, Sri Lanka, and Thailand, among many others.

Many utilities have developed and adopted a grid code and a distribution code,[1] which describe the entire process of connecting a generating plant from first application through to on-grid operation. Depending on the voltage of interconnection allowed, the SPP interconnection may be described in either the grid code or the distribution code, in both codes, or in a separate document. In Sri Lanka, Thailand, and Tanzania, the energy ministry, regulator, or utility has developed and adopted a separate document containing guidelines for the grid interconnection of embedded generators.[2] Regardless of where the guidelines or rules are located, it is important that those relevant to SPPs cover the (a) application process, (b) the locus of responsibility for analysis and approval, (c) payment and construction responsibilities, (d) protection requirements, (e) testing and commissioning procedures, and (f) the data exchange process and follow-up activities.

Key Recommendation

Regulators should require a standardized process for SPPs that wish to interconnect to a regional or national grid. The standardized process should include specific guidelines for the application process, responsibility for analysis and approval, payment and construction responsibilities, protection requirements, testing and commissioning procedures, and a data-exchange process and follow-up activities.

Scope of the Engineering Standards for Interconnection

Engineering standards for interconnection should ensure safe and reliable operation of the grid as well as of the SPP. The interconnection should be designed and built with good-quality equipment of the correct rating and should be protected with the necessary relays.[3] The interconnection of any generator should have standard protection facilities to prevent damage from (a) overvoltage, (b) undervoltage, and (c) overcurrent. In addition, other safety-related protection (such as protection against lightning damage) should be implemented. To meet these operating requirements, the engineering standards should spell out the relays (over- or underfrequency, over- or undervoltage, overcurrent, reverse power, and so on) that should be required for different generator types (synchronous, induction, or asynchronous[4]) and over what size range.[5]

Since most SPPs are small and are embedded in the distribution network, they will generally be nondispatchable. This means that the generation dispatcher on the main grid cannot (remotely or through an oral instruction) switch the SPP on or off or control how much output each SPP should produce at any given time, because (a) the resource is intermittent (such as in run-of-river hydro and wind), or (b) contractual obligations require the grid to purchase the SPP's full electrical output at all times under "must-take" contracts. Under such contracts, if the SPP produces it, the grid must take it (see chapter 6).

SPP generators that are connected to the main grid are usually not designed to operate to serve an isolated grid that may become disconnected from the main grid, and SPPs that are embedded and nondispatchable require *protection against islanding*. This means that in the event of a grid failure, the SPP must quickly shut down without attempting to serve the section of the grid and its customers in the vicinity of the SPP (that is, not operating as an isolated island electrical system). This is a basic safety requirement to prevent dangerous overvoltages and other abnormal operating conditions that may damage the equipment of SPPs, the utility, and customers. Such protection against islanding can be implemented using a number of engineering techniques, including detection of over- and underfrequency, the rate of change of frequency, reverse VAr,[6] or voltage vector shift. These relays and their application in interconnecting SPPs to a main grid are explored in greater detail in Greacen, Engel, and Quetchenbach (2013).

All the above techniques to prevent islanding depend on measuring the indirect effects of the onset of islanding, such as abnormal changes in voltages and frequency, that will then automatically cut the SPP off from the grid. An SPP with its own mini-grid or other captive load[7] may, in rare cases, remain stable when the grid fails. This can happen only when the mini-grid load or the captive load reasonably match the power output that was delivered by the SPP when the grid failed. This is a dangerous situation, as voltages higher than the normal voltages may appear on customer supplies, and when the grid supply returns, the reconnection will most likely be out of synchronism, causing equipment damage.

A simple and definitive—but relatively expensive—technique to prevent islanding is to provide intertripping. This ensures that when a designated substation circuit breaker is tripped[8] by the grid operator or automatically as a result of a fault on the grid—or when a main grid line has no power—the SPP will trip out and isolate itself from the main grid, so that no power will flow to the grid from the SPP. Because of the higher costs of direct communication links between the SPP and one or several points on the grid, intertripping is rarely used among SPPs.[9]

The measurement and detection of abnormal voltages, currents, and frequency are all available now in a single digital protection relay unit. In most cases, the cost of this single relay unit will be lower than the traditional approach of using separate relays to receive measurements on each parameter to detect every abnormal condition. Such an integrated relay is likely to cost about $5,000, excluding the cost of the circuit breakers. For small induction generators (IGs), cheaper options (under $1,000) are available.

Key Recommendation

Engineering standards for interconnection should mandate the use of good-quality equipment with the correct ratings; protection to prevent damage from overvoltage, undervoltage, and overcurrent; protection against islanding; and protection against lightning damage and other safety hazards.

Paying for Interconnection Costs

Two general approaches are used in charging large and small generators for the cost of connecting to an existing grid. The first is to charge for a "shallow" connection. This approach is based on the view that any generator, whether large or small, that wishes to connect to an existing grid should pay only for the section of line and other equipment from the generator up to the POI. But if the grid does not exist in the area, a shallow connection may extend beyond the POI, and even beyond the PCC. The second approach is to charge for a "deep" connection. It presumes that the generator, large or small, should pay the shallow connection charge plus any upgrades to the upstream network that need to be made to accommodate the SPP's output. A "deep" connection may also constitute a direct connection from the generator to a point upstream on the grid where the required capacity is available, bypassing any locally available lower-capacity lines.

If the regulator allows SPPs to pay only shallow connection charges, the remaining deep connection costs are borne by the utility. But since these remaining capital costs will be rolled into the utility's overall capital costs at the time of a tariff application, they will ultimately be paid for by all customers on the utility's system. This approach is sometimes referred to as the "socialization of connection costs." The rationale for sharing the costs of the interconnection is that upstream capacity is shared among many customers and SPPs. In addition, in some countries regulators have established mechanisms to subsequently compensate the first newly connected customer or SPP for payments made for a network connection if the connection is later used to serve other customers or SPPs.[10]

If a good renewable energy resource for an SPP is far away from the closest possible POI on the grid with adequate capacity, the financial viability of the SPP project may be compromised by requiring it to pay the full cost of constructing an interconnection to the main grid. The central regulatory question then becomes who should pay the capital costs of the connection and any upgrades required upstream (if those costs are not socialized)? In Sri Lanka, more than a dozen mini-hydro SPPs in the Central and Sabaragamuwa provinces have been waiting more than five years for a grid connection because of the absence of grid capacity upstream from their closest POI. The Ceylon Electricity Board (CEB), the national utility, initially offered a cost-sharing arrangement (50 percent by the utility, 50 percent shared by the proposed SPPs) that would have required up-front payments by the SPPs. But no agreement could be reached for several years because of varying levels of commitment among the SPPs and lack of financial capacity among several. Finally, the CEB applied for and received a concessionary loan from the Asian Development Bank (a multilateral development bank that funds new infrastructure in developing countries in Asia) and built the upstream network capacity on its own. The nonsubsidized cost of this investment will eventually be paid by the CEB's electricity customers (through tariffs). SPPs were required to pay the connection costs only up to the closest POI (that is, the shallow connection charge). It appears that a similar approach will soon be taken in Kenya. The cost of a long radial line to connect the 300 megawatt (MW) Lake Turkana wind project in northern Kenya will be paid by all customers of KETRACO, the utility that owns and operates Kenya's transmission facilities.

In another example from Sri Lanka, four permits for wind power were issued at the same time for sites in the same area. The grid interconnection required 15–20 km lines from each power plant to reach the grid. But it was also recognized that it would be wasteful (in terms of investment and energy losses) for each SPP to build a separate 33 kV line to the closest grid point. The CEB therefore proposed a cost-sharing arrangement: The CEB would finance and build a grid substation in the vicinity of the wind resource area to step up the wind power SPP outputs to 132 kV, if the four SPPs would jointly finance the 132 kV transmission line to reach the grid. However, the four SPPs could not reach a joint agreement, so this less costly option could not be implemented. As a consequence, one SPP went ahead of the others. It financed and built a new 15 km 33 kV line to connect its 10 MW wind power plant to the grid. This 33 kV interconnection was adequate to serve only 10 MW, with no capacity for others to share the line. The other three wind power plants were built subsequently and connected to a new grid substation jointly financed and built in the vicinity of the wind resource area, while the first power plant continues to operate on the long connection, having now completed two years of operation.

Should an SPP Be Allowed to Construct the Interconnection?

As noted earlier, an interconnection is a general term that refers to all the physical equipment needed to connect a new generator to an existing grid. An interconnection will typically consist of (a) the transformers, switchgear, and protection equipment of the SPP; (b) new lines (or upgrades) and other switchgear and protection equipment to be installed farther away from the SPP (toward the POI); and (c) lines and equipment farther upstream from the POI. It is usually the case that the equipment referred to in (a) will be built and paid for by the SPP, whereas the equipment in (b) and (c) will be paid for by the SPP but built and commissioned by the utility (see table 8.1).

In expanding networks, most utilities give highest priority to expanding the transmission and distribution network to serve more customers and to connecting new, large generating plants. These are given highest priority because they are very visible and are driven by social and political pressure on the utility. Under these circumstances, which are common in many African countries, even if SPP interconnection guidelines have been issued, SPP-requested interconnections often get pushed to the end of the queue. And even if the utility wants to make a connection, it may simply not have the money to pay for upgrades farther upstream of the SPP's POI. Therefore, it has become increasingly common for utilities to allow SPPs to build their own interconnection—consisting of lines, transformers, and switchgear—even beyond the POI. When utilities allow construction by SPPs, they typically

Table 8.1 Cost Allocation of Interconnection Equipment Generally Observed in Asia and Africa

Equipment	Purpose	Paid by	Cost sharing	Built by
Transformer(s), switchgear, and line up to the POI	To supply power to the grid	SPP	None	SPP
Protection equipment	To protect the grid from adverse effects of the SPP and vice versa	SPP	None	SPP
Energy meter and metering equipment at the point of supply	Invoicing	SPP	None	Utility
Line upstream of the POI up to a designated point in the grid	To deliver power from the POI to the grid	SPP	May be possible, with another SPP	Utility (SPP may be allowed to build)
Lines and equipment farther upstream of the designated point in the grid	Enhance capacity of the lines and other equipment to deliver the output of the SPP to the grid	SPP	May be possible, with another SPP, a customer, or the utility	Utility (SPP may be allowed to build)

Note: POI = point of interconnection; SPP = small power producer.

From the Bottom Up • http://dx.doi.org/10.1596/978-1-4648-0093-1

impose one or more of the following conditions: (a) the interconnection should be based on the standard costs of the utility; (b) the material should be provided by the utility[11] and paid for up-front by the SPP or purchased from a short list of approved suppliers of the utility; and (c) construction labor and management should be provided by the SPP, but supervised by the utility.[12] Upon completion, the utility will test, commission, and take over the line and other equipment. The SPP's interconnection equipment described in item (a) above will also be tested and commissioned in the presence of a utility representative, but will be maintained by the SPP.

Transfer of Interconnection Facilities to the National or Regional Utility

All SPP assets up to the POI remain under the SPP's ownership. This ensures that the equipment is maintained by the SPP to deliver power safely to the grid and with the required quality. In most instances legal ownership of the SPP's power output is transferred from the SPP to the utility at the POS, which in most cases is the same as the POI or adjacent to it.

It is common practice that customer-paid network assets or SPP-paid assets upstream from the POI are transferred to the utility at zero cost. This seems reasonable because the lines upstream carry power that is now owned by the utility. From an SPP's perspective, this convention has both positive and negative aspects. On the positive side, the utility assumes the obligation to maintain the upstream assets and to replace them in the future when required—and no further obligations are imposed on the SPP. On the negative side, the SPP will rightly point out that an asset that it paid for is now fully owned by the utility, without the SPP receiving any direct compensation for its investment (unless the investment costs are later shared with other neighboring SPPs). As SPPs are relatively permanent facilities (compared with customers, who change locations), the mere fact that the utility takes over the ownership of upstream assets should not be of major concern to the SPP. An important advantage is that it relieves the SPP of the burden of maintenance and replacement of upstream assets beyond the POS.

Utility Equity Returns and Depreciation on SPP-Built Interconnection Facilities Transferred to the Utility

How should regulators treat these "gifted assets" when deciding tariffs that utilities may charge? Utilities tariffs typically earn a return on capital assets they have invested in, and tariffs they are allowed to charge also account for depreciation of these assets. As discussed in chapter 5, the utility should be allowed to take depreciation for gifted assets because they will have to be replaced at the end of their economic life, but should not be allowed to earn an equity return on these assets because they were not investments paid for by the utility. Therefore, the regulatory agency should require clear separation of assets funded by the utility from gifted assets funded by SPPs or other entities when determining investment/depreciation on network assets in order to establish the revenue levels that will be recovered through the utility's tariffs.

Compensating the SPP for Later Use of an Interconnection

Sometimes an SPP (or even a new customer) will apply to use an interconnection that was previously paid for by another SPP. It is generally accepted that it is unfair for a later SPP or customer to get a "free ride" on assets paid for by another SPP or customer. Hence, it has become increasingly common for regulators to establish a system that requires new SPPs or users who wish to connect during a specified period of time (say, within five years of the establishment of the interconnection) to reimburse the SPP(s) or customer(s) who paid for interconnection facilities that they now seek to use. For such a reimbursement system to be successful, the utility must maintain accurate records, charge the new customers and SPPs a *pro rata* share of the initial capital cost, and provide reimbursement to the first customer or SPP.[13]

Such situations are especially likely to arise in the case of new small hydro and wind power developments. It is often the case that SPPs in the same area will secure approvals and reach financial closure on different time schedules. The first SPP may not be able to wait until all other SPPs in the area are ready to pay jointly for an interconnection. If the first SPP were forced to wait, it might run the risk of losing access to financing or other approvals having a defined expiration date. The first SPP may be willing to pay for the entire interconnection if it knows that it will be compensated when others need to use some of the interconnection capacity in the future.

A related issue is the sizing of the interconnection. For example, a 2 MW SPP can be connected using a conductor with a smaller cross-section, but it may be the policy of the utility that all medium-voltage distribution lines should be of a certain minimum size. That minimum size may be much larger than would be required for the 2 MW SPP. In this case, the SPP will overpay for capacity that it does not need. In this situation, we think that it is both fair and efficient that the initial SPP should be compensated when more customers (or SPPs) are connected over the higher capacity line.

Minimum Interconnection Voltage for Generators of Different Sizes

SPPs are usually embedded in the distribution network and connected to common distribution lines. The main reason for embedding SPPs into the distribution network is to minimize the cost of interconnection and relieve the SPP of the large capital costs that would be required to reach the high-voltage transmission network through expensive high-voltage step-up transformers and switchgear.

In deciding on an appropriate interconnection voltage for SPPs, the two primary considerations are (a) to minimize the impact on customers, and (b) to lower the cost of connections to SPPs. *The general rule is to limit the size of the SPP's generating capacity to match the current carrying capacity of the typical conductor used by the utility at that voltage level.* But this cannot be considered a hard-and-fast rule. For example, a long 33 kV radial distribution line serving a town 30 km from a grid substation in Tanzania may not be able to accommodate a 10 MW SPP at the town end of the line. This is because when the SPP is generating, the voltage at the SPP (town) end may increase above specified levels unless the customers in the town exert a demand of 10 MW or more. This voltage rise at the town end of the line may be unacceptable and could damage appliances and other electrical equipment used by customers in the town. As such situations occur frequently in Africa and Asia, where many communities and resources are located far from the grid, there are two options for utilities and regulators endeavoring to establish a policy on the interconnection voltage and the SPP capacity limit. The first is to predefine the allowed connection capacity at each voltage level (for example, 5 MW at 11 kV, 10 MW at 33 kV), to allow the utility to conduct studies on each application for interconnection from an SPP, and then to request the SPP to pay for network strengthening to enable the voltage standards to be maintained. The second is to specify only the upper limit of generating capacity allowed for SPPs anywhere in the grid, and to allow the utility to decide the connection voltage on a case-by-case basis.[14] Sri Lanka and Tanzania generally use 33 kV as the interconnection voltage for SPPs, but 11 kV lines exist as well. The SPP capacity in both countries is limited to 10 MW per installation.[15] Vietnam allows SPPs on 35 kV and 110 kV lines, and the SPP capacity limit is 30 MW.

Key Recommendation

When an SPP wishes to connect to the main grid, the following interconnection and cost-recovering policies are recommended.

1. SPPs should pay for and construct the transformers, switchgear, and protection equipment required for interconnection up to the point of interconnection (POI). They should also pay all so-called shallow interconnection costs, even beyond the POI. The regulator should decide whether other (deep) connection costs should be paid by the SPP or "socialized"—that is, paid for initially by the connecting utility and ultimately by its customers through transmission or distribution tariffs.

2. The utility should, in principle, construct the new lines and other upstream equipment on the grid beyond the POI and install the meters and metering equipment at the point of supply (POS). As a standard policy, to ensure timely construction, the SPP should be allowed the option to construct facilities beyond the POI under construction and engineering standards established by the utility and subject to review by the regulator.

3. SPPs should retain ownership of their assets up to the POI but transfer ownership of SPP-paid assets upstream of the POI to the utility at zero cost. The utility can claim depreciation on these assets but should not earn a profit on such "gifted" facilities.

4. New SPPs or users who wish to connect to interconnection facilities paid for by another SPP or user should reimburse the initial SPP(s) or customer(s). In these cases, the utility must maintain accurate records, charge the new customers and SPPs a *pro rata* share of the initial capital cost, and provide reimbursement to the first customer(s) or SPP(s).

Successful Integration of SPPs into the Grid: Technical and Commercial Requirements

Once SPPs come online, whether by being connected to the main grid or by operating on an isolated grid, they must satisfy certain operating practices and maintain several key electrical parameters in order not to cause physical harm to others and themselves. Regulators, who are often not electrical engineers, cannot be expected to develop the required operational technical codes. Instead, the job of the regulator should be to make sure that such a code is created (if it does not exist) and is agreed to by both the main-grid operator and the SPP operators. If there are disputes, the regulator, with the support of technical advisors, must quickly resolve them. In this section, we provide a brief primer on what should be included in such a technical code, as well as a checklist for preparation of the code.[16]

Do SPPs Need to Be Dispatchable?

Grid-connected SPPs that are powered by renewable energy or are part of a cogeneration scheme where both electricity and steam are produced generally need not receive dispatch instructions from the system operator (also known as the dispatcher). In other words, these SPPs are generally not dispatchable. This is because renewable energy is intermittent and may not be available for dispatch at any given time. Similarly, a generator in an industrial cogeneration facility generates electricity according to the industrial heating (steam) needs of the industrial process. For these reasons, and because SPPs are small compared to other generators on the grid, most SPPs have "must-take" PPAs that allow the SPP to generate at the SPP's convenience and require the utility to purchase all of the electricity generated (see chapter 6).

But there are exceptions to this general rule, especially when the number of SPPs in the network grows and begins to have a significant impact on the grid. For example, in some rare cases in the United States where large concentrations of wind power are found and at times when the transmission system is congested or there is an excess of water in hydropower reservoirs that must be released, wind power generators are ordered to curtail power generation (that is, not generate electricity). Also, when a system has a lot of wind power generation in the same geographic area, some dispatch (or control) will be required to ensure that a sudden increase or decrease in wind speeds that would lead to increases or decreases in the electrical output of wind generators does not cause instability in the grid.[17] Small hydropower plants, even if they are run-of-river, have a more stable and predictable pattern of production. Normally, small hydropower systems provide adequate time for other generators in the system to respond to changes in hydropower production.

In such cases, the contract between the SPP and the main-grid operator may allow for dispatch, even when the SPP resource is renewable and intermittent. In these cases the generation dispatcher will have the ability and authority to decide when to operate the power plant and the level of output at which it should operate. To perform these functions, the generation dispatcher will require information on the status of the resource (water availability, rainfall, wind speed, and so on) and the degree of readiness of the power plant to start up (ready to start, on standby, under long-term maintenance). This information may be obtained from the SPP either manually (by telephone or fax) or online (through an automatic data acquisition system). The contract permits and dispatch instructions (stand by, start up, shut down, raise or lower power output) may be issued either manually (by telephone or fax) or remotely (through a supervisory control system). For this to be possible, hardware must be installed at the SPP. The hardware at an SPP for an automatic data-acquisition system may cost about $20,000, in addition to which will be the cost of the monthly or annual fee for the communication link. A supervisory control system may cost about $30,000 in addition to the fees for the communication link. For both a data-acquisition system and a supervisory control system to be functional, the dispatch center will need to be equipped with the necessary hardware and software to analyze the acquired data and optimize the grid operating costs. Such supervisory control and data-acquisition systems will generally be too expensive for small SPPs. Hence, the norm in most countries is for SPPs to be nondispatchable.

As an alternative to dispatch, SPPs can be incentivized to generate electricity at different times through pricing mechanisms. For example, Thailand and Vietnam pay renewable energy generators according to a time-of-day (TOD) rate,[18] with higher tariffs during work-week daytime hours and lower payments at night, on weekends, and on holidays. To respond to such TOD tariffs, it may be advantageous for some SPPs to have storage, such as a pond in a mini-hydro power plant or stockpiled biomass fuel. For intermittent renewable energy like solar and wind power, higher TOD rates may coincide with their periods of peak production: the sun shines and the wind blows most strongly in the daytime—typically the peak rate or a daytime rate in a TOD tariff schedule. In Sri Lanka and Tanzania, SPPs receive a lower tariff during the rainy season and higher payments during the dry season. In Tanzania, this incentivizes maximum production to help alleviate strain on hydropower resources. (For more on this, see the discussion of feed-in tariffs in chapter 7.) Even if storage were limited or not available, such TOD tariffs would encourage SPPs to schedule maintenance during the off-peak period (Thailand and Vietnam) or the rainy season (Sri Lanka and Tanzania).

Essential Electrical Parameters to Be Specified and Controlled

Four major electrical parameters need to be specified and controlled for successful integration of SPPs into the grid: voltage, frequency, harmonic distortion, and the power factor.

Voltage

The output voltage of the SPP facility should be equal to the nominal (usual) voltage of the grid. Most countries require the same voltage standard on the main grid and mini-grids so that machinery and appliances need satisfy only a single uniform voltage standard throughout the country. For a grid-connected SPP, the voltage regulation (the variation in voltage between full power and zero power output) at the POI should be within specified limits. For example, most countries establish a voltage regulation standard of ±6 percent on the low-voltage network and ±5 percent on the medium- or high-voltage network. Voltages that are outside this specified band are harmful to the SPPs, to utility equipment, and to customers' equipment. Under emergency conditions, such as the temporary outage of a transmission line that shares power at normal times, a voltage regulation of ±10 percent is allowed in most countries for a short period of prespecified duration.

Frequency

The normal operating frequency of the SPP is the same as the frequency of the national grid (either 50 hertz [Hz] or 60 Hz). SPPs will follow the grid frequency, and because they are small, the presence or absence of a single SPP at any given time will not make any impact on the frequency of the main grid. In contrast, the loss of a large generator feeding the grid will cause the grid frequency to drop. If not balanced quickly with additional generation or a reduction in customer loads, or both, the loss of a large generator may lead to large frequency drops, causing other remaining generators to trip out (for their safety). If this happens, there will be a cascading failure in which all the remaining generators stop generating electricity one by one, and the entire system will experience a blackout.

SPPs are affected by changing grid frequency in three principal ways. First, if the value of the frequency or the rate of change of frequency passes the preset threshold, based on the protection systems installed, the SPPs will trip themselves off the grid. Second, prolonged operation at off-nominal frequency may not be permissible for certain SPPs, especially those generating electricity from a steam cycle or from combustion turbines, because it could damage the turbines. Third, when the grid undergoes a frequency variation because of a disturbance (for example, the loss of a large generator or of a transmission line), even if the grid has not yet failed, automatic or semi-automatic protection for underfrequency and rate-of-change of frequency may cause SPPs to trip out, thus worsening the crisis on the grid (if the grid frequency is decreasing) or helping it to recover (if the grid frequency is increasing). Therefore, the threshold below the nominal frequency is set at a larger percentage shift compared with the upper threshold: for example, the setting's high frequency threshold may be nominal frequency plus 4 percent, with a low frequency threshold of nominal frequency –6 percent. The rate-of-change-of-frequency protection is typically set to operate at 2.5 Hz per second, to ensure that it responds only to very severe frequency changes, such as with the onset of islanding.[19] Frequency protection should not respond to normal changes in grid frequency from which the grid is likely to recover through other means.

From the Bottom Up • http://dx.doi.org/10.1596/978-1-4648-0093-1

In the mini-grid case, the SPP represents a large portion (sometimes 100 percent) of the generation serving the mini-grid. Here the SPP must control the rotational speed of the generator's prime mover (steam turbine, reciprocating engine, or other) to keep frequency within limits. In small grids, frequency variations are generally allowed and are expected to be larger than the trip settings on SPPs connected to the national grid, for example ±10 percent.

Harmonic Distortion

The voltage of an alternating current (AC) electricity supply should be a perfect, smooth, sinusoidal waveform. Ripples and distortions in this waveform are referred to as harmonic distortion or "harmonics." Harmonic distortion may be caused by SPPs that use power electronic devices such as inverters (in solar photovoltaic [PV] systems) or sometimes from small rotating generators.[20] Harmonic distortions, if injected onto the grid in large quantities, may damage the electrical equipment of customers. Therefore, grid operators must specify maximum limits for harmonic generation by SPPs. Harmonic-related specifications are standard on any well-run electrical system, and grid or distribution codes generally provide comprehensive guidelines on the allowable levels of harmonic distortion and the methods of measurement.

Power Factor

The power factor is a measure of the degree to which current and voltage at a point in an AC electrical system rise and fall in phase. Devices that contain coils of wire (for example, most motors or all transformers) cause the current to lag behind the voltage (called "lagging power factor"). Lagging power factor lowers voltages and consumes "reactive power."[21] Conversely, devices that have significant capacitance (rarer in SPPs) cause current to lead voltage (called "leading power factor"). Leading power factor increases voltage and creates reactive power. When current flows out of phase with voltage, losses in the system increase.

When a synchronous generator (SG) operates at a power factor close to 1 (phase angle zero degrees), it provides the minimum or zero "reactive power" to the grid. SPPs using SGs can control their operating power factor by controlling the field current in the generator. By increasing the field current, an SPP using an SG can deliver more reactive power to the grid in order to regulate the voltage at the node at which it connects to the grid.

Grid operators or dispatchers often complain about the absence, in PPAs with SPPs, of adequate provisions governing reactive power support during day-to-day operation. They contend that in the absence of such provisions all of the grid's reactive power requirements must be provided by other generators. But this ignores the fact that producing reactive power in generators, especially in generators (including SPPs) serving a grid through long lines, can be a wasteful exercise. This is because providing reactive power requires currents larger than the currents required to deliver the useful output of an SPP to the grid. These larger currents cause additional losses in SPP generators, as well as in the utility's own generators, both before and after the POS. Reactive power can be produced by other means, the cheapest being the installation of capacitors closer to the locations in the grid where such reactive power is required (for example, at transformers and on customers' premises). A good system for managing reactive power would enable all generators (not only SPPs) to produce the minimum currents, thus minimizing heating losses across the network. Therefore, while SPPs may be required to be *capable* of operating at a power factor of up to 0.8 to serve any reactive power requirements of the grid under emergency conditions, the correct practice in most cases will be to operate the SPP at or close to a power factor of 1.

A special case related to the power factor involves induction (that is, asynchronous) generators, which are increasingly used in SPPs. Unlike SGs, asynchronous generators cannot produce reactive power. In fact, they require magnetizing power (that is, reactive power) to be provided externally, from the grid or by other means, similar to a customer's induction motor.

But an SPP using an asynchronous generator can, by installing capacitors, "generate" the required reactive power to provide magnetization to the SPP's own IG without drawing power from the grid. If the SPP does not install capacitors, however, it must draw reactive power from the grid. Grid codes or distribution codes typically specify a limitation—and penalties for such use of reactive power from the grid. These may be in the form of a limit on the operating power factor (such as a limit of 0.98 lagging to 0.98 leading), a penalty for violating the operating power factor limits, a reactive energy charge (measured in kVArh), or a combination of all the above.

Operational Communications between Utility and Operator

As described earlier, SPPs are mostly embedded and nondispatchable. In such cases, in principle, there is no need for any communication between the grid dispatcher and the SPP during operations. Many SPPs operate on this basis: neither party has real-time information about the operating status of the other party's system. In Sri Lanka, where more than 120 SPPs (mostly mini-hydro) are in operation, with a total capacity exceeding 230 MW on a grid that has a peak demand of about 2,200 MW and a minimum nighttime demand of about 900 MW, there is currently no communication (on- or offline) between the SPP and the main grid system operator. At present, not even the SPPs' day-ahead or week-ahead plans or their power plant maintenance schedules are exchanged with the dispatcher. Outages for line maintenance are coordinated at the distribution level. While no serious problems have occurred because of the absence of online communication systems on operational status, the addition of wind power into the system has introduced the requirement that specific online status information be provided to the dispatch center. As wind flow varies significantly at shorter intervals when compared with variations of water flow in a river, frequent changes in wind power output and voltage flicker (frequent ups and downs) have been observed along the distribution lines in the area where the first few wind SPPs were located in Sri Lanka. Online communication and data acquisition from wind power plants have since been specified as a requirement.

With mobile telephone coverage now extending to many parts of Africa and Asia, and with the availability of communication interfaces in the current generation of automatic meter reading (AMR) meters used to measure SPP energy inputs to the grid (for invoicing purposes), it has become simpler and less expensive to provide online status and output information. As the SPP contribution to total output grows, especially in wind and solar, the grid operator will benefit from knowing the operational status of distributed generation (DG) in order to make forecasts and ensure that other large generators on the grid will be used in an optimal manner. Future SPP agreements are likely to move away from "must-take" regimes to some limited form of control and dispatchability. The availability of online operational information to both the dispatcher and the SPP will enable the two parties to move toward better use of resources and investments.

Grid-Connected SPPs: Key Elements of a Good Billing and Payment System

A meter fixed at the POS measures the energy sent to the grid by the SPP and, ideally, power imported from it. Newer meters are digital and record all important electrical parameters (current, voltage, real and reactive power, and maximum demand, at intervals of a few minutes). Now it is common to use four-quadrant meters[22] that can measure and record the direction and combinations of two key parameters: real and reactive power and imports and exports. Meters must satisfy the appropriate standards applied by the utility to its customers. The accuracy of meters is usually specified as class 0.2 (meaning the accuracy is ±0.2 percent).

Most meter manufacturers provide a communication interface that allows for remote reading by the utility. Utilities are moving away from reading meters at the site, using AMR meters instead, which can be queried from utility offices. In developing countries, this trend is presently limited to bulk customers. Some SPP agreements may require joint reading of meters by the utility and the SPP, which might require a physical visit to the site.

Both the import and export registers of the meter are jointly read by the utility and the SPP, and the invoice for export is prepared by the SPP accordingly. The utility prepares an invoice for imports by the SPP (that is, backup power) from the grid.[23] Thereafter, the usual payment systems follow, as in the case of any other independent power producer (IPP) (export) or bulk customer (import) of the utility—unless the utility has created a special backup tariff for SPPs (see chapter 6).

Key Recommendation

When SPPs prepare to connect to the main grid, the following technical and commercial requirements are recommended:

1. Grid-connected SPPs should not be required to be dispatchable, except in the rare case when the share of renewables on a grid reaches a high enough threshold to require dispatchability. In lieu of dispatchability, regulators and utilities can incentivize SPPs to generate electricity at certain times of the day or in certain seasons.
2. The regulator or grid operators should specify standards for voltage, frequency, harmonic distortion, and power factor.
3. Grid operators do not necessarily need to be able to contact SPPs that are embedded in the grid at the distribution level and that are nondispatchable, but as SPP penetration grows to high levels the grid operator will find it increasingly useful to know the SPP's operational status, make forecasts, and use other generators in the grid in an optimal manner.

Factors to Consider When Connecting to an Isolated Mini-Grid with Existing Diesel Generators

SPPs connected to mini-grids with existing diesel generators present a separate set of technical and financial issues that are generally more challenging than those for SPPs connected to the main grid. At the core is the question of how the timing of the availability of electricity from the SPP coincides or overlaps with the timing of loads on the mini-grid, and how the SPP operation can be coordinated with the operation of existing diesel generator(s).

The Case of an SPP Connected to the Main Grid

To conceptualize these issues, first recall that in the case of SPPs connected to the main grid, the SPP is a small fraction of the total installed capacity of the grid. The SPP is essentially injecting current into a grid whose frequency is controlled moment to moment by much larger generators. Typically this "frequency control" function is handled by load-following generators—often large hydropower or natural-gas-fired turbines that can adjust their power output instantaneously in response to fluctuations in demand for electricity. The grid is able (except in rare instances) to absorb all the power that the SPP is able to generate. The SPP coming online or going offline causes only very small changes in power input requirements for larger generators and is generally lost in the "noise" of minute-to-minute or hour-to-hour variations in national load.

The Case of an SPP Connected to an Existing Isolated Mini-Grid

This situation should be contrasted with an isolated mini-grid where the capacity of an SPP generator is significant with respect to the total load on the grid.

From the Bottom Up · http://dx.doi.org/10.1596/978-1-4648-0093-1

Here we run into several difficult issues. We will start with the financial issue, and then move onto technical issues (which have financial implications).

When Demand Is Less than the Small Power Producer's Capacity

The fundamental financial issue is perhaps easiest to conceptualize: If there are times of day when load on the mini-grid is lower than the amount of electricity that the SPP can generate, then the SPP cannot generate to its full production capacity. Because the SPP's capacity will not be fully utilized, it will not be receiving full revenue. This represents a shift from the "must-take" contractual arrangement of SPPs on the main grid (described in more detail in chapter 6) to one in which electrical energy can be sold only if someone on the mini-grid is demanding it at that moment. This temporal discrepancy between the ability to supply and the availability of demand is one reason why tariffs for SPPs selling to existing utility-owned isolated mini-grids in Tanzania are much higher (about three times higher) than for SPPs selling to the main grid.

PPAs for SPPs connected to a utility's isolated mini-grid should take this into account. One possible way to do this is to include in the PPA general language such as the following:

> The Buyer has no obligation to purchase and accept the portion of electric energy from the Seller when said energy would exceed the amount of power that the Buyer's mini-grid system can safely accept while maintaining overall power quality to customers.

Complicating Technical Factors

The engineering realities of keeping electrical systems stable and power quality acceptable (alluded to in the sample PPA text above) present further challenges and constraints that complicate and exacerbate the fundamental financial issue raised above.

The main issue is that in mini-grids it is necessary to determine who is responsible for frequency and voltage regulation.[24] The situation is analogous to that of a musical group: someone needs to keep time and set the beat. If there are multiple generators in charge they may end up working against one another—and the music will not be pretty.

Another constraint that arises specifically with diesel generators is "wet stacking." Diesel generators, whether powered by diesel fuel or some other fuel such as biogas, build up residues inside the cylinders and condensation in the exhaust if operated at low loads. For these reasons, vendors of diesel generators do not like to see their products run at low load—because doing so may have implications for maintenance and warranty service. A related issue is that diesel generators are less efficient at low loads. An SPP that injects electricity into the grid may force the diesel generator to operate at levels at which efficiencies are low, and the diesel generator will be forced to consume more fuel per kWh than if it were operating at higher loads. This has the effect of reducing the diesel fuel-saving benefit of the renewable energy SPP generator. In cases in which the mini-grid diesel is owned by an entity separate from the SPP (a utility, for example, as is the case for 16 mini-grids in Tanzania), these operating issues can complicate the relationship between the SPP and diesel mini-grid operator.

Views of Engineers

To further complicate all of this, experts disagree on the question of how much DG can be integrated into a diesel mini-grid. Here are responses from several experts:

British Columbia. "As far as I can tell PV can be added with relatively constant incremental benefit until there is spillage. And then the spillage starts to affect the economics of additional increments of PV. I'm not sure there are any obvious technical limitations." —Brett Garret, power system engineer, BC Hydro

Thailand. "The [diesel] generator maker and synchronization panel for the DG integrator told me that they do not want their generator operating below 20 percent of [the diesel generators'] rated capacity." —Dr. Wuthipong Suponthana, mini-grid designer

Hawaii. "The bigger issue with high penetrations of PV is frequency control. That requires some detailed modeling with models that can look at transient responses. It depends on the rotating inertia of the generator, the type of generator (electronic control, turbo, etc.), and the physical layout of the grid and the PV system. It also depends on the frequency tolerances of the system. Lanai, Hawaii, put 1.4 MW of PV in a single array on their 5 MW system and can't run it at more than half capacity while maintaining their desired frequency stability. They are putting in a big battery bank now. It would have helped a lot if they had spread the PV out, but Hawaii has a lot of fast-moving patchy clouds." —Dr. Peter Liliental, president and CEO of HOMER Energy

Tanzania. "To prevent [power quality problems] total export capacity ceilings have been established for certain types of SPPs connected to TANESCO Mini Grids:

Synchronous generator (SG): Export capacity must not exceed 75 percent of the TANESCO's Mini Grid's total installed generating capacity;

Wind farm/turbine must not exceed 50 percent of the TANESCO's Mini Grid's total generating capacity;

Induction generator (IG): Must not exceed 50 percent of the TANESCO's Mini Grid's total generating capacity; The project's generating capacity must not exceed 1 MW;

Inverter-based DG: May not exceed 50 percent of the TANESCO Mini Grid's total generating capacity." (TANESCO 2011)

Greece. A simulation study of adding solar arrays to a diesel-powered mini-grid on the island of Kythnos assumed five 450 kilovolt-ampere (kVA) generation sets. The simulation finds "the system is generally stable for penetration levels lower than 50 percent while it becomes unstable at higher levels of penetration due to insufficient spinning reserve of the diesel generators" (Rikos, Tselepis, and Neris 2008, section 3).

Key Recommendation

When an SPP prepares to connect to an existing isolated mini-grid that has been powered by one or more diesel generators, the following technical and commercial requirements are recommended:

1. The receiving grid should have no obligation to purchase and accept electric energy from the SPP if the energy exceeds the amount of power that the receiving mini-grid system can safely accept, while maintaining overall power quality to customers.
2. Only one of the two entities should be responsible for maintaining the nominal voltage and frequency of the isolated grid.
3. The existing diesel generator and incoming SPP must reach an agreement about how to compensate the diesel generator if the incoming SPP electricity forces the diesel generator to operate at a lower load, thereby reducing its efficiency and incurring higher costs, and potentially causing damage. A better solution would be to encourage one of the two parties—the SPP or the entity that owns the diesel generator—to assume ownership and operating responsibility for the entire isolated mini-grid system.

Two Technology-Specific Examples

Let us develop these ideas a bit further through consideration of two examples.

Example 1: Integration of a Solar PV Array with a Diesel-Powered Mini-Grid

Consider a mini-grid with a 1 MW diesel generator. Imagine that current peak load is 700 kW (figure 8.2), and this peak load occurs mostly in the evening time as people light their homes, cook dinner, and watch TV or listen to the radio.

Figure 8.2 Hourly Load Profile for Example Village Mini-Grid System

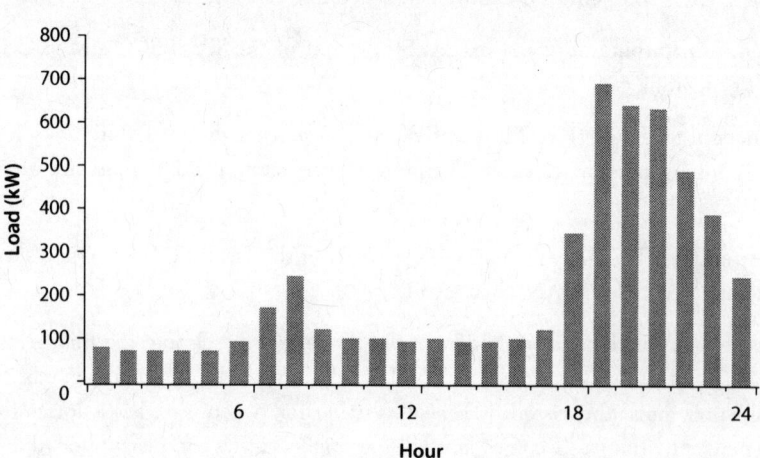

The minimum load in this hypothetical case might be only 80 kW, occurring between 2 am and 4 am, but the daytime load is also quite low, with a minimum of only 100 kW because most people are outside working. Let us assume that this village does not include large intermittent loads such as arc welding equipment. Assume that an SPP developer wishes to add a solar electric system that interconnects to the mini-grid through an inverter. How big a system could be accommodated by the mini-grid?

A solar PV system up to 50 percent of minimum daylight hour load (50 kW in this example) would probably have minimal integration issues. The maximum output of the solar PV system is less than minimum daytime customer load, and most diesel generators can provide frequency and voltage control in this regime. The existence of arc welding or other large intermittent daytime loads could complicate this arrangement as would fast-moving patchy clouds, which would cause rapid ramp-up/ramp-down in power production from the PVs. This is because such intermittent loads would force the diesel generator to make rapid relative changes in power output to keep electricity within frequency specifications.

Approaching 100 percent of minimum daytime load (100 kW in this example), however, we run into the issue that the diesel generator is now operating with zero load but still trying to modulate frequency and voltage. This is a recipe for instability. If the customer load were to decrease even momentarily below PV power output, the PV panel would need equipment that allows it to take over the role of frequency and voltage regulation and to spill power generated in excess of load. Otherwise, these conditions could cause voltage or frequency disturbances that could cause the generator to shut off, inverter equipment (which converts direct current [DC] electricity to AC electricity) to shut off, or cause problems with customer equipment.

A solar PV installation with capacity above 100 percent of minimum daytime load would generally include a battery bank to absorb peaks when generation exceeds load. If a battery was put in place, typically it would be accompanied by an inverter that has stand-alone capability: providing frequency and voltage regulation so that the PV/battery/inverter system can carry the load entirely, allowing the diesel generator to shut off entirely at times.

With the addition of batteries it is also possible to store a portion of electricity produced by the diesel generator, allowing the diesel generator to operate at full capacity for a shorter duration producing electricity for load and to charge the battery simultaneously. This increases diesel efficiency by operating at closer to full capacity and reduces the diesel generator's run time. If one is interested in optimizing the cost-efficiency of the entire hybrid mini-grid system (and not just the SPP generator as a private sector merchant generator selling electricity to a utility owned mini-grid) it makes sense to optimize the sizing of the battery bank capacity and renewable energy source to minimize overall levelized cost of energy—taking into account the temporal characteristics of the load profile and renewable energy resource. Free software such as HOMER[25] can facilitate these calculations by modeling economic dispatch of different-size generators under many different candidate system configurations.

Example 2: Micro-Hydro SPP Added to an Existing Diesel Mini-Grid

Somewhat similar issues arise in the case of an SPP micro-hydropower[26] generator added to an existing diesel grid. Again, the key issue is which generator is going to be controlling frequency regulation. Using the same load profile (figure 8.2), we can imagine the following:

A micro-hydro system providing up to 80 percent of minimum load would be able to run at full power continuously, injecting current to the diesel mini-grid and reducing diesel usage. System-interconnection equipment need only comprise equipment to synchronize the micro-hydropower project with the grid (if a SG) and basic over/under frequency and over/under voltage relays.

Beyond this level, but still below peak load, the micro-hydro generator would need two sets of controls: one that allows it to operate in stand-alone mode with the diesel generator off, and one in which it can synchronize to the diesel. The stand-alone controls include a regulator that controls frequency either through modulating the volume of water flow to the turbine (electro-mechanical controls) or an electronic diversion load controller that keeps frequency constant through dumping excess electricity generation to a "ballast load"—essentially a heater.

In this regime, at certain times (for example, the middle of the night or morning time), the micro-hydro is sufficient to carry the load, but the diesel generator is turned on to supplement the hydro to meet the peak for evening loads. The coordination between these two modes would need to be worked out.

If the micro-hydro is larger size, for example, to meet peak loads, then the predominant mode for the micro-hydro would be stand-alone operation. But the ability for a diesel generator to synchronize would be useful—for example to meet peak loads in the dry season.

The Need for Operating Protocols

Clearly, many of these cases require specialized equipment, as well as coordinated operation protocols between the renewable energy operator and the diesel operator. These complicate a financial picture already compromised by energy sales that are less than optimal because of hourly and daily variability in the load. Whether or not higher tariffs in a mini-grid case (as is the case in Tanzania) are sufficient to make up for these factors requires careful consideration on a case-by-case basis.

Coordinating voltage and frequency regulation presents challenges for which there is no one-size-fits-all technical solution. It depends on the intermittency of the renewable energy source, the presence or absence of storage (batteries, and so on), the technical characteristics of the generator, and the control system that monitors and supervises different elements in the system. These various factors have to be studied by competent electrical engineers in the context of the specific project, and the solution will require both technology as well as an operations protocol. It is unrealistic to expect that a regulator can or should write regulations to cover these many site-specific technical coordination issues.

These technical issues arise whenever there is a sufficiently large intermittent energy source combined with a diesel generator. If the diesel generator is owned and operated by the same party as the intermittent renewable energy source, then these issues are internal to the developer. It is more difficult to resolve these problems when the SPP is owned by a different party than the owner of the diesel generator and mini-grid. *Therefore, the best solution would be to encourage one of the two parties to take over ownership and operation of both the renewable and diesel generators.*

The challenges of dispatch, frequency, and voltage regulations would further multiply in the event that more than one company is operating SPPs on a single mini-grid. It may be most practical to allocate full rights of the market in a particular mini-grid to the first project on that mini-grid that passes certain milestones, such as signing a PPA or obtaining a letter of intent.

Notes

1. In general, the grid code covers the transmission network and associated generators, equipment, and customers, while the distribution code covers the network and equipment downstream of the transmission network. The "grid" refers to the entire network, both transmission and distribution. Specific definitions of transmission and distribution depend on the standard voltages used in a given country. In Tanzania and Sri Lanka, the network above 33 kV is the transmission network, and the balance is the distribution network. In Thailand and Vietnam, 110 kV and above is considered to be the transmission network.

2. A proposed Tanzania Guide for Grid Interconnection of Embedded Generation, which was developed by the regulator, is available in three parts here: http://www .ewura.go.tz/sppselectricity.html. Tanzania's utility, TANESCO, also has a grid code specifically addressing SPPs (available from TANESCO) that is expected to be issued in 2014. Sri Lanka's guidelines (in two parts) are available in English upon request from the Ceylon Electricity Board. Key features of Thai regulations (in English), are available at: http://www.eppo.go.th/power/vspp-eng/index.html.

3. A relay is a device that receives and processes a measured quantity, such as the voltage on a line, and then issues a signal to activate a device such as a switch, a warning indicator, or an alarm. A relay can be programmed to make certain calculations with the measured quantity and to issue the output signal based on the results of the calculation. A relay can also be programmed to issue the output after a time delay.

4. Synchronous generators (SGs) rotate in phase with the rest of the SGs on a grid. Asynchronous generators rotate typically between 0.5 to 2 percent faster than the SGs on a grid. An induction generator is effectively the same as the induction motor very widely used in industry to convert electricity to motive power; when working as a motor, it rotates below the synchronous speed, and as a generator, it rotates above the synchronous speed.

5. These issues are discussed in sections B7 to B9 in the Tanzanian guidelines mentioned in footnote 2.

6. Reverse flow of reactive power from the grid into the SPP.

7. Captive load refers to load that is on the SPP's side of the meter.

8. *Tripped:* switched open so that no current can flow.

9. Depending on the distance between the utility and SPP breakers, an intertripping system may cost about $10,000 or more to install, plus a monthly fee for the communication link.

10. A good survey analysis of the issues surrounding transmission charges for new renewable generators can be found in Madrigal and E3 (2010). For the most recent statement of the Sri Lankan regulator's policy on sharing of transmission investment costs between a first and later SPP, see http://www.pucsl.gov.lk/english/wp-content/themes/pucsl/pdfs/methodology_for_charges.pdf.

11. This provides assurance to the utility that the material used will meet its technical standards. This is an important utility concern because the interconnection (or a portion of it) would subsequently be owned and operated by the utility.

12. Such decisions are usually taken on a case-by-case basis.

13. Sri Lanka has recently approved such a reimbursement to the first customer or SPP (PUCSL 2010). Up to nine customers/SPPs pay 10 percent of the first customer's shallow interconnection cost if they request connection within the first five years of operation of the interconnection asset.

14. For example, on a short 11 kV line in Tanzania, it may be possible to deliver 10 MW to the grid, without being bound by a predefined regulation (for example, a regulation that capacities over 5 MW should be connected to 33 kV lines).

15. In Tanzania, the 10 MW limit is for exported capacity, while the installed capacity and the actual generation may be higher than 10 MW.

16. Examples of such technical codes include EPPO and Ministry of Energy (2010) and TANESCO (2012).

17. Instability may be caused by a shortage or surplus of power on a grid. Unless arrested, a shortage or a surplus of power may lead to a cascading failure of other generators on the grid.

18. May also be referred to as time of use or TOU rates, which are applicable to electricity customers.

19. Islanding is a situation in which an SPP and a segment of the local distribution network would isolate from the main grid and continue to operate as an "island" within a grid.

20. Solar PVs produce direct current (DC), whereas all grids operate with alternating current (AC). DC is converted to AC using an inverter. Smaller generators typically used in SPPs may also cause harmonic distortion when compared with large, utility-scale generators that have more stringent specifications.

21. Reactive power (measured in kVAr or MVAr) is the portion of power flow in a circuit that is temporarily stored in the form of electric or magnetic fields in a circuit. Reactive power cannot contribute to useful work, but it does contribute to increased current flow on wires and transformers. Utilities try to minimize reactive power flow and measure and charge industrial customers for the reactive power consumed by their loads. Devices with large coils of wires such as transformers or motors consume reactive power. Reactive energy (measured in kVArh or MVArh) is reactive power used over a period of time.

22. Import and export of both real and reactive power flow across the POS provide a combination of four "quadrants."

23. An SPP may import power from the grid, for use in the power plant during periods of low generation or during a maintenance shutdown. An SPP with a captive load

(such as a cogeneration SPP in a sugar industry or a mini-hydro SPP in a tea factory) may also import when the SPP generation is inadequate to meet the requirements of the industry. The issues for pricing of backup power are discussed in chapter 6.

24. Without proper frequency and voltage regulation, equipment powered by electricity either fails to function properly or is damaged. In Africa frequency is typically regulated at 50 Hz, with household voltage regulated at 220, 230, or 240 volts.

25. Available here: www.homerenergy.com.

26. Biomass or biogas generators would have essentially the same characteristics as this micro-hydro example.

References

Daunghom, Kitsanapol. 2010. "Thailand Grid Code for VSPP." Presentation at Solar Business Bangkok 2010, Bangkok, Thailand, March 23.

EPPO (Energy Policy and Planning Office), and Ministry of Energy. 2010. *Distribution Utilities' Regulations for Synchronization of Generators with Net Output under 10 MW to the Distribution Utility System.* http://www.eppo.go.th/power/vspp-eng/VSPP%20 Synchronization%2010%20MW-eng.pdf.

Greacen, Chris, Richard Engel, and Thomas Quetchenbach. 2013. "A Guidebook on Grid Interconnection and Island Operation of Mini-Grid Power Systems up to 200 kW." LBNL-6224E, Schatz Energy Research Center, Humboldt State University, Arcata, CA, and Lawrence Berkeley National Laboratory, Berkeley, CA, April. http:// palangthai.files.wordpress.com/2013/04/a-guidebook-for-minigrids-serc-lbnl-march -2013.pdf.

Madrigal, Marcelino, and E3 (Energy and Environmental Economics). 2010. *Creating Renewable Energy Ready Transmission Networks: A Survey of 14 Jurisdictions Highlights Emerging Lessons.* World Bank and E3, Washington, DC.

PUCSL (Public Utilities Commission of Sri Lanka). 2010. *Cost Reflective Methodology for Tariffs and Charges: Methodology for Charges.* September 30. http://www.pucsl.gov.lk /english/wp-content/themes/pucsl/pdfs/methodology_for_charges.pdf.

Rikos, Evangelos, Stathis Tselepis, and Aristomenis Neris. 2008. "Stability in Mini-Grids with Large PV Penetration under Weather Disturbances: Implementation to the Power System of Kythnos." 4th European PV-Hybrid and Mini-Grid Conference, Center for Renewable Energy Sources, Glyfada, Greece, May 30. http://www.cres.gr /kape/publications/photovol/new/Stability%20in%20Mini-Grids%20with%20 Large%20PV%20Penetration%20under%20Weather%20Disturbances-Im_.pdf.

TANESCO (Tanzania Electric Supply Company). 2011. *TANESCO Manual for SPP Application Approval and Interconnection.* Dar es Salaam, Tanzania.

———. 2012. *Draft—TANESCO Grid Code for Embedded Generation.* Dar es Salaam, Tanzania.

Regulatory Decisions for Small Power Producers Serving Retail Customers: Tariffs and Quality of Service

The most expensive electricity is no electricity at all.

—PARTICIPANT AT AFRICA ELECTRIFICATION INITIATIVE WORKSHOP, MAPUTO, MOZAMBIQUE, 2009

Remember that the end user is both a consumer and a voter.

—MORGAN LANDY, IFC OFFICIAL, WORLD BANK ENERGY DAY, 2012

Mind the gap.

—LONDON UNDERGROUND ANNOUNCEMENT

I think I'll stay out of the retail business. If I sell electricity to hundreds of individual customers, I will have no time to pray.

—CATHOLIC NUN IN CHARGE OF DEVELOPING A NEW HYDROPOWER PROJECT IN EAST AFRICA, IN CONSIDERING WHETHER TO PARTNER WITH A SEPARATE DISTRIBUTION OPERATOR

Abstract

Chapter 9 examines different approaches to setting retail tariffs for isolated and connected mini-grids. It presents a Microsoft Excel–based financial spreadsheet tool for analyzing the effect of grants and different tariff structures on the financial viability of a representative isolated mini-grid. The chapter also discusses service quality and safety standards for mini-grids.

The two key regulatory concerns for small power producer (SPP)–owned mini-grids serving retail customers are setting maximum tariffs and establishing minimum quality-of-service standards. This chapter addresses both concerns, with a particular focus on setting tariff levels and structures. Tariff levels determine how much total or overall revenue an SPP operator will be able to collect. Tariff structures refer to the individual elements of the tariff (for example, per kilowatt-hour [kWh] charges, flat charges, and prepaid and postpaid charges). Ideally, tariffs should be cost-reflective, which simply means that the SPP operator can reasonably expect that the overall revenues received from the tariffs paid by its customers will recover total operating and capital costs. If cost-reflective tariffs are not allowed because the SPP operator's tariffs are capped at a lower level (either because of informal political pressures or formal legal requirements), there will be a financial gap and the SPP will not be commercially sustainable. In this situation, if the SPP is going to survive as a viable supply option, the central question for both regulators and policy makers is: how can the revenue-cost gap be closed?

Setting Retail Tariff Levels: Concepts and Cases

The principal job of the regulator, whether it be the national electricity regulator or the rural electrification agency (REA) acting as a *de facto* regulator, is to set tariffs that are neither too low nor too high. Tariffs must be high enough that they will, after a transition period of several years, recover operating costs and depreciation on all capital (whether supplied by the operator or others) as well as any debt payments (if any), and provide for reserves to deal with emergency repairs and replacements. And if the SPP developer is a private entity, it must be allowed to earn a return on the equity capital that it has invested in the project that is commensurate with the risks that it faces. But the regulator must also balance the need for commercial sustainability with its legal obligation to protect customers. This means that the regulator must protect consumers from SPPs that try to exercise monopoly power after receiving a license or permit issued by the regulator.

At the outset, it is useful to flag several different tariff principles that are frequently encountered in Africa (and in developing countries in other regions):

- *Uniform national tariffs.* All citizens in the same tariff categories pay the same tariff for electricity regardless of where they live in the country.
- *Avoided-cost tariffs.*[1] An SPP operator is allowed to set tariffs that produce monthly bills to consumers that are equal to or below what the consumers would have been paying on other energy purchases (for example, kerosene, cell-phone charging) that are now replaced by electricity supplied from a mini-grid.
- *Cost-reflective tariffs.* Tariffs that produce enough revenues to recover the overall capital and operating costs likely to be incurred by an actual or hypothetical SPP operator. The tariffs for individual customer categories such as households and businesses are elements that help to recover the overall cost of supply.

Each of these principles for defining and setting tariffs, used alone, can lead to considerably different tariff levels for electricity and can be in conflict with one another. For example, if a country requires a uniform national tariff, then it would be virtually impossible at the same time to satisfy the requirement that the tariffs for an isolated mini-grid operator should be cost reflective. Yet each of these tariff-setting principles plays a role in tariff setting in contemporary Africa, shaping the financial terrain in which electricity retailers, including SPPs, operate. Contemporary practice in retail tariff setting, both for the main grid or isolated mini-grids, requires that regulators make judgments on how to apply these different principles to SPP tariffs.

These tariff principles are applied in different SPP retail sales contexts:

- Sales of electricity to retail customers on isolated mini-grids owned and operated by the national utility
- Sales of electricity to retail customers from SPP-owned mini-grids (Case 1 in table 2.1, chapter 2: an isolated SPP that sells at retail)
- Small power distributors (SPDs), who purchase electricity at wholesale for distribution and resell to households and businesses (discussed in chapter 10)

Within these settings there may be considerable variation in technology, scale, geographic and historic factors, and customer types. As a result, the structure and level of retail tariffs vary widely. For example, at one end of the scale a 2 megawatt (MW) hydropower plant serving tens of thousands of customers in a town with a good road network will generally have much lower energy costs and per-customer costs than a 20 kilowatt (kW) solar-diesel hybrid micro-grid system operating in a remote region accessible only by seasonal roads.

In this section we will consider different tariff-setting options, focusing on reasons why SPP developers might choose (or regulators might encourage) one tariff-setting option over another depending on circumstances. To illustrate practical applications, we include examples of retail tariff arrangements in several African and Asian countries.

Key Observation

The three tariff principles most frequently encountered in developing countries are: uniform national tariffs, avoided-cost tariffs, and cost-reflective tariffs. These principles can be applied to three types of retail sales of electricity by SPPs: sales to retail customers by isolated mini-grids owned and operated by the national utility, sales to retail customers by SPP-owned mini-grids, and sales to retail customers by SPDs that purchase electricity at wholesale from the utility.

Setting Tariffs: The Case of Isolated Mini-Grids Operated by National Utilities

Many national utilities in Africa operate isolated mini-grids in some rural areas in addition to the communities that they serve from the main national grid. In most instances, the national utility (which is usually government owned) did not actively seek to build and operate these isolated grids, but was probably forced to accept this obligation because of political pressure from a prime minister or a president to provide electricity to isolated communities that were unlikely to be reached by the main grid in the near or medium term. Once the isolated mini-grid was built, the political presumption (or legal requirement) was that the customers on the isolated grid would pay the same tariffs as electricity customers served by the national utility who take electricity from distribution facilities connected to the national grid. In other words, it is assumed that all customers in the country, whether rural or urban or served from the main national grid or an isolated mini-grid, will pay the same uniform national tariff (see box 9.1).

Key Observation

While many of the newer national electricity laws in Sub-Saharan Africa require the regulator to set cost-reflective tariffs rather than a uniform national tariff, the current reality is that most retail tariffs are both uniform and too low, particularly for isolated mini-grids operated by the national utility. This results in significant financial losses for national utilities that serve rural communities. It also creates a major problem for any independent mini-grid operator because the national utility's retail tariff establishes a *de facto* fair price in the eyes of rural customers that will be too low to recover the mini-grid operator's higher costs.

Setting Tariffs: The Case of Isolated and Main-Grid-Connected Mini-Grids Operated by SPPs

Tanzania is typical of many low-income countries in Africa that are pursuing rural electrification. The government, with the help of donors, is able to provide initial capital cost subsidies, but it does not have the money to provide ongoing operating subsidies to isolated mini-grids. Recognizing this reality, the Tanzanian electricity regulator has decided that isolated mini-grids will simply not be commercially viable if project developers are limited to charging their customers no more than the uniform national tariff. In rules proposed in 2012, the Tanzanian regulator stated that it would allow SPP operators to charge tariffs that exceed the national utility's uniform retail tariffs. The proposed regulatory rule states that: "SPP and SPD tariffs will be allowed to exceed the TANESCO [Tanzania Electric Supply Company] National Uniform Tariff if this is necessary for the SPP or SPD to recover its efficient operating and capital costs" (EWURA 2012c, Section 41 [5]).

In contrast, in India the central government has taken the opposite approach. The national government has stated that the retail tariffs of decentralized distributed generation, or DDG (that is, isolated mini-grids), must be comparable to the tariffs of nearby grid-connected villages.[2] Obviously, this policy mandate is only feasible because the Indian central government has committed to providing significant capital and operating subsidies to isolated mini-grids. The government has stated that it will provide subsidies equal to 90 percent of capital costs (including project preparation costs), the cost of five years of spare parts, and any shortfalls in operating costs for five years (ABPS 2011, 42).[3] As of this date, there is no clear evidence that the government has followed through on its commitment, and there have been reports of potential DDG developers expressing concern about getting the promised subsidies if they are routed through the SEB, the local state-owned utility.

Box 9.1 Uniform National Tariffs and Rural Electrification

A uniform national tariff (sometimes referred to as a pan-territorial tariff) requires that all retail electricity consumers, or at least residential or domestic consumers, be charged exactly the same tariff regardless of whether they live in the capital served by the main grid, or in an isolated village 650 kilometers from the capital served by a mini-grid operator.

The stated or unstated rationale for a uniform national tariff is that electricity is seen as a basic right to which all citizens are entitled. *For most elected officials, who have their eyes on the next election, fairness is much more important than cost recovery.* And fairness is usually defined by these elected officials to mean that every customer should pay the same retail tariff. A uniform national retail tariff is viewed as fair for two reasons: (a) consumers are being treated equally, since all households throughout the country are charged the same tariff, and (b) poor households in rural areas are receiving the help they need by being charged the same tariff as urban customers, even though the cost of serving rural households is almost always much higher.

While the principle of fairness is noble, the on-the-ground outcome is often financial hemorrhaging: the national utility will lose money on every kilowatt-hour sold to customers served on isolated mini-grids. For example, the Tanzania Electric Supply Company (TANESCO, the national utility) currently operates more than 10 isolated mini-grids that are powered by diesel generators. TANESCO currently sells electricity to isolated mini-grid customers at a uniform national tariff, which produces an average revenue of about 9 cents/kWh even though its actual production costs are about 40 cents/kWh or higher (EWURA 2012b).[a] Under a uniform national tariff that does not differentiate between rural and urban customers, most state-owned utilities lose money even on rural customers that are connected to the main grid. In India, where tariffs are usually the same for both urban and rural customers in a state, it has been estimated that the government-owned state electricity boards (SEBs) lose, on average, 8 cents for every kilowatt-hour that they sell in rural areas to customers connected to the main grid (Dixit 2012).

Very few of the newer national electricity laws in Africa mandate a uniform national tariff (CORE International 2008). Instead, African electricity laws usually require that the regulator set cost-reflective tariffs. This, in turn, would imply geographic differentials to reflect the varying costs of supplying customers at different locations. But this does not happen in practice. One Tanzanian government official observed that despite the absence of any legal requirement, a uniform national tariff has become a national habit and one that is hard to change.

A uniform national tariff ignores the fact that the real cost of providing electricity in rural areas is almost always higher than providing the same electricity in urban areas. This is true for three reasons. First, more capital is required in a grid-connected rural system than in an urban system to deliver the same amount of electricity. For example, a study of Senegal found the average capital cost of an urban electrical connection was $409, while the average capital cost per rural household (including the medium-voltage [MV] line extension and transformers to extend the grid) was $1,140 (ASER and Columbia Earth Institute Energy Group 2007). Second, rural operating costs will be higher because technical losses[b] increase when electricity has to be delivered to a more distant rural location.

box continues next page

Box 9.1   **Uniform National Tariffs and Rural Electrification** *(continued)*

Third, operating costs on the rural system will be higher especially if the power has to be generated from a diesel, heavy-fuel-oil, solar, or wind-power unit. As mentioned, in Tanzania diesel mini-grid electricity costs about 40 cents/kWh (EWURA 2012b); main-grid power, on the other hand, costs about 12.2 cents/kWh (EWURA 2012a).[c]

In most Sub-Saharan African countries, the economic distortions caused by a uniform national tariff are exacerbated by two other features that almost always accompany this type of tariff. The first is that the existing uniform national tariff will rarely cover the national utility's actual costs of production (Eberhard and others 2008, 29).[d] This is true because political authorities are afraid of the general public's reaction if tariffs were allowed to rise to cover the costs that are actually incurred by the national utility. One high-level African energy ministry official referred to immediate implementation of an overall cost-recovering tariff for the national utility as the equivalent of political suicide, especially when the general public thinks that the national utility is inefficiently run and filled with corrupt employees. Consequently, uniform national tariffs are rarely allowed to rise to cost-reflective levels (even when this is a legal requirement under the national energy law). *The current reality for most national utilities in Sub-Saharan Africa is that tariffs are both uniform and too low.* A second common feature of uniform national tariffs is that they almost always include a social tariff or lifeline tariff component that provides for subsidized, low-cost (Tanzania) or even free (South Africa) electricity for poor customers with low levels of monthly consumption (usually 50 kWh or less). The social tariff block recovers an even smaller percentage of actual costs than the rest of the national tariff.

A uniform national tariff is also a barrier to rural electrification if there is a mixed, two-track rural electrification strategy in which rural electricity service is provided by both the national utility and independent SPP operators. The SPP operator may get a formal waiver from the regulator or government to charge a price higher than the uniform national tariff. *But while the waiver may be legal, it may not be politically sustainable.* As a longtime Tanzanian mini-hydro specialist observed: "It costs TANESCO at least 500 shillings/kWh [about $0.33] for operational cost alone, and add management and distribution costs, then power from TANESCO's isolated mini-grids costs something like 800 [about $0.50] shillings. It is crazy that TANESCO turns around and sells it at 130 shillings [about $0.08]. No investor can possibly build projects with this situation. And the problem is that everyone now expects electricity at 130 shillings. If someone gets permission to charge a higher price than this, some villagers will go to the power plant and break equipment saying, 'you make too much money.'" The basic problem, then, is that a uniform national tariff establishes a "fair" price among rural consumers, and any price above it, even if allowed by law, is viewed with suspicion and anger.

a. T Sh 629.55/kWh.

b. Technical losses refer to losses not due to theft.

c. Adjusted avoided cost at MV distribution level is T Sh 192.37/kWh.

d. Pervasive underpricing was found in the surveys of the Africa Infrastructure Country Diagnostic (AICD), which found that only 10 of 21 national utilities in Sub-Saharan Africa were allowed to charge tariffs that covered their historic operating costs, and that only 6 of 21 national utilities covered historic operating and capital costs (Eberhard and others 2008, 29).

Key Observation

It is not feasible to require SPPs to charge a national utility's retail tariff if they are operating an isolated grid. The electricity regulator in Tanzania has proposed allowing SPPs and SPDs to charge tariffs that exceed the uniform national tariff if it is necessary for their commercial viability. The central government of India has committed to subsidizing capital and operating costs of SPPs and SPDs to keep their tariffs equal to the tariffs of grid-connected customers of the state-owned utility. It remains to be seen whether the subsidies will be delivered as promised.

Cost-Reflective Tariffs for Isolated SPPs: How Can the Cost-Revenue Gap Be Closed?

> To measure is to know … if you cannot measure it, you cannot improve it.
>
> —LORD KELVIN

Rural electrification is expensive, as customers are often dispersed over a wide geographic area. Consumption by the average customer is generally low, and often with a low load factor.[4] Low energy sales limit the number of kilowatt-hours over which fixed costs of lines, poles, and generating equipment can be recovered. Generation costs may be higher if the electricity must be generated on-site—either with diesel generators, renewable energy, or some combination of the two. High costs and low sales lead to a gap between the costs of supplying electricity and the revenues that can be collected.

Full cost-reflective tariffs cover the SPP's overall cost of generating and distributing the electricity plus a reasonable profit. If a donor can be found, options for less-than-full-cost-reflective tariffs are possible. For example, in many off-grid projects it is common to set tariffs to cover all operating expenses and some depreciation, but the initial capital cost is paid for in whole or in part by grants.

The central question is: how can the gap be closed between the SPP operator's costs and its customers' ability to pay? In principle, eliminating the gap involves some combination of three basic approaches: charging higher tariffs, cutting costs, and obtaining subsidies. All three approaches have their limits. Raising tariffs depresses consumption and raises the risk of a disgruntled customer base. Cutting the investment or operations budgets can lead to more breakdowns or lower quality of service—which ultimately increases costs or lowers revenues, and creates more disgruntled customers. And external subsidies are not always delivered as promised.

Measuring the Gap

The first step toward eliminating the gap is to know how large it is. This is important for the SPP operator and those who provide it with equity or loans. In addition, both regulators and REAs need estimates of an SPP's costs and its likely revenues. The regulator needs this information to set tariffs and the REA to decide how much grant money to provide.

Table 9.1 shows the output of a simple one-page spreadsheet that was created for the Tanzanian REA to help SPPs develop realistic business plans and as a tool to determine SPP subsidies.[5] Its structure is general enough to be used by REAs and regulators in other Sub-Saharan African countries. The advantage of using a single common spreadsheet is that it allows all interested parties to work from a common financial framework. It can also be used to perform what-if calculations—for example, to estimate the effect of lowering tariffs or increasing grant levels.

The spreadsheet includes the following inputs:

- Capital costs, including plant costs and costs of extending the local mini-grid
- Grants, including grants received from the REA for providing connections to new customers, as well as other grants from donors
- Operating costs, including salaries, maintenance, and fuel costs
- Construction schedules
- Portions of financing coming from debt and equity
- Interest rate (for debt) and expected returns (for equity)
- Loan grace period and term

On the basis of these inputs, the spreadsheet can be used to calculate standard financial indicators, including:

- The project's internal rate of return (project IRR)[6]
- The after-tax IRR on equity (equity IRR)[7]
- Project net present value (project NPV)
- Equity net present value (equity NPV)
- Annual project cash flows
- Debt service coverage ratio (DSCR)
- Tariffs needed to make the project financially viable

The spreadsheet used in table 9.1 could be used to model the finances of the different SPP cases described in chapter 2. As shown in table 9.1, it is populated with data from a hypothetical, isolated mini-grid SPP project located in a rural area of Tanzania, which is assumed to be powered by a 300 kW mini-hydro facility. This is the case of a pure isolated mini-grid (Case 1 in table 2.1, chapter 2: an isolated SPP that sells at retail) that has no connections to the national grid or to some existing isolated mini-grid. Hence, its only revenues are sales of electricity to businesses and households in villages, plus a small amount of additional revenues that it expects to earn from carbon credits. As is the case in most small villages, the load is intermittent so that the capacity factor[8] of the generator will reach only about 40 percent.

The total project cost is $1.35 million. This is equivalent to a capital cost of $4,500/kW of installed capacity, which is consistent with recent projects in operation or under development in East Africa. It is assumed that the SPP operator will obtain a loan with an interest rate of 12 percent and a tenure of 10 years. Construction will take three years and operations begin at the start of year four.

The project receives $0.6 million in REA grants of $500 for each of the 1,200 households it serves. Typically, it will take several years before the mini-grid actually reaches its full market potential of 1,200 households, but assuming that all households will come online as soon as the project begins operations makes the project easier to model.[9] Half of the electricity produced is sold to business customers (for example, mills, carpentry shops, welding shops, and barber shops) and the remaining amount to residential customers.

Banks in Tanzania allow the REA connection grant to be treated as equity for the purpose of meeting the bank's minimum equity requirements when securing a loan with a bank. This is neutered equity because the SPP operator is not allowed to earn an equity return on this gift of equity.[10] In this example, the REA grant alone provides 44 percent of the project investment cost, which easily passes the bank equity threshold requirement of 30 percent. In all three scenarios described below, we assume that the project developers provide an additional 15 percent equity stake in the project, and it is this 15 percent on which they earn equity returns.

Key Recommendation

Regulators, rural energy agencies, and SPPs should use a common financial analysis tool similar to the one presented in table 9.1 to measure the gap between the high costs of supplying electricity to rural communities and the revenues that can be collected.

The Financial Viability of Isolated Mini-Grids under Three Scenarios

What can be learned about the revenue-cost gap that can arise for rural isolated mini-grid SPPs under different scenarios? The discussion that follows is specifically focused on the financial viability of what is arguably the most difficult SPP case: a mini-grid SPP that only sells directly at retail and does not make any sales of electricity to the national utility (Case 1: an isolated SPP that sells at retail). If commercial sustainability can be achieved for this case, then it can also be achieved for other SPPs that will have the benefit of being able to sell to the national utility.

Consider three scenarios:

Scenario 1: No Donor Grants Other than a Connection Grant and Tariffs Constrained to Uniform National Tariffs

This scenario represents a business-as-usual case for a private sector developer if the SPP is required to charge the uniform national tariff and must obtain significant private sector debt and equity financing. The only outside grant is the REA grant of $500 per connection. Tariffs for businesses fall under TANESCO's T1 (general usage) tariff[11] of approximately T Sh 230/kWh (15.3 U.S. cents).

Table 9.1 Simple Project Financial Analysis Spreadsheet

Inputs

Project inputs

Plant capacity	300 kW
Capacity factor	40%
Annual generation	1,051 MWh
Number of connections	1,200

Generation costs

	USD 000s	T Sh Mn	
Generation	600	900	44%
Soft costs	150	225	11%
Grid connection	400	600	30%
Grid extension	200	300	15%
Total investment costs	1,350	2,025	100%
REA grants	600	900	44%
Other donor grants	160	240	12%
Total grants	760	1,140	56%
Investment (excluding grants)	590	1,125	44%

Other inputs

Construction time	3 years
Collection efficiency	90%
Distribution losses	5%
Tariff inflation	4.0%
Consumer price index	6.0%
Terminal value	0.0 times exit FCF
Exchange rate	1,500 T Sh/USD

Revenues

		T Sh/kWh	US c/kWh
Residential <50 kWh/month	40%	100.0	6.7
Residential >50 kWh/month	10%	273.0	18.2
Business	50%	360.0	24.0
Average tariff		247.3	16.5
Carbon revenues		9.0	0.6

Operating costs

	USD 000s	T Sh Mn	US c/kWh
Salaries	25	38	2.38
Maintenance	14	20	1.28
Fuel costs	0	0	0.00
Other	15	23	1.43
Total	53.5	80.3	5.1

Capital structure

			USD 000s	T Sh Mn
Equity share	15%	Equity	203	304
Debt share	29%	REA grant	600	900
Required returns	14%	Other grants	160	240
Equity including REA	59%	Loan	388	581
Loan rate	12%			
Loan grace period	3			
Loan term	10			
WACC[c]	9.2%			

Outputs

Project returns

Project IRR[a]	4.8%
Equity IRR[b]	19.7%
Project NPV	($286)
Equity NPV	$73

Cost recovery tariff

Capital investment	4.50 $/W
Operating costs	5.1 c/kWh
Output per watt	3.50 kWh/W
Capital recovery	20.22 c/kWh
Total	25.3 c/kWh
	380 T Sh/kWh

Schedule

	Year			
Schedule	1	2	3	4
Construction	33%	33%	33%	0%
Operations	0%	0%	0%	100%

Other assumptions

Depreciation period	25 years
Corporate tax rate	30%
Tax holiday	0 years

table continues next page

248

Table 9.1 **Simple Project Financial Analysis Spreadsheet** *(continued)*

Cash flow USD 000s							Year									Residual value[d]
	1	2	3	4	5	6	7	8	9	10	11	12	13	14	15	
Investment	$450	$450	$450	$0	$0											
Increase in working capital																
Grants	$253	$253	$253	$0	$0											
Equity	$68	$68	$68	$0	$0											
Debt	$129	$129	$129	$0	$0											
Revenue (c/kWh)	17.1	17.7	18.4	19.1	19.9	20.7	21.5	22.3	23.2	24.1	25.0	26.0	27.0	28.1	29.1	
Total revenues	$0	$0	$0	$201	$209	$217	$226	$234	$243	$253	$263	$273	$284	$295	$306	
Operating costs	$0	$0	$0	$64	$68	$72	$76	$80	$85	$90	$96	$102	$108	$114	$121	
Operating cash flow	($197)	($197)	($197)	$138	$142	$146	$150	$154	$158	$163	$167	$172	$176	$181	$185	
Cash flow if no grant	($450)	($450)	($450)	$138	$142	$146	$150	$154	$158	$163	$167	$172	$176	$181	$185	
Depreciation[e]	$0	$0	$54	$54	$54	$54	$54	$54	$54	$54	$54	$54	$54	$54	$54	
Loan payments	$0	$0	$0	$96	$96	$96	$96	$96	$96	$96	$0	$0	$0	$0	$0	$0
Interest	$0	$16	$33	$52	$47	$41	$35	$28	$19	$10	$0	$0	$0	$0	$0	$0
Principal	$0	($16)	($33)	$43	$48	$54	$61	$68	$76	$85	$0	$0	$0	$0	$0	$0
Loan balance	$129	$274	$436	$393	$344	$290	$229	$161	$85	$0	$0	$0	$0	$0	$0	$0
Effective tax[f]	$0	$0	$0	$9	$12	$15	$18	$22	$25	$30	$34	$35	$37	$38	$39	$0
SPP's total costs	$197	$212	$230	$125	$127	$128	$129	$130	$130	$130	$130	$137	$144	$152	$160	$0
Equity cash flow	($68)	($68)	($68)	$33	$34	$35	$36	$37	$37	$38	$133	$136	$139	$143	$146	
Debt service coverage ratio	0.00	0.00	0.00	1.44	1.48	1.52	1.57	1.61	1.66	1.70	0.00	0.00	0.00	0.00	0.00	0.00

Source: http://tinyurl.com/SPPevaluator.

Note: c/kWh = cents per kilowatt-hour; FCF = free cash flow; IRR = internal rate of return; kW = kilowatt; kWh = kilowatt-hour; MWh = megawatt-hour; NPV = net present value; REA = rural electrification agency; SPP = small power producer; T Sh Mn = million Tanzanian shillings; WACC = weighted average cost of capital.

a. The project IRR is low, but this is mitigated somewhat by the presence of REA grants, which do not need a return or repayment.

b. An equity return is allowed only on the equity supplied by the developer (not on the REA and donor grants).

c. WACC = equity share × return on equity + debt share × loan rate × (1 – tax rate).

d. Present value of residual value at year 15 is for revenues from year 16–25.

e. Depreciation is allowed on the total investment (SPP developer equity plus donor funds).

f. A project like this would typically collect value added tax (VAT) and perhaps other duties (for example, in Tanzania VAT on power sales is 18 percent, and there is an additional 3 percent duty collected to fund the REA and 1 percent collected to fund the Energy and Water Utilities Regulatory Authority). These charges are not included in this spreadsheet because, while they affect ratepayers' finances, they are passed through directly to the government and do not affect cash flows from the developer's perspective.

Business customers are estimated to consume 50 percent of the electricity sold. Residential customers are charged a tariff of T Sh 100/kWh (6.7 U.S. cents) for usage less than 50 kWh/month (estimated to consume 40 percent of electricity sold) and T Sh 273/kWh (18.2 U.S. cents) for usage above 50 kWh/month (accounting for the remaining 10 percent of electricity sold). These tariffs are roughly consistent with TANESCO's current tariff schedule (TANESCO 2012).

Scenario 2: REA and Donor Grants That Cover 56 Percent of Project Capital Costs, but with Tariffs Still Constrained to Uniform National Tariffs

In this scenario it is assumed that the project benefits from additional foreign donor assistance in the form of a $160,000 grant that covers about 12 percent of capital costs. Together with the REA grants for individual connections, total grants cover 56 percent of initial costs. Tariffs to businesses and households continue to be capped at the uniform national tariff.

Scenario 3: Donor and REA Grants (as in Scenario 2), but Tariffs Are No Longer Constrained to Uniform National Tariffs, and Cross-Subsidies Are Allowed from Commercial/Industrial Customers to Residential Customers

This scenario includes the same grants as in scenario 2. The two differences between scenarios 2 and 3 are: first, the SPP is allowed to charge tariffs that exceed the uniform national tariff (T Sh 360/kWh [24 U.S. cents]) for commercial customers and T Sh 100/kWh (7 U.S. cents) for low-usage residential customers and second, the tariffs for commercial customers cross-subsidize the tariff for household customers. *Table 9.1 shows the spreadsheet results for scenario 3.*

How do these scenarios compare on key financial parameters? Banks are particularly concerned about whether the project will generate sufficient revenues to meet loan payments. The ability of a project to meet debt payments is captured in the DSCR, which is defined as the ratio of revenues to debt payments. A DSCR of less than 1, say 0.9, would mean that there is only enough net operating income to cover 90 percent of annual debt payments. Typically a DSCR of at least 1.4 is required for Tanzanian banks to feel comfortable providing a project loan.

The DSCRs under the three scenarios are shown in figure 9.1.

- Scenario 1 generates only enough cash flow to meet 55 percent of yearly debt payments. No bank would lend to this project.
- In scenario 2 the addition of the $160,000 grant improves the DSCR, bringing it to 0.77—a level still unacceptable to banks.
- Scenario 3 achieves a DSCR of 1.44 through a combination of donor grants and higher tariffs.

If, however, the SPP developer is able to connect only 600 residential customers instead of 1,200 (and, in fact, it is not uncommon for a project to take four to five years to sign up most or all households in its geographic area), then it would struggle to meet DSCR requirements. If households are slow in connecting, more donor funding (which would lower the amount of loans taken out by the operator) or higher tariffs would be needed.

Figure 9.1 Comparison of Debt Service Coverage Ratio (DSCR) in Three Scenarios

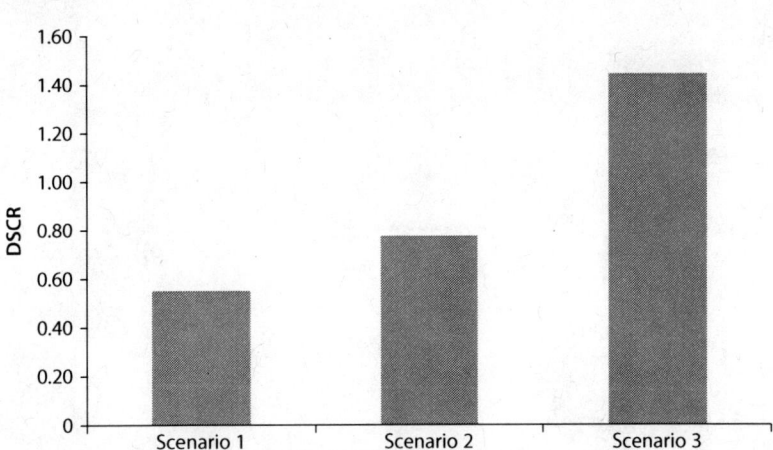

Let us look at these projects from the investor's perspective. Once the yearly bank loans and operation and maintenance (O&M) costs have been paid, the remaining revenues can be used to pay dividends to equity investors. The ability of the project to provide returns to equity shareholders is measured by the project's equity IRR. The equity IRRs for the three scenarios are shown in figure 9.2.

Again, scenario 1 looks dreadful: the project earns a negative return (that is, loss of money) for equity investors. Scenario 2, with a foreign donor grant of $160,000, has a very low equity IRR: only 1 percent. No rational equity investor would invest in this project given that other investments are likely to be available with higher equity returns. The equity return could be increased to a more reasonable (but hardly lucrative) equity IRR of 11 percent if the SPP is allowed to charge higher tariffs, which means that more money would be available to increase dividends to equity investors.

Scenario 3 is the only scenario that manages to provide attractive investor returns (an equity IRR of 20 percent).

The take-home message for regulators and REA officials is that given current technology and technology prices, isolated mini-grid SPPs that receive no outside capital grants and that are required to charge the uniform national tariff on retail sales are unlikely to be able to earn enough money to cover loan payments under loan tenors and interest rates found in much of Africa. But these same projects can become commercially viable if they:

- Are allowed to charge tariffs that are higher than the often politically suppressed uniform national tariff
- Are allowed to charge higher tariffs to business customers to cross-subsidize the tariffs of households
- Receive grants to lower capital costs

From the Bottom Up • http://dx.doi.org/10.1596/978-1-4648-0093-1

Figure 9.2 Equity Internal Rate of Return (IRR) for the Example Project across Three Scenarios

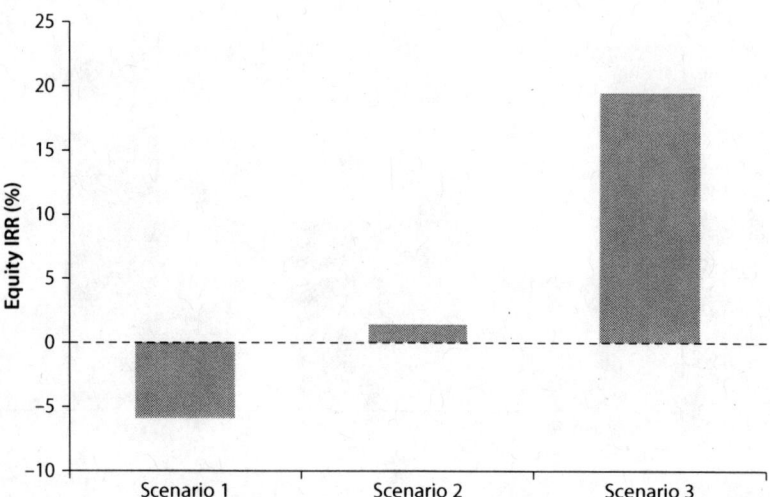

Key Recommendation

SPPs that operate isolated mini-grids must be allowed to charge tariffs that are higher than the uniform national tariff if they are going to succeed as commercially viable entities. SPPs must be allowed to charge higher tariffs to business customers to cross-subsidize household customer tariffs. SPPs will also generally need to receive outside donor grants to lower capital costs.

Are Cost-Reflective Tariffs Affordable for Rural Customers?

It is all well and good to perform detailed spreadsheet calculations of the tariffs necessary for SPPS to achieve commercial sustainability. Without these what-if calculations, anyone who is promoting, financing, or regulating an SPP project would be flying blind. *But these calculations, by themselves, do not tell us whether cost-reflective tariffs are affordable for rural customers.* The best way to address the affordability question is to take a close look at the amount of money that rural customers currently spend on sources of energy that could be replaced by electricity from mini-grids.

Households

In the absence of electricity, people make do with some combination of kerosene, candles, cooking fires, or dry-cell batteries. Using data from typical kerosene lamps in Kenya, Peon and others (2005) and Radecsky (2009) find that 50,000 hours of operation of a typical kerosene wick lamp costs $1,250 at kerosene prices of

$0.50/liter.[12] Since Peon's study in 2005, current kerosene prices have more than doubled—for example, kerosene currently sells for T Sh 2,300 ($1.45)/liter in rural Tanzania—pushing the cost of 50,000 hours of kerosene lamp operation to above $3,600 (Reuters 2012). By contrast, Peon and others (2005) calculate that 50,000 hours of light from a 7 watt fluorescent bulb using electricity from a mini-grid at $1.00/kWh costs $350—which is about ten times less expensive.[13]

Because an electric compact fluorescent light (CFL)[14] is much brighter than a typical kerosene lamp, in terms of the actual light output per dollar, the results are much more striking. Peon and his coauthors found that a typical kerosene wick lamp light provides 0.008 lumens/dollar while an electric fluorescent light provides 0.66 lumens/dollar. Adjusted to current kerosene prices in Tanzania, light from kerosene is almost 240 times more expensive than light from a CFL, even if the household were paying $1.00/kWh for the mini-grid electricity.

It should be noted, however, that the price of electricity in this example, $1.00/kWh, would be unrealistically high even for a remote mini-grid. The hydroelectric-powered mini-grid example discussed earlier (in table 9.1) was based on more likely residential electricity prices of T Sh 100–T Sh 273/kWh ($0.06/kWh–0.17/kWh). These electricity tariffs put the cost of light (on a lumen basis) from a kerosene wick lamp 1,400–4,000 times higher than that of a CFL powered by electricity from a mini-grid.

Household expenditures for kerosene and candles add up. A 2010 study of 97 rural households in Malawi found that all sources of spending on lighting—including kerosene, candles, dry-cell batteries, battery charging, rental fees, and other sources—consumed 19.7 percent of the household income. Among survey respondents, kerosene was the primary lighting source in 96.3 percent of homes (Adkins and others 2010).

Within the Malawi study, monthly kerosene expenditures ranged from $2 to $12/month, with an average of $2.91/month. In these homes, kerosene lamps were used an average of 2.9 hours/day. Candles were a secondary light source in 82.5 percent of the households and were used an average of 2.2 hours/day. The 97 respondents reported a mean monthly expenditure on candles of $1.51. The average for kerosene and candles together is $4.42/month. Similar findings were reported in an unpublished market study for several possible isolated SPPs in rural Tanzania. The survey found costs ranging from $1.40 to $17/month for kerosene, candles, and charging for small batteries. A market study for the Mwenga hydropower project in Tanzania also found that households pay an average of T Sh 12,000/month for kerosene ($8/month) (Gratwicke 2012).

Table 9.2 compares monthly noncooking energy expenditures in nonelectrified houses (relying on kerosene, candles, and so on) to monthly expenditures of electricity at two indicative monthly levels of consumption (16 kWh/month and 30 kWh/month). (See table 9.3 for lights and other devices that could be powered at these two levels of monthly electricity consumption.) *Even when electricity is relatively expensive on a per kWh basis ($0.50/kWh), the total monthly electricity costs would be comparable to current monthly expenditures on kerosene and candles.*

Table 9.2 Rural Household Monthly Expenditures on Energy Services That Can Be Supplied by Electricity (Pre- and Postelectrification)

	Supply source	$/month
Preelectrification total energy costs (not including cooking)	Kerosene (liters)	$2–5
	Candles	$1–4
	Dry-cell batteries	$2–7
	Cell phone charging	$2–7
	Total	$7–23
	$/kWh	16 kWh from mini-grid
Postelectrification 16 kWh from mini-grid	$0.04	$0.64
	$0.10	$1.60
	$0.15	$2.40
	$0.20	$3.20
	$0.50	$8.00
	$/kWh	30 kWh from mini-grid
Postelectrification 30 kWh from mini-grid	$0.04	$1.20
	$0.10	$3.00
	$0.15	$4.50
	$0.20	$6.00
	$0.50	$15.00

Sources: Unpublished World Bank survey, Kenya 2011; authors' calculations.
Note: kWh = kilowatt-hour.

Table 9.3 Typical Lights and Appliances and Duration per Day That Could Be Powered with 16 kWh/Month and 30 kWh/Month

Quantity	Appliance	Watts	Hours/day	Days/month	kWh/month
3	CFL	15	5	30	6.75
1	Radio	10	4	30	1.2
1	TV	90	3	30	8.1
				Total	16.1
8	CFL	15	5	30	18
1	Radio	10	5	30	1.5
1	TV	90	3	30	8.1
1	DVD player	30	3	30	2.7
				Total	30.3

Note: CFL = compact fluorescent lamp; kWh = kilowatt-hour.

The conclusion that can be drawn from these data is that if connection charges can be made affordable by being subsidized through grants and paid for in small monthly payments over time, electricity can provide power for lighting and information services at a total cost comparable to what rural families currently spend for kerosene and candles. While the costs can be made comparable, the quality of the services is higher with electricity. Once a household switches to electric lighting, the dangers of fires and indoor air pollution disappear. More important, even though the total monthly costs may be the same, lighting provided by electricity is hundreds of times brighter than lighting provided by kerosene lanterns and candles. Household electrification offers cleaner lights, better safety, and more convenience, and an ability to use a variety of communication and electromechanical appliances compared to the preelectrification options of kerosene, candles, dry-cell batteries, and traveling to charge cell phones. So comparing just the energy costs of pre- and postelectrification would be an apples-to-oranges comparison, because the postelectrification energy services are often of much higher quality.

Businesses

Businesses are likely save even more. Small shops use kerosene lamps or candles to illuminate their premises at night, and thus face situations similar to residences. Many businesses in rural areas rely on their own small diesel generators to provide electricity, typically at a cost of T Sh 800/kWh ($0.53/kWh) or more. Even in the case in which a village mini-grid charges businesses considerably higher tariffs than residences (for example, T Sh 360/kWh for businesses and T Sh 100 or T Sh 273/kWh for residences, as shown in table 9.1), businesses will still save compared to the cost of generating their own electricity. If a milling business is using its own diesel generator to generate 3,500 kWh/month at T Sh 800/kWh, it will save more than $1,000/month once it is able to purchase the same kilowatt-hours at T Sh 360 from the operator of an isolated hydro mini-grid.

Key Observation

Rural household customers can afford cost-reflective tariffs if the initial connection charge is not too high and can be paid in small monthly payments over time. Once rural households get over the connection charge hurdle, they can afford to pay electricity tariffs that will produce monthly expenditures equal to or less than their prior expenditures on nongrid energy sources (kerosene, candles, batteries). Electricity has the added benefit of producing better energy services: higher-quality lighting, better access to information, and health benefits.

What Can a Regulator Do to Promote the Commercial Viability of Isolated Mini-Grids?

A regulator can choose to promote isolated SPPs either *implicitly* or *explicitly*. In the implicit approach, the regulator does not issue any specific tariff-setting guidance for SPPs. Instead, he states that he will rely on the general tariff-setting principles found in the national electricity law. For example, if the regulator were to take this approach in Tanzania, he would simply announce that SPP tariffs would be based on the national electricity law's general principle that tariffs should reflect the costs of efficient business operations. The SPP would then have to depend on the regulator to interpret this general principle to mean that if an isolated SPP's costs are higher than the national utility's costs, the SPP would be allowed to charge higher tariffs to cover its higher costs. The problem with this implicit approach is that it does *not* provide much clarity or certainty for a potential SPP developer. It is risky because the general principles can be interpreted in a variety of ways.

An alternative explicit approach is for the regulator to state up front the specific regulatory rules that will be applied to SPPs. This would provide more up-front certainty for potential SPP operators. This is the approach that has been proposed by the Tanzanian electricity regulator. In the second-generation SPP rules issued for public consultation in June 2012, the Tanzanian Energy and Water Utilities Regulatory Authority (EWURA) proposed a set of specific retail-pricing rules designed to promote the commercial viability of SPPs for both connected and isolated SPPs. Five of EWURA's proposed rules have direct implications for SPP cost recovery:

1. **SPP operators are not required to charge the uniform national tariff.**
   EWURA proposed that:

   SPP and SPD tariffs will be allowed to exceed the TANESCO National Uniform Tariff if this is necessary for the SPP or SPD to recover its efficient operating and capital costs. (EWURA 2012c, Sections 41–45)

   In fact, EWURA went beyond just stating that it would allow deviations from the national uniform tariff. It also said that SPPs would be granted permission to charge the national uniform tariff only if they "can demonstrate that it [the national uniform tariff] will lead to commercial sustainability" (EWURA 2012c, Sections 41–44). Since most SPPs that operate isolated mini-grids will incur costs that exceed the national uniform tariff, EWURA's proposed rule would effectively preclude SPPs from charging the national uniform tariff.

2. **SPP operators are allowed to cross-subsidize among their customers.**
   EWURA states that it will allow SPPs to cross-subsidize between customer groups. EWURA's specific proposal was:

   To facilitate commercial sustainability, an SPP or SPD may propose tariffs for specific customer categories or for customers within a single category, subject to the Authority's approval, that take account of the ability to pay of these customers. (EWURA 2012c, Section 39.c)

   By stating that "ability to pay" can be considered in setting tariffs, the proposed tariff rule would allow the SPP operator to charge tariffs to businesses at a level just below the per-kWh costs that the businesses incur to run their own on-site diesel generators. If it costs a mill operator T Sh 800/kWh ($0.53/kWh) to operate its diesel generator, this would be strong evidence that the mill operator has the financial ability to pay as much as T Sh 800/kWh to an SPP operator and definitely the ability to pay T Sh 360/kWh ($0.24/kWh). By taking account of "ability to pay," EWURA is saying, in effect, that it will allow for cross-subsidies.

   Cross-subsidies already exist in the Tanzanian electricity sector. The national utility TANESCO is allowed to charge businesses taking service off the national grid a much higher price (T Sh 132/kWh) than the price charged to low-consumption household customers (T Sh 60/kWh), even though the costs of serving the business customer are likely to be lower than the costs of serving a household customer (since businesses tend to be more centrally located than rural households). Hence, this proposed tariff policy for SPPs would offer the same pricing flexibility to SPP operators that already exists for TANESCO.

3. **SPPs are mandated to take depreciation on equipment that was financed through grants.**

EWURA proposed that an SPP or SPD shall charge a tariff that:

> at a minimum, after a transition period of 3 to 5 years, recovers operating costs and depreciation on all capital, whether supplied by the operator or others, as well as any debt payments (if any), and provide for reserves to deal with emergency repairs and replacements. (EWURA 2012c, Section 44 [a])

This provision would allow the SPP operator to recover depreciation but not an equity return on equipment that was financed by outside grants. As discussed in chapter 5, the rationale is that a piece of equipment will eventually have to be replaced regardless of how it was originally financed. Depreciation generates funds that can be used to pay for the costs of replacement.

4. **SPPs can enter into power sales contracts with business customers without obtaining prior or after-the-fact regulatory approval of the price and nonprice terms of the contract.**

The Tanzanian electricity law establishes a category of electricity customers that it describes as eligible customers. All electricity sales to eligible customers are deregulated, which means that the selling entity does not have to obtain the approval of the regulator for the price of electricity sales to an eligible customer.

EWURA proposes to define an eligible customer in the following way:

> "Eligible customer" means, for customers connected to, or seeking to connect to, an SPP or SPD, any entity with a peak load of 250 kVA or higher or any communication tower (such as a cell phone tower). (EWURA 2012c, Section 3)

EWURA's specific proposal is that:

> If an SPP developer reaches an agreement with a business or commercial entity to sell electricity to that entity under a power purchase agreement such agreement shall be deemed to constitute a sale of power to an eligible customer. (EWURA 2012c, Section 45)

This provision allows an SPP to enter into a negotiated sales contract with some businesses in its village. And the terms and conditions of the power sales contract need not be reviewed and approved by the regulator. In effect, it goes one step beyond the pricing flexibility given to the SPP developer in the previously cited rule that allows it to take account of the ability to pay. The rule would eliminate any requirement for regulatory approval for sales to businesses that are deemed to be eligible customers. This could, for example, cover SPP sales to a mobile-phone tower operator or a mine. In essence, it is a form of price deregulation.

5. **SPPs should be allowed to recover the administrative and financing costs of providing loans to actual or potential customers that will allow the customers to connect to the SPP system and to facilitate productive uses of electricity. The loans could be repaid through extended payment plans implemented through on-bill financing.**[15]

The regulatory language that has been suggested to implement this proposal would be:

The retail tariff structure may include as an allowed component of tariffs any interest subsidies and administrative costs for on-bill financing such as financing of connection charges, financing of internal wiring, construction of upgrades to dwellings necessary to meet minimum electrification requirements, and the costs of purchasing electrically powered equipment for productive uses.

If implemented, this could lead to a more rapid increase in the number of customers and their average electricity usage. By providing a source of financing that might not otherwise be available, it would provide a way to increase the SPP's sales revenues and allow it to become commercially viable at an earlier time. So it would be a win-win outcome for both the SPP and its customers. However, simply changing the regulatory rules by itself is not likely to accomplish very much. The problem is that most SPPs are not likely to have the funds needed to establish such a line of credit for their customers. Therefore, it has been recommended that bilateral and multilateral donors in Tanzania should provide funding to SPPs through loans or grants that SPPs can use to establish a line of credit for their customers. The customers would be allowed to use this line of credit to finance expenditures that would allow potential household customers to get connected and potential and existing business customers to purchase electrical machinery that would increase their productivity.

Key Recommendation

To ensure the commercial viability of SPPs that operate isolated mini-grids, regulators should explicitly:

- Allow SPPs to charge tariffs above the uniform national tariff if it is required to recover efficient operating and capital costs.
- Allow SPPs to cross-subsidize among their customers.
- Mandate SPPs to take depreciation on equipment financed through grants.
- Allow SPPs to enter into power sales contracts with businesses without requiring prior regulatory approval of the contract terms.
- Allow SPPs to recover in tariffs the administrative and financing costs incurred to provide on-bill financing to customers for uses such as connection charges, internal wiring, dwelling upgrades, and the purchase of electric-powered appliances and machinery.

Tariff Levels for Community-Owned SPPs

Community- or local-government-owned SPPs (as opposed to privately owned SPPs) often charge tariffs that are too low. Typically, community-based SPPs will, at best, set retail tariffs to cover basic operating costs (for example, the salaries of the operator and staff) with little or no provision for covering maintenance or replacement of equipment that wears out. And if this continues for several years, blackouts will become more frequent and the system may eventually collapse. When this happens, it means that the government's or donor's up-front capital cost grant has been wasted on a project that will fail to achieve commercial sustainability. And equally important, villagers will probably be reluctant to pursue other community-based projects in the future.

The tariffs may be low and unable to recover operating costs and depreciation, even though the community-based operator might have signed an agreement with a government agency or a rural electrification fund (REF) in which he has committed to charging cost-recovering tariffs in return for receiving an initial capital cost grant. For example, community-based mini-hydro systems in Nepal are required to charge tariffs that "must be able to meet operational cost (staff salary, regular maintenance, and so on) along with a provision of setting aside 20 percent of revenue collection for major repairs and maintenance" (NEA 2003). But what is written on paper is often ignored in practice. And, unfortunately, national regulators have limited leverage to do very much when legal requirements are ignored.

Consider the situation of a regulator who observes or is informed that a community-based SPP is charging too low a tariff. It is clear to the regulator that if the SPP continues charging low tariffs that do not support even basic maintenance, the SPP will limp along for a few years but eventually the system will collapse. In this situation, the regulator could take away the community-based SPP's license or permit to persuade the SPP to raise its tariffs, but the regulator would most likely face considerable political pressure to give the license back if he did.

The head of the regulatory agency will probably receive an urgent telephone call from a member of parliament or even the president's office, saying something like this:

> How dare you take away the operating license of the SPP in this poor community? Don't you realize that there are businesses in this community that will collapse if there is no electricity? And the health clinic in the village will no longer be able to store medicines that need to be refrigerated. OK, I understand that the community may not have done a good job in setting and collecting tariffs but certainly the solution is not to shut them down completely. Let's be reasonable. You need to give them time and I am sure that things will improve.

Faced with the common situation of non–cost recovering tariffs being charged by community-based SPPs, what can be done? We recommend four possible solutions:

- *Encourage bank loans.* The first is to encourage the SPPs to take out loans. The need to make periodic loan payments and to have collateral at risk creates an incentive to raise tariffs to cost-recovering levels. The loan will be especially effective in creating tariff discipline, if the collateral at risk is the property of individual community members. The regulator can piggyback on the existence of such a loan. For example, as a condition for granting a license or permit, the regulator could require that a community-based SPP applicant have a provision in the loan agreement that gives the lender the right to call in the remaining principal on the loan (or any monies in an escrow account or assets offered in collateral) if the community-based SPP borrower loses its license or permit.

- *Offer future grants.* The second is to persuade the government or REA to offer future rehabilitation grants, or grants to connect new customers only if the applicant provides evidence that it is charging cost-recovering tariffs. A weakness of many current grant-giving systems is that grants are only given for the initial capital costs and then there is no serious follow-up to ensure that the community-based organization will charge cost-recovering tariffs once the system is in place. This could be described as the hit-and-run approach to grant giving and unfortunately it is typical of many donor grant programs. But if the grant-giving organization also offers the possibility of follow-up grants, it would have the leverage that it does not have under the current prevailing system of one-off, up-front grants.

- *Convert grants to loans.* The third is to threaten to convert grants to loans if the community fails to set tariffs at cost-recovering levels. This last approach exists (at least on paper) in India. India's national Rural Electrification Policy states that "if conditionalities of the scheme are not implemented satisfactorily, the capital subsidy could be converted into interest bearing loans" (Government of India, Ministry of Power 2006, Section 7.3). It is unclear whether this central government threat has actually been implemented. This action would, however, need to be taken by the granting agency (acting as a *de facto regulator*), rather than by the designated national regulator.

- *Use private operators.* The fourth is to encourage the community to use private operators. Private operators create financial discipline because they will walk away from the project if they are not paid, which creates pressure for the community to set tariffs that will generate at least enough revenue to pay the private operator. Under one type of private involvement, the community owns all plant and equipment and simply hires one or more private individuals (either as employees of the community owner or independent private contractors) to operate the system. This arrangement, which is the lowest level of private sector involvement, helps ensure that tariffs will at least cover operating costs (that is, the costs of paying these individuals to operate the system). A higher level of private involvement would be for the community to hire a private operator to both own and operate the system. If the agreement between the community and the private operator also specifies performance standards, then no rational private operator would sign the agreement unless tariffs cover all operating expenses and have a provision for maintenance. If the contract is for a longer time period, it would also need to provide the private operator with funds for future replacements (that is, depreciation). The existence of a contract between the community and a private owner/operator creates pressure on the community to charge tariffs that will recover some or all of the operator's costs. The contract can thus be thought of as the functional equivalent of a locally granted license. See Mahé and Chanthan (2005, annex 2) for an example of a well-specified agreement between a private operator and a community-owned SPP in Cambodia.

Our general view is that positive or negative financial incentives are more likely to be effective in encouraging community-based SPPs to charge cost-recovering tariffs than the threat of license or permit removal by the regulator. It has also been proposed that village women's groups collect and safeguard or deposit monies that have been collected. For example in Bulelavata, a small remote village in the Western Solomons, the Bulelavata Women's Committee took over the job of collecting and depositing monthly electricity tariff payments from the male-dominated Village Hydro Management Committee. The community's decision was based on the belief that the women's group would be more transparent, reliable, and honest in handling the tariff payments. While this may help in stopping leakages in payments, it does not solve the problem of tariffs being set at too low a level.

Key Recommendation

Community-owned SPPs often charge tariffs that are too low. To overcome this tendency, regulators should:

- Encourage these SPPs to take out bank loans.
- Persuade the government or REA to offer future grants for rehabilitation or new connections, only if the applicant provides evidence that it is charging cost-recovering tariffs.
- Threaten to convert grants to loans, if the community fails to set tariffs at cost-recovering levels.
- Encourage or require communities to hire private operators to run the generation and distribution system.

Setting Tariff Structures: Concepts and Cases

Even the most basic design elements of SPP retail tariffs may differ from project to project. Some SPP retail tariffs are based on energy (kWh), while others are based on power (watts or kilowatts) or the number of lights or appliances in each household. Tariffs can be pre- or postpaid. In some countries, retail tariffs are different for each project depending on the developer's costs or customer's ability to pay; in others, they are set according to a standardized schedule. There is clearly no one-size-fits-all tariff that is appropriate in all circumstances. In this section, we examine some factors that should be considered in designing retail tariffs for SPP customers. Our general recommendation is that regulators should allow SPP developers wide latitude to propose tariff structures that work best for their technology and rural electrification context. But regulators also require evidence that the tariffs, when combined with any subsidies, will achieve commercial sustainability within a few years.

Energy Tariffs (per kWh)

Energy (kWh) electricity tariffs are familiar to most people who purchase electricity from a national grid. Under traditional practice, when a customer uses electricity a disk spins in a meter installed on the customer's premises, recording the customer's cumulative energy usage measured in kilowatt-hours. The total energy consumption measured by the meter is read once a month by a meter reader. In some countries, the latest generation of kWh meters are digital, and can be queried or read automatically by way of electrical pulses sent over the electricity wires. The electricity bill paid by the customer is, in the simplest case, the recorded kilowatt-hours consumed that month, multiplied by the allowed price per kilowatt-hour. This is a postpaid energy tariff.

kWh energy tariffs can also be charged on a prepaid basis, where the customer prepays for his energy consumption. And if the money paid in advance is insufficient, the customer has the option of topping up the available balance in the same way that mobile-phone customers purchase more minutes in advance of expected usage.

Power Tariffs (per Watt)

In some cases, tariffs are based on peak power (watts) consumed rather than energy (kWh).[16] These peak-power tariffs are sometimes referred to as flat-rate tariffs, or subscription tariffs because the tariff charge remains the same (or is "flat") regardless of the energy consumed by the customer, as long as the maximum demand at any given moment does not exceed the subscription amount. For example, a customer with a 50-watt flat tariff can turn on two 25-watt lightbulbs for as many hours as she wishes, but cannot turn on three bulbs simultaneously (since three 25-watt bulbs would exceed 50 watts). Flat-rate customers are unmetered customers. Flat-rate tariffs are usually offered to customers whose consumption is so small (for example, just one or two small lights) that the expected collected revenue would not warrant the costs of installing a meter and the labor costs of sending the reader out to read the meter. On mini-grid systems with flat-rate tariffs in several West African countries, the customer usually has a choice among several monthly maximum demand levels (such as 50 watts, 100 watts, 150 watts, and so forth).

Flat-rate customers are typically subject to maximum or peak consumption controls. If a flat-rate customer simultaneously turns on loads that exceed the maximum subscription amount he has purchased, then an overcurrent device (a miniature circuit breaker [MCB] or similar device) trips, cutting off power to the customer's house. To use electricity again, the customer's load must be reduced to a level below the maximum demand ceiling and the device reset. Sometimes, the overcurrent device is installed in a locked box, as a disincentive for users to repeatedly push their consumption beyond the subscribed ceiling. In Nepal basic load-limiting devices cost around $10/household, not including installation cost (Smith 1995).

A variation of the flat-rate tariff is to charge per lightbulb, or per outlet, with no overcurrent device. This produces a small amount of initial cost savings because there is no need to install an overcurrent device. A major disadvantage, however, is that without an overcurrent device, it becomes necessary to physically inspect usage in the household at unannounced times to ensure that the customer has not added additional lights or appliances, exceeding the wattages assumed in the original subscription agreement.

For the operator, flat-rate tariffs are easier to administer than kWh tariffs because there are no meters to read and no monthly energy levels to calculate and bill. A flat-rate tariff structure keyed to subscribed demand is particularly suited for micro-hydropower or other energy sources where the fuel cost is zero (box 9.2). In most village micro-hydropower plants, peak power production is limited (by the capacity of the turbine/generator) but the plant can generate at this maximum electrical output 24 hours a day at little or no additional cost. Indeed, at every moment when electricity is produced in excess of total consumption on the mini-grid, the surplus electricity must be dissipated as heat in a ballast load.[17] It is truly a use-it-or-lose-it situation. A flat-rate tariff with no per-kWh charge works well in this context because it provides a price signal that encourages consumption with a high load factor (Greacen 2004).

Economists often talk about the importance of sending good price signals. This means that the price customers are charged should measure the incremental or marginal cost of an additional kilowatt-hour consumed at any given moment. The underlying rationale is that the consumer should be informed through price signals of the additional costs that are incurred to provide him with an additional unit of consumption (that is, a kilowatt-hour) at any given time. On a mini-hydro system, the marginal cost is essentially zero at any time of the day other than at the time when total consumption of all users reaches the system's peak capacity. Hence, there is no need to install a meter that measures energy consumption because the additional cost of consumption for most of the day is zero. But there is a need to install load-limiting or overcurrent devices that prevent customers from consuming more than their

---

Box 9.2 Retail Tariffs for Mini-Hydro SPPs: Some Examples

Thailand: kWh tariffs without load limiters. In dozens of Thai villages, mini-grid micro-hydropower systems have been installed with kilowatt-hour (kWh) meters but with no load limiters. In a typical village, users had no incentive to spread their loads out throughout the day. As populations grew and villagers acquired more appliances, the evening-time peak load grew, driven by the use of rice cookers, water heaters for coffee and tea, lights, and televisions. Eventually power consumption during this evening peak period exceeded the capacity of the system, resulting in chronic low voltage (brownout) throughout the village. When voltage is low, "fluorescent lights will not start, television pictures become distorted and motors run hot or stall" and in the worst case, appliances are damaged (Greacen 2004, 56). Therefore, a Thai villager remarked, "Yes, we have electricity 24 hours a day … except when we need it." Brownouts were often followed by blackouts and (not infrequently) equipment failure at the micro-hydropower plant. Power plant repairs would often take weeks, compounded by the remote location of the installation.

Nepal: Flat-rate tariffs with load limiters. In contrast, most village mini-hydro schemes in Nepal have combined subscription tariffs with load-limiting devices so users can only get the maximum power that they paid for. For example, a guesthouse in Ghandruk village in Nepal with a load limiter installed purchases electricity on a subscription tariff from a community hydropower mini-grid. The guesthouse carefully schedules loads to keep consumption balanced throughout the day at a level just below the subscription limit: lights at night until guests go to sleep, then electric water heaters which heat water throughout the night to provide hot showers for guests. In the daytime, the electricity in the guesthouse is used in bakery ovens and to cook rice and other meals. This arrangement helps avoid the brownout situations encountered in the Thai villages because the total wattage of subscriptions available for sale is (in principle) no greater than the power plant capacity. While very useful, the subscription tariff is no silver bullet: it does not solve problems of seasonal variations in hydropower output (due to water supply constraints), or ensure that theft by bypassing the subscription device does not occur.

Bhutan: kWh tariffs with a sophisticated load limiter. A promising high-tech version of load limiters called GridShare has recently been piloted in a 40 kW micro-hydropower project in Rukubji, Bhutan, a village of approximately 90 households (Dorji 2012). As is the case in many villages served by micro-hydropower, the power supply is sufficient during off-peak times; however, during preparation of morning and evening meals, the use of high-power kitchen appliances regularly caused brownouts. GridShare communicates the state of the grid to its users (a green light indicates the grid is fine; a red light indicates the grid is close to brownout conditions) and regulates usage before severe brownouts occur. When the red light is lit, users cannot turn on large loads such as rice cookers, but if a user is already operating a rice cooker and the grid slides into brownout-approaching mode, the rice cooker is allowed to continue operation for one hour. This demand-side solution encourages users to distribute the use of large appliances more evenly throughout the day, allowing power-limited systems to provide reliable, long-term renewable electricity. While GridShare could be used for either flat or kWh-based tariffs, Bhutan uses kWh tariffs.

---

box continues next page

Box 9.2 Retail Tariffs for Mini-Hydro SPPs: Some Examples *(continued)*

In the summer of 2011, GridShares were installed in every household and business connected to the Rukubji mini-grid. Installations were accompanied by an extensive education program. Following the installation of the GridShares and the training, the occurrence of severe brownouts has dropped by approximately 85 percent and the average length of brownouts has substantially decreased. Additionally, residents stated that their rice has cooked more consistently and that they would recommend installing GridShares in other villages facing similar problems. The cost per household for the GridShares device is $80, but could be expected to fall to $40 or less if mass manufactured. The electronic circuit itself (not including the box and installation) costs $10–15 (Quetchenbach and others 2013).

---

subscribed amount. This physical device is, in effect, sending a price signal—if you exceed your prespecified allowed limit, you will pay the very high price of receiving no electricity at all.

Hiding the kWh Charge

Sometimes, flat-rate tariffs are favored by private mini-grid operators and REAs because it is difficult for flat-rate customers to compare their tariffs with the uniform national tariffs paid by friends and relatives connected to the national utility. Flat-rate tariffs help operators of mini-grids achieve cost recovery because they obscure the effective per-kWh price that the flat-rate customers are actually paying. For example, is a tariff of T Sh 1,000 a month for a 50-watt connection in a mini-grid operated four hours a day a bargain, compared to a tariff of T Sh 60/kWh paid by urban consumers? The answer to this apples-to-oranges comparison requires a bit of mathematics, and therefore hides the fact that the flat-rate customer on the mini-grid is usually paying a much higher tariff on a per-kWh basis than comparable customers taking service off the national grid under a uniform per kWh national tariff.[18] (See appendix B for an example from Tanzania of calculating the estimated per-kWh charge associated with an existing flat monthly charge.)

Private operators of isolated rural mini-grids who charge a flat monthly charge are generally uncomfortable with calculations that convert their flat charges to an implied per kWh tariff. They argue that the relevant comparison is not their implicit per-kWh charge versus the national utility's per-kWh charge. From the operators' perspective, the relevant comparison is the customers' prior energy expenditures on kerosene for their lanterns and for mobile-phone charging versus the lower monthly costs the customers now incur on purchases of electricity from the operators' mini-grids. Private operators assert that their household and commercial customers do not care what electricity costs on a per kWh basis. Instead, they contend that their customers' only concern is whether the same or more energy services can now be acquired from the mini-grid operator at a lower monthly cost. In other words, they argue that the only relevant comparison is the monthly cost of purchasing the energy services and not the cost per kWh of electricity supplied. In addition, they point out that they will be able to persuade households to sign up as customers only if they can provide the households with better quality energy services at a monthly cost to households that is at least 20–30 percent lower than its prior monthly energy expenditures from other sources. Although these are valid points, it becomes politically more difficult to make these arguments if the villagers are charged on a per kWh tariff rather than a flat tariff and if there are friends and relatives in nearby villages who are purchasing grid-based electricity from a national or regional electric utility on a tariff with a much lower kWh charge.

Two Major Disadvantages of Flat-Rate Tariffs
Disadvantage 1: No Incentive to Be Energy Efficient
Flat-rate tariffs do not encourage people to turn off lights or other appliances when they are not using them, since there is no additional charge for higher electricity usage. The customer pays the same amount whether his lights are left on or turned off. There are some mitigating situations as discussed earlier: the customer's behavior is not a problem if the generation source is a micro-hydropower project with zero fuel costs, or a mini-grid that provides power only a few hours a night (as is common with many diesel-fired mini-grids). In this latter case, excess cumulative energy consumption is limited by virtue of the power plant being on only during those hours when most people would want lights anyway. But flat-rate tariffs are undesirable in mini-grid systems in African villages where a fossil-fuel generator is used all day long and households are not being sent a price signal that shows the cost of the additional fuel that is being burned every hour. Similarly, flat-rate tariffs are undesirable in cases where the customer is connected to the main grid and is supplied with electricity produced from fossil-fuel generators. In this situation, it costs money to generate each additional kilowatt-hour but the customer does not see that additional cost in the price that he or she is charged.

Disadvantage 2: Discourages Productive Uses of Electricity in a Village Where Both Flat and kWh Tariffs Are Offered
In Sub-Saharan Africa and elsewhere, a principal motivation for providing electricity to rural communities is that it offers the possibility of undertaking economic activities that would not be possible in the absence of grid electricity. But flat-rate tariff structures can also be an impediment to promoting productive activities in a village. For example, in Senegal many of the operators in isolated villages that are supplied by hybrid generating systems (typically 5 kWp[19] of solar and 15 kVA[20] of diesel) offer both flat-rate and per-kWh tariffs (shown in table 9.4). For the S1 tariff customers, the fixed or flat monthly charge is roughly equivalent to 55–73 cents/kWh depending on the customer's total monthly consumption.

In addition, most of these mini-grid operators also offer a per-kWh tariff of $0.23/kWh for metered businesses combined with a fixed monthly customer charge. This is below the mini-grid's actual operating costs per kilowatt-hour,

Table 9.4 Mini-Grid Tariffs (S1, S2, S3, and S4) in Senegal

	Peak (watts)	Fixed monthly charge ($)	$/kWh
S1	50	4.68	n.a.
S2	90	8.62	n.a.
S3	180	16.16	n.a.
S4	180+	n.a.	$0.23

Source: Assani 2011.
Note: In addition to the charges shown in the table, the customer has the option of paying a separate monthly charge if he or she decides to pay for indoor wiring over time. n.a. = not applicable.

estimated to be about $0.27. The capping of the per-kWh tariff at $0.23 reflects the fact that the government does not want mini-grid operators to sell electricity at a per-kWh charge that exceeds the per-kWh charge paid by customers connected to the main grid. The effect of this cap is that the mini-grid operator loses about 4 cents for every kilowatt-hour that it sells to S4 customers. Hence, no rational mini-grid operator will want to take on more S4 customers.

Senegal's mixed system of flat-rate and kWh tariffs for mini-grid operators with the per-kWh charge capped at the main grid's per-kWh charge produces two unintended (and presumably undesired) consequences. First, the poorer customers who are taking service under the S1, S2, and S3 tariffs will pay more on a per-kWh basis than the more well-to-do S4 customers. Consequently, it should not be surprising that S1, S2, and S3 customers will always seek to pay a lower per-kWh charge by becoming S4 customers. But as one Electricité de France (EDF) official involved in deploying mini-grids in Senegal observed, "Small customers will ask for the S4 tariff, and the financial viability of the company is lost" (Marboeuf 2011). Second, it creates a strong financial incentive for the mini-grid operator to refuse to offer service to potential new S4 customers because the more electricity that the operator sells, the more money it loses. So a mixed-tariff structure of flat and kWh charges, with the per-kWh charges set at less than cost-recovering levels, creates a major barrier to the promotion of village-level productive businesses—exactly the opposite of what a government wants.

Tariffs That Combine Energy and Power Charges: An Example from Senegal

Advances and cost reductions in electronic metering technology using microprocessors have enabled mini-grid operators to create tariff structures that combine energy and power charges and temporal factors that better match the technical characteristics of mini-grids.

For example, a solar photovoltaic (PV)/diesel hybrid mini-grid has low marginal electricity costs as long as it is sunny and overall consumption in the village is low, but very high marginal costs when a diesel generator must be dispatched, or when new capacity must be added to accommodate growth in power demand. To address this, Integrated Energy Supply Systems (INENSUS), a German company, has proposed an electronic metering system in which

Senegalese villagers must purchase one or more energy blocks every six months. The blocks impose two limits: total energy consumption (kWh) and peak load (watts). These blocks, typically of 50 watts maximum demand and 6 kWh/week cumulative energy consumption, can be added together to meet the varying electricity demand of different customers. The purchase of these blocks informs the SPP mini-grid operator of the capacity of resources (solar panels, batteries, diesel generator, and so on) to deploy. These blocks can be bought and sold among villagers and if surplus electricity is needed, it can be purchased on the spot at higher prices (reflecting the higher marginal cost of electricity when the diesel generator is dispatched, or batteries are deeply discharged) (INENSUS GmbH 2011). The cost per household for INENSUS meters is €168–220 depending on the quantity ordered, but is expected to decrease if there is mass production.

Lifeline Tariffs and Increasing Block Tariffs

Another dimension to consider is how to design tariffs to apportion costs among and within different customer classes. A lifeline tariff or social tariff is a lower tariff charged to customers who consume below a certain amount of electricity per period. Lifeline tariffs are often part of a progressive block rate or increasing block tariff (IBT) structure in which a customer whose electricity usage is greater pays progressively higher rates for that usage. As such, this typically represents a cross-subsidy from high- (generally wealthier) to low-consumption customers.

Under an IBT tariff all residential customers receive the benefit of the subsidy on the first block, regardless of their total monthly consumption. In contrast, under a volume-based tariff (VBT), higher-volume customers pay higher prices for each successive block but they lose the subsidy on the first block (hence VBT tariffs are sometimes referred to as tariffs with "the disappearing first block"). IBT tariffs are much more common than VBT tariffs.

Sometimes IBTs are justified on grounds that they mirror the mini-grid's underlying cost structure. But this is a misconception. A household consumer does not impose increasingly higher unit costs on the systems with each unit of electricity consumed. The reality is that: "It is a customer's load profile rather than total volume of consumption that affects his or her contribution to system costs" (Komives and others 2005, 13). The effect of a customer's additional consumption on an SPP's unit costs will depend on the time at which the electricity is consumed and not on the customer's total volume of consumption.

Both energy and power tariffs can be structured to incorporate lifeline tariffs in mini-grids. For example, the 250 kW Khandbari village micro-hydropower mini-grid in Nepal charges 4.43 rupees/kWh to domestic users, and a higher rate of 5.84 rupees to tourist lodges and other commercial enterprises (Vaidya n.d., table 2.5). A typical micro-hydro in Nepal might charge NPR 0.75/watt/month for low-usage domestic users, and NPR 1/watt/month to commercial enterprises and high-use domestic consumers.

A ratcheted lifeline approach has been proposed by the private developers of the Mwenga Hydro Limited project in the Mufindi district of Tanzania. Like the national utility, Mwenga's distribution subsidiary will charge TANESCO's lifeline tariff of T Sh 60 (4 cents)/kWh to any household customer who consumes 50 kWh or less per month. But if the household's monthly electricity usage exceeds more than 50 kWh for more than three months of a calendar year, then the customer will be automatically switched to another tariff category where the energy charge is T Sh 234/kWh (15.6 cents). And equally important the customer loses the right to return to the lifeline tariff even if his consumption later drops back down to less than 50 kWh/month (Mwenga Hydro Limited 2012). EdM (Electricidade de Moçambique, the Mozambican national utility) uses a similar ratcheted tariff structure, with about 80 percent of its household customers being on prepaid meters. If they consume 100 kWh or less per month, they pay about 3.45 cents/kWh on the company's social tariff. But if their consumption goes above 100 kWh/month, they are automatically moved onto a higher domestic tariff and they pay 8.05 cents/kWh on all consumption. More important, like the customers of Mwenga Hydro, EdM's customers can never return to the social tariff. At present, less than 1 percent of EdM's household customers are on the social tariff (Mills 2013).

Key Observation

Regulators usually approve both the level and structure of tariffs. Tariff structures can take many different forms. Customers can be charged: an energy tariff, where they pay for the kilowatt-hours that they use (that is, a per unit of energy used charge); a power or flat-rate tariff, where customers pay a fixed monthly fee based on their peak power usage or maximum allowed power demand rather than their unit consumption; or a combination of the two. In the case of energy tariffs, the tariff may also contain (a) lifeline or social components, where customers who use very little energy pay a low, flat monthly rate, and (b) increasing block tariffs (IBTs), which divide the overall tariff into consumption blocks with higher prices paid for consumption in each succeeding block of consumption. Our general recommendation is that regulators allow SPP developers wide latitude to propose tariff structures that they believe will work best for their technology and customer base, rather than imposing a one-size-fits-all approach.

Should Tariffs Be Prepaid or Postpaid?

Twenty years ago, virtually all electricity tariffs were postpaid. A customer used electricity, and at the end of the month received a bill for the electricity consumed. But electronic innovations and cost reductions, combined with the huge success of prepaid cell-phone services, have brought prepayment into the mainstream in retail electricity sales. In first-generation prepaid metering systems, customers purchased special number-bearing cards or cards with magnetic strips that could be physically inserted onto the prepaid electricity meters, or whose number could be keyed in, providing the customer with additional kilowatt-hours equal to the monetary value of the purchased cards. In second-generation prepaid metering systems, a customer can use her mobile phone to pay the electricity supplier and she receives an SMS (text message) with a confirmation number that can be keyed into the prepaid meter. The customer then uses the electricity, and tops up her account (by purchasing another card or making another electronic transfer via her mobile phone) when the meter indicates that the balance is running low and she is at risk of running out of electricity.[21]

Pre- and postpaid electricity supply systems have upsides and downsides. To the consumer, the advantages of the prepaid electricity supply system are:

- The ability to control expenditures
- No unpleasant surprise of a big electricity bill at the end of the month
- Immunity from penalties arising from late bill payment, including penalty payments or disconnection
- Typically no deposits required because there is no risk of nonpayment

To the SPP, the advantages are:

- No risk that users will use electricity without paying for it
- Low or zero meter-reading costs

In Tanzania the Mwenga hydro system SPP developer is using second-generation prepaid meters as a key element in electrifying 16 rural villages. Since the villages are widely disbursed over an area of 1,000 square kilometers it would be prohibitively expensive to read traditional postpaid meters and then deliver paper bills to each customer. Also, customers would find it very difficult to pay their bills at locations outside their villages, especially during the rainy seasons when roads become almost impassable. But a reliable mobile-phone service is now available in each of these villages, therefore, the developer has established a second-generation prepaid metering system called M-Luku,[22] which allows the SPP's customers to recharge their electricity meters using their mobile phones (Mwenga Hydro Limited 2012). The customer purchases a top-up card at local village stores and sends an SMS with the number on the card and the customer's meter number to the M-Luku server. He immediately receives a return SMS that includes a number that, if keyed into the prepaid meter, will add the purchased amount to his account.

Key Observation

New technologies, most notably in metering and telecommunications, are giving SPPs more options to offer pay-as-you-go or prepaid energy plans to their customers, which offers benefits both to the operator and to its customers.

Alternatives to Setting Retail Tariffs on a Case-by-Case Basis

The standard textbook discussion of regulation assumes that the regulator will establish tariffs on a case-by-case basis for each individual enterprise under its jurisdiction. But in fact there are alternatives to case-by-case-basis tariff setting, which are often used when the regulator is confronted with numerous small electricity providers. The three most common alternatives are:

- Retail tariff setting by category
- Periodic case-by-case tariff setting, combined with automatic adjustment clauses (AACs)
- Full or partial deregulation of retail tariffs

Two general justifications are usually given in support of these alternatives. The first is that it is too costly and time consuming to set SPP and SPD tariffs on an individual case-by-case basis. If SPPs and SPDs become widespread in a country, a regulatory commission or an REA would be quickly overwhelmed if it is required to perform separate cost-of-service determinations with periodic updates on each of possibly 100 or more individual SPPs or SPDs. The regulator would quickly run out of time and resources to regulate the national utility or other large utilities. Therefore, it is rational for the regulator to use his limited resources to set tariffs for the national utility and any other large utility whose performance will affect many more people, and employ simpler and less-time-consuming tariff-setting mechanisms for SPPs and SPDs.

A second justification for avoiding or limiting case-by-case tariff setting for SPPs is that this has been the norm in countries that have successfully electrified rural areas. For example, in Cambodia, more than 200 small, privately owned diesel-fired mini-grid systems were created and successfully served thousands of rural customers before the national regulator and the REA came into existence. The Cambodian SPPs were created spontaneously and from the bottom up, without support from any Cambodian government program. Proponents of partial and total deregulation of SPPs argue: what is the benefit of imposing tight case-by-case tariff regulation on SPPs if they were successfully providing electricity service to poor and isolated rural customers before either the regulator or the REA existed?

Setting Tariffs by Category

In 2005 India initiated a major program (known as the Rajiv Gandhi Grameen Vidyutikaran Yojana, RGGVY) to increase the number of grid-connected and off-grid villages that would be electrified. To promote the program, India's national government announced that it would provide grants to pay for up to 90 percent of the capital costs of electrification (that is, transformers and distribution lines) and a 100 percent subsidy for the connection costs (that is, poles, meters, droplines, and internal wiring) for any household below a prespecified poverty line. In addition to these grants, there was also a legal requirement that mandated retail tariff setting by categories, for any entity receiving grants under the RGGVY program. Specifically, the Rural Energy Plan policy document issued by the Central Government stated that: "The Appropriate Commission would lay down guidelines for this purpose for various types of projects (for different fuels, technology and size) receiving subsidy *as opposed to tariff determination on case by case basis*" (italics added) (WESCO 2011, 4).[23] At the time of this writing, it is unclear the extent to which this mandated policy has actually been implemented.

Automatic Adjustment Clauses (AACs)

Another option is to combine periodic individual cost-of-service reviews with AACs between reviews. AACs allow for automatic adjustments in tariffs without performing a new tariff review. The adjustments are made according to a prespecified formula and can be partial or total. If it is partial, then the retail tariff is adjusted for changes to one or more underlying cost components according to a prespecified formula. If it is total, then the overall tariff (rather than individual cost components) is adjusted according to a prespecified formula. AACs, whether partial or total, are typically used by regulators when a high proportion of the underlying costs are variable, and the changes in the variable costs are hard to predict and largely beyond the SPP's or SPD's control. AACs are usually combined with periodic full tariff reviews to recalibrate tariffs. But in the period between full tariff reviews, adjustments to tariffs are essentially on autopilot and determined by a prespecified adjustment formula.

Cambodia

The proportions of variable and fixed costs in the generation component of SPPs vary considerably by technology and fuel source. As shown in table 9.5, diesel-fired generators have the highest proportion of variable costs while mini-hydro generators have the lowest. In addition, the costs of diesel generation change constantly with fluctuations in diesel fuel costs, which in turn are directly affected by world oil prices. When faced with this situation for more than 100 SPPs, the Electricity Authority of Cambodia (EAC, the Cambodian electricity regulator), decided to create a fuel-adjustment mechanism for isolated, diesel-fired mini-grids.

Table 9.5 Split Between Fixed and Variable Costs in Mini-Grids (by Generation Technology)

	Share of total cost (percent)				
	Mini-hydro	Wind	Biomass grown	Biomass waste	Diesel
Fixed cost	96.7	97.5	68.1	81.0	5.9
Variable cost including fuel (if any)	3.3	2.5	31.9	19.0	94.1

Note: The level of fixed costs will depend heavily on the capacity factor of the generator. A higher capacity factor causes the fixed costs to decrease, and vice versa. For example, table 9.5 assumes a capacity factor of 30 percent for a hydropower plant serving a mini-grid. If the capacity factor reduces by 20 percent, then the fixed cost share increases from 96.7 to 97.8 percent.

In an April 2012 order for one isolated mini-grid operator, the EAC included a table that prespecified the retail tariff that the SPP operator would be allowed to charge based on the price of diesel oil (EAC 2012). For example, the table specifies that if the price per ton of diesel oil is between $620 and $680 (approximately $87.32–95.77/barrel), the SPP operator would be allowed to charge a retail price of $0.52/kWh. But if the price per ton of diesel oil goes up to $1,010–1,075 (approximately $142.25–151.41/barrel), then the SPP operator is allowed to charge its retail customers a price of $0.67/kWh.

The Cambodian fuel-adjustment mechanism is semiautomatic rather than fully automatic. Under a fully automatic adjustment, the operator makes the adjustment subject to possible after-the-fact auditing by the regulator. In other words, the operator does not need prior approval to make the adjustment. But in Cambodia, the isolated mini-grid operator is required to "seek approval from the EAC before increasing the retail tariff, so that the EAC can review and monitor whether the fuel price variation is to the level requiring change in the tariff or not" by providing "details of fuel price variation including the bills of fuel purchased from sources" (EAC 2012). So the price change is not automatic, but requires a before-the-fact review by the regulator before it can go into effect. But this is less burdensome to an operator than a regular tariff hearing because public consultation is not needed. The EAC commits to making its decision within 30 days of receiving the complete requested information. In fact, once the operator submits invoices from the authorized dealers, the EAC will usually approve the tariff adjustment within two to five business days.

The Philippines

In 2011 the Philippine Energy Regulatory Commission (ERC) instituted an automatic tariff adjustment system for 119 on-grid electricity cooperatives. These cooperatives, scattered across several islands, are distribution entities that generally buy 100 percent of their power needs from larger utilities on their island. The ERC decided that the cooperatives' retail tariffs should incorporate two automatic adjustment mechanisms. The first one would cover power purchase costs, typically about 85 percent of total operating costs for these distribution utilities. Once a cooperative's power purchase contract (usually a long-term bilateral contract) is approved by the ERC, the cooperative is entitled to automatically and fully pass-through the power purchase costs in the retail tariffs that it charges its members.

The second automatic adjustment mechanism covers the cooperative's total distribution costs (about 15 percent of total costs). This is based on a formula that allows overall distribution costs to be adjusted based on a regional consumer price index (CPI), with additional positive or negative adjustments based on measures of the cooperative's efficiency and service performance. For administrative ease of implementation, the ERC grouped the on-grid cooperatives into three groups based largely on geographic proximity. *While the adjustment formula is common to all cooperatives in a particular group, there is no requirement that the tariff levels must be the same for each member of the group.*

From the Bottom Up • http://dx.doi.org/10.1596/978-1-4648-0093-1

The ERC has stated that tariff levels of individual members of each group will be reviewed every three years along with each cooperative's capital expansion plan. So unlike most AACs, which allow for annual or quarterly adjustments, the distribution charge AAC allows for adjustments only once every three years. This was done to gain political acceptance. Under Philippine law, the ERC is required to hold a public hearing whenever it performs a new tariff review. But under this new system, it is expected that it will not be a full cost-of-service review, but a review of just the distribution charge. In about 90 percent of past tariff cases, there has been no opposing intervention by customers. If this holds true in the future, the ERC should be able to make its review fairly quickly. Since the system has only recently been initiated, it is too early to know whether it will produce significant benefits for the ERC and the cooperatives (Tan 2012).

Full or Partial Deregulation of Retail Tariffs and Elimination of Other Regulatory Reviews

Full Deregulation—The Indian Approach

In 2003 India deregulated SPPs operating in rural areas. The 2003 Electricity Act states that "where a person intends to generate and distribute electricity in a rural area to be notified by the State Government, such person shall not require any license for such generation and distribution of electricity but he shall comply with the measures which may be specified by the Authority under Section 53" (Government of India 2003, Section 14, Proviso 8). In the rural electrification guidelines that were issued soon after the act became law, the Ministry of Power stated that this provision would exempt rural SPPs and SPDs from both licensing and tariff regulations, but the exempted SPPs would still be subject to safety and technical regulations under Section 53 of the 2003 Act.

The ministry went on to explain that tariffs for these exempted entities would be "based on mutual agreement between such person [the exempted entity] and the consumers." The justification for this exemption was that "[s]ince these would be micro enterprises with low capital expenditure, short gestation periods and no entry barriers, competitive market forces would ensure reasonable prices reflecting actual costs" (Government of India and Ministry of Power 2006). In effect, the ministry was predicting that the village electricity markets would be contestable because potential competitors would prevent the mini-grid operator from charging monopoly prices.

On paper, the Indian approach deregulates isolated mini-grid systems by eliminating both the requirement for a license and prior approval of the retail tariffs to be charged. But deregulation in theory may not translate into deregulation in practice. The traditional regulatory function of setting maximum tariffs appears to have been transferred from the state electricity regulatory authority to the state-level grant-giving agency. This seems to be implied in the guidelines issued by the Indian central government that direct the granting agency to calculate maximum tariffs (which would require something like a regulator's cost-of-service study) as a key step in deciding how much of a capital and operating subsidy should be granted to the project operator. In addition, the grant-giving agency is required to ensure that the benefits of the grants are passed on to the consumers, which is difficult to do without first calculating tariffs with the grant and then without the grant. Finally, the Central Government's 2006 Rural Electrification Policy states that: "[the implementing agency] shall have right to intervene by scrutinizing tariffs if these guidelines are not implemented in any particular case" (Government of India and Ministry of Power 2006, Section 8.6). Arguably, the combined effect of these several requirements is a *de facto* regulation, even though the rural operator has ostensibly been exempted from any tariff or licensing regulation by the 2003 Electricity Act.

Another concern raised by several developers is that it is difficult to obtain financing from commercial banks if they do not have a license, that is, some official piece of paper from a government entity that shows that they have a legal right to generate and distribute electricity. Since they have been exempted from the usual requirement of having a license, they are operating in a legal limbo that makes commercial banks hesitant to provide them with loans.

Partial Deregulation—The Tanzanian Approach

In contrast to India, Tanzania has proposed what might be called partial deregulation. The Tanzanian regulator's proposal is that very small power producers (VSPPs, generators with an installed capacity of 100 kW or less) need not apply for a license, but instead are required to register and provide periodic reports as specified by the regulator (EWURA 2013, section 59). Presumably, these two lesser requirements are intended to allow the regulator and the government to know where the VSPPs are located, the technology and fuel that is being used, and the amount of the electricity that is being produced. A VSPP might apply for a license, even though it is not a legal requirement, to obtain a recognized legal identity that would presumably help if the VSPP were to seek a bank loan.

Like India the Tanzanian regulator does not require prior regulatory review and approval of retail tariffs, but unlike India, it reserves the right to review a VSPP's retail tariffs if it receives complaints from the VSPP's retail customers and it can lower tariffs on an after-the-fact basis. In reviewing these complaints, the regulator uses the same cost-of-service model employed to determine maximum allowed revenues for larger SPPs (that is, greater than 100 kW of installed capacity) or some alternative tariff benchmark that is yet to be specified. This is different from India, where the government has taken the position that neither prior nor after-the-fact tariff reviews are required because of the expectation that potential competitors will keep tariffs down to reasonable levels.

Partial Deregulation—The Cambodian Approach

The Cambodian regulator has used yet another approach in regulating the pre-existing rural electrification enterprises (REEs)—the term used to describe isolated, diesel-fired mini-grids. By the end of 2010 it had issued more than 180 licenses to REEs, but in the earlier years the regulator took a light-handed approach to regulating their retail tariffs. Since virtually all of the REEs used either diesel oil or heavy fuel oil, the tariffs in these early years were often quite high—ranging up to 90 cents/kWh (currently as high as $1.25/kWh). In contrast, the regulator took a much stricter approach in setting the duration of licenses. Several consultants recommended that licenses be granted for relatively long periods of at least seven or more years to increase the chances of REEs being able to get loans from Cambodian banks—the rationale being that banks would provide loans to REEs if the borrower had a license to operate for a time period longer than the possible loan period.

But the Cambodian regulator rejected this advice and initially gave licenses only for relatively short periods such as two years. It also stated its willingness to increase the duration of the license if the REE showed evidence of investing in the technical and operational quality of its system. This strategy worked—the benefit of being able to get a longer license incentivized REE operators to make significant investments in their physical facilities through the informal funding channels of family and friends. Over the past several years the average duration of awarded REE licenses has increased, as REEs have been able to demonstrate substantial improvements in their physical plant infrastructure to the regulator. The hope is that these longer licenses will now allow some REEs to borrow from local banks (Rekhani 2012).

Key Observation

To reduce their burden of having to set retail tariffs on a case-by-case basis, regulators can instead set tariffs based on categories of SPPs, implement automatic adjustment clauses, or fully or partially deregulate retail tariffs.

Setting Quality-of-Service Standards

Don't let the perfect be the enemy of the good.

—UNKNOWN

The power goes out four times every night.... In the daytime, we hardly get three hours of electricity, and when we do, it is such a low voltage we can't run any appliances at home. Our refrigerator and television are in comas. The fan moves but doesn't throw any breeze at us. All we can do is charge our mobile phones.

—VILLAGER IN INDIA (LAKSHMI AND DENYER 2012)

If poor service is economically the equivalent of high price, why is there not just as great a danger that monopoly power will involve the one as the other?

—ALFRED E. KAHN (1970, 24)

Quality of service is especially important for any SPP that sells to retail customers. Its three principal components are: quality of the product, quality of supply, and quality of commercial service.[24] *Quality of product* refers to the technical parameters of supply such as whether the frequency and voltage of the electricity are at or near the target levels. *Quality of supply* refers to the availability and continuity of supply. For example, how many hours of the day does the SPP operator provide electricity? How frequent are unexpected blackouts, and how long do the outages last for each incident and in total over the course of a year? *Quality of commercial service* refers to the quality of service provided in numerous commercial interactions with customers. Some key dimensions of commercial quality include: number of days to connect a new customer, time to respond to and resolve a complaint about billing and metering, and number of days to reconnect a customer who has paid the balance due on an account. Appendix C summarizes the quality-of-service standards established for rural service providers in Peru.

For each of the three quality-of-service areas, five basic design questions have to be answered:

- What dimensions of quality of service will be regulated?
- What minimum levels of service will be required for each quality-of-service dimension?
- Who sets the standards?
- How are the standards monitored?
- How are the standards enforced?

In answering these questions, the regulator, whether it is the national regulator or an REA that is effectively acting as a regulator, must always keep in mind three overriding considerations.

Three Design Considerations

The Cost of Quality

The first consideration is that customers will always prefer higher-quality goods and services, all other things being equal. But the reality is that all other things are not equal: higher quality involves higher costs and these costs must ultimately be borne by the consumer. Therefore, the regulator must always ask: Can the customer realistically afford this level of service?

OSINERGMIN (El Organismo Supervisor de la Inversión en Energía y Minería), the Peruvian electricity regulator, has decided that quality-of-service standards should be lower for electricity service providers in rural areas. Its rationale is that it is more difficult and costly to provide comparable service in rural areas at a price that is affordable to the generally poorer rural customers. Table 9.6 shows the standards set by OSINERGMIN for the System Average Interruption Frequency Index (SAIFI) and the System Average Interruption Duration Index (SAIDI) for different categories of urban and rural service providers. SAIFI is a standard measure of the number of outages during a specified

Table 9.6 Targeted SAIFI and SAIDI Standards in Peru

Types of service areas	SAIFI (number of interruptions per year)	SAIDI (hours per year)
Urban high density	12	7
Urban medium density	16	9
Rural concentrated	25	10
Rural dispersed	40	10

Source: Revolo Acevedo 2011.
Note: SAIDI = System Average Interruption Duration Index; SAIFI = System Average Interruption Frequency Index.

calendar period and SAIDI is the total duration of these outages, measured in hours per year.

The maximum number of allowed interruptions per year for customers in an urban, high-density area (for example, Lima) is 12. In contrast, an electricity service provider serving in a rural dispersed area (most isolated mini-grids) is allowed up to 40 interruptions per year without incurring a penalty. Similarly, OSINERGMIN has established lower standards for quality of product and quality of commercial service for rural providers (see appendix C).

Monitoring and Enforcement

The second consideration is that the regulator (or the regulator's agent) must be able to monitor and enforce the service standards that are established. When a village gets grid-based electricity for the first time, there is a honeymoon period during which villagers will overlook poor performance because of the novelty of having real electricity. But this period will not last very long if service deficiencies persist. While the initial complaints and anger will be directed against the SPP, later it will be directed against the regulator for failing to enforce its rules or regulations. As one villager observed: "I want more from the regulator than just pretty poetry." Hence, regulators should not expect more from SPPs than they can realistically produce, and should also have a workable system in place to monitor the SPP's quality-of-service performance.

Inputs versus Outputs

Regulators are not the only government entities that regulate quality of service. REAs, also, establish minimum quality-of-service standards as the *quid pro quo* for providing connection grants to rural service providers. The REA requirements typically focus on the technical specifications of the equipment that will be purchased with the grants that they provide. For example, the grant agreement between an REA and an SPP will typically specify the height of the distribution poles, the materials that can be used in the distribution poles, and the minimum height of the wires above the ground—which can be thought of as input rather than output standards. From the perspective of an REA, it is easier to specify input standards as they need only be specified and validated once. By contrast, output standards require ongoing and more costly monitoring.[25] The downside of regulating quality by specifying inputs is that the regulator becomes a micromanager. What ultimately matters is the quality of the electric service that is provided rather than the particular inputs used to achieve that quality of service. But that requires a sophisticated and ongoing monitoring system that may not be initially feasible in isolated rural areas.

One example of a standards development process underway that focuses on inputs is an initiative by the International Electrotechnical Commission (IEC) to specify recommendations for rural electrification mini-grids. The IEC's *82-62257-9: Recommendations for Small Renewable Energy and Hybrid Systems for Rural Electrification* (http://webstore.iec.ch/webstore/webstore.nsf/Artnum_PK/41866) is a document that provides guidance on micropower and mini-grid design from power plant to the power outlets in users' homes. It covers safety; erection of equipment; power generation; system voltage selection; operation, maintenance, and replacement; marking; and documentation. The goal of the IEC initiative is to provide a clear and well-written set of practical recommendations relevant for mini-grids in developing countries. These IEC recommendations, in turn, could be adopted as standards by regulators or as *de facto* standards by REAs, or by the donor community through requiring that these recommendations be followed as preconditions for receiving grants.

Key Recommendation

Regulators should set minimum quality-of-service standards for the quality of product, quality of supply, and quality of commercial service. The standards should not be cost prohibitive for SPPs and should be relatively easy to monitor and enforce. Initially, it is easier for regulators to establish standards for inputs (equipment, materials, and so on) rather than for outputs (quality of service), because the input standards typically need to be specified and validated only on a one-time basis. But over time it is preferable that regulators move to output-based quality-of-service standards, so as to avoid micromanaging the mini-grid's equipment choices and operations.

Notes

1. In Tanzania, the avoided-cost tariff principle is also used in setting tariffs for an SPP's wholesale sales to the national utility. In this instance, the avoided cost is based on an estimate of the financial costs that the national utility would have incurred to obtain the same electricity supply in the absence of the purchase from the SPP.

2. Section 7.6 of India's 2006 Rural Electrification Policy states that to achieve the "objective of parity," the consumer tariffs should be comparable between "remote villages yet to be electrified and adjoining grid connected villages."

3. In addition, the Indian central government has committed to providing grants that would cover 100 percent of the cost of connecting any rural household whose income is below a designated poverty line.

4. Load factor is the ratio of the average electric load to the peak load over a period of time.

5. The spreadsheet (http://tinyurl.com/SPPevaluator) has also been recommended for use by the Tanzanian Energy and Water Utilities Regulatory Authority (EWURA) in evaluating retail tariffs submitted by SPPs for approval.

6. IRR is a measure of the rate of return of an investment over a period of years. The project IRR considers the project cost and project revenues without taking into consideration where this money comes from (debt, equity, grants). Project IRR is thus a measure of the economic viability of the project itself, specifically, the economic returns the project provides over its lifetime, weighed against the project costs.

7. Equity IRR is a measure of the rate of return of an investment to *equity investors*, taking into consideration the equity investments (generally at the start of the project) and the returns to the equity shareholder over the years of the project lifetime. It is similar to return on investment (ROI), but a more nuanced and complicated calculation because it considers cash flows throughout the duration of the project, not just the initial and ending values of the investment.

8. Capacity factor is the ratio of the average power production by an electric generator to its nameplate capacity.

9. A private developer in Tanzania found that of 2,600 households that expressed interest in connecting immediately, only 900 signed up. He expects that within several billing cycles there will be a flood of new signups as people realize that electricity is less expensive and has a higher level of service than kerosene for lighting.

10. See the discussion of advance payments in chapter 5.

11. The energy charge for T1 customers is T Sh 221/kWh, accompanied by a service charge of T Sh 3,841/month. We make the simplifying assumption that this service charge is spread over a typical usage of 400 kWh/month, yielding an effective tariff of T Sh 230/kWh (TANESCO 2012).

12. With kerosene consumption of 0.05 liter/hour and kerosene price of $0.50/liter, the kerosene lamp lasts 5,000 hours with a replacement cost of $1.00.

13. CFL lifetime of 6,000 hours at a replacement cost of $3.00.

14. Compact fluorescent lamp.

15. This proposal is also discussed in chapter 5. On-bill financing means that loan and interest payments are made through separate payments incorporated into a pre- or postpaid billing system.

16. The distinction between power and energy is important here. Power is a measure of the instantaneous electricity consumed. Energy is a cumulative measure of electricity consumed over time. The distinction is analogous to the difference between speed (instantaneous measure) and distance traveled (cumulative measure of travel over time).

17. A ballast load (or ballast heater) is an electrical resistance heater typically used in isolated village hydropower mini-grid projects. An electronic load controller carefully monitors generator frequency and diverts electricity to the ballast load keeping total electrical load constant on the generator regardless of real-time variations of the load in the village.

18. T Sh 1,000 for 50 watts × 4 hours/day × 30 days/month = 6 kWh/month, which would cost only T Sh 360 at an energy charge of T Sh 60/kWh. In Senegal it is estimated that in 2011 flat-rate customers on hybrid mini-grids were paying about 66–93 cents/kWh, whereas comparable residential customers of SENELEC (Société National d'Éléctricité du Sénégal, the state-owned national utility) were paying a

maximum price of about 25 cents. In Guinea in late 2011 it was estimated that flat-rate customers on 24 isolated small diesel-fired mini-grids paid an effective price of 55–73 cents/kWh consumed (BERD 2011).

19. Kilowatt peak (kWp) is the peak power rating of a solar array, meaning the maximum amount of electricity it can generate under ideal conditions (temperature 25 degrees centigrade and insolation of 1,000 watts/square meter).

20. Kilovolt-amps.

21. In Kenya, for example, electricity prepayment to the national utility, the Kenya Power and Lighting Company (KPLC), can be done using the MPESA, a mobile phone money transfer system (Bert and Rich 2011). This system is also now available in Tanzania, where most of TANESCO's household customers in Dar es Salaam are on prepaid meters.

22. Luku is an acronym for Lipa Umeme Kadri Unavyotumia, which when translated from Kiswahili means "pay for electricity as you use it."

23. India has a two-level electricity regulatory system. There is a single regulator at the national level, and each state has a state regulatory commission. In most instances, the "appropriate commission" that will implement this policy is the state electricity regulatory commission.

24. Quality of product and quality of supply are sometimes grouped together and described as technical quality of service.

25. As a condition for awarding a concession and giving a grant, AMADER, the Malian REA, specifies output standards for the minimum number of hours per day that electricity must be supplied and the deviation from targeted voltage and frequency standards. It is unclear how tightly these standards are actually monitored (AMADER, n.d., Articles 6 and 10).

References

ABPS Infrastructure Advisory Pvt. Ltd. 2011. *Policy and Regulatory Interventions to Support Community-Level Off-Grid Projects*. Final Report. http://www.forumofregulators.gov .in/Data/Reports/CWF%20Off-grid%20final%20report%20nov%202011_Latest _feb2012.pdf.

Adkins, Edwin, Sandy Eapen, Flora Kaluwile, Guatam Nair, and Vijay Modi. 2010. "Off-Grid Energy Services for the Poor: Introducing LED Lighting in the Millennium Villages Project in Malawi." *Energy Policy* 38: 1087–97.

AMADER (Malian Agency for the Development of Household Energy and Rural Electrification). n.d. "Specifications Annexed to Concession Order." Unofficial English translation. http://ppp.worldbank.org/public-private-partnership/sites/ppp.worldbank .org/files/documents/Mali0Specifications.pdf.

ASER (Agence Sénégalaise d'Electrification Rurale) and Columbia Earth Institute Energy Group. 2007. *Costing for National Electricity Interventions to Increase Access to Energy, Health Services, and Education: Senegal Final Report*. Report prepared for the World Bank. http://modi.mech.columbia.edu/wp-content/uploads/2013/04/Senegal _WorldBank_Report_8-07.pdf.

Assani, Mansour. 2011. "Regulatory and Technical Issues in Operating Hybrid Mini-Grids." AEI Practitioner Workshop, Dakar, Senegal, November 15. http://siteresources .worldbank.org/EXTAFRREGTOPENERGY/Resources/717305-1327690230600

/8397692-1327691245128/Regulatory_TechnicalIssues_Operating_Hybrid_MiniGrids.pdf.

BERD (Bureau d'Électrication Rurale Décentralisée). 2011. Personal communication. November 10.

Bert and Rich (blog). 2011. "Buy Prepaid Electricity by MPESA." May 9. http://www.bertandrich.com/blog/how-to/buy-prepaid-electricity-by-mpesa/.

CORE International. 2008. "Study on Tariff Setting Principles and Issues Surrounding Tariffs and Electricity Pricing in Southern Africa." Submitted to the Southern African Power Pool, Houston, Texas, April 23.

Dixit, Shantanu. 2012. "Powering 1.2 Billion People: Case of India's Access Efforts." Presentation at World Bank Energy Days 2012, Washington, DC, February 23.

Dorji, Chhimi. 2012. *Smart Grid Technology: GridShare Project in Rukubji, Bhutan.* http://www.sari-energy.org/PageFiles/What_We_Do/activities/BhutanCrossBorder WorkshopAug2012/PResentations/GridShare_SARIE_CD_Final.pdf.

EAC (Electricity Authority of Cambodia). 2012. "On Determination of Electricity Tariff Based on Fuel Adjustment Mechanism for Consumers in the Distribution Area of Mr. Chet Layhim." Decision 098-SR-12-EAC. Phnom Penh, Cambodia.

Eberhard, Anton, Vivien Foster, Cecilia Briceño-Garmendia, Fatimata Ouedraogo, Daniel Camos, and Maria Shkaratan. 2008. "Underpowered: The State of the Power Sector in Sub-Saharan Africa." Background Paper, Africa Infrastructure Country Diagnostic, World Bank, Washington, DC. https://openknowledge.worldbank.org/handle/10986/7833.

EWURA (Tanzanian Energy and Water Utilities Regulatory Authority). 2012a. *Detailed Tariff Calculations for Year 2012 for the Sale of Electricity to the Main Grid in Tanzania under Standardized Small Power Purchase Agreements in Tanzania.* http://www.ewura.go.tz/pdf/SPPT/2012/2012%20SPPT%20Calculation%20for%20Main%20Grid.pdf.

———. 2012b. *Detailed Tariff Calculations for Year 2012 for the Sale of Electricity to the Mini-Grids in Tanzania under Standardized Small Power Purchase Agreements in Tanzania.* http://www.ewura.go.tz/pdf/SPPT/2012%20SPPT%20Calculation%20for%20Mini-Grid.pdf.

———. 2012c. *The Electricity (Development of Small Power Projects) Rules.* Draft circulated for public consultation.

———. 2013. *The Electricity (Development of Small Power Projects) Rules, 2013.*

Proposed for Public Consultation, June. Dar es Salaam, Tanzania.

Government of India. 2003. *The Electricity Act, 2003.* http://guj-epd.gov.in/extra_no_46.pdf.

Government of India, Ministry of Power. 2006. *Rural Electrification Policy, 2006.* http://www.powermin.nic.in/whats_new/pdf/RE%20Policy.pdf.

Gratwicke, Michael. 2012. Personal communication. June.

Greacen, Chris. 2004. "The Marginalization of 'Small Is Beautiful': Micro-hydroelectricity, Common Property, and the Politics of Rural Electricity Provision in Thailand." PhD dissertation, University of California, Berkeley. http://palangthai.org/docs/Greacen Dissertation.pdf.

INENSUS GmbH (Integrated Energy Supply Systems). 2011. *The Business Model of Micro Power Economy.* http://www.inensus.com/download/MicroPowerEconomy.pdf.

Kahn, Alfred E. (1970) 1988. *The Economics of Regulation: Principles and Institutions.* Vol. 1. Cambridge, MA: MIT Press.

Komives, Kristin, Vivien Foster, Jonathan Halpern, and Quentin Wodon. 2005. *Water, Electricity and the Poor: Who Benefits from Utility Subsidies?* Directions in Development Series. Washington, DC: World Bank. http://siteresources.worldbank.org/INTWSS /Resources/Figures.pdf.

Lakshmi, Rama, and Simon Denyer. 2012. "Lack of Power Symbolizes India's Inequalities." *Washington Post*, August 6. http://www.washingtonpost.com/world/asia_pacific/lack -of-power-symbolizes-indias-inequalities/2012/08/06/ecdbef64-df20-11e1-a19c -fcfa365396c8_print.html.

Mahé, Jean Pierre, and Ky Chanthan. 2005. *Rehabilitation of a Rural Electricity System.* GRET and Kosan Engineering, Phnom Penh, Cambodia, August. http://www.gret.org /wp-content/uploads/07409.pdf.

Marboeuf, Guy. 2011. "Mini-Grids and Regulatory Issues: EDF's Experience in Mali." Presentation at the Practitioner Workshop, Dakar, Senegal, November 14. http:// siteresources.worldbank.org/EXTAFRREGTOPENERGY/Resources/717305 -1327690230600/8397692-1327691245128/Mini_grids_And_RegulatoryIssues _Guy_Marboeuf.pdf.

Mills, Rob. 2013. Personal communication. March 11.

Mwenga Hydro Limited. 2012. "Application for Tariff Approval by Mwenga Hydro, Ltd. (MHL)." Submitted to EWURA by Mwenga Hydro Limited, Dar es Salaam, Tanzania.

NEA (Nepal Electricity Authority). 2003. *Nepal Electricity Authority Community Electricity Distribution Bye Laws, 2060.* http://www.nea.org.np/images/supportive_docs /Community%20Electricity%20Distribution%20Bylaw.pdf.

Peon, Rodolfo, Ganesh Doluweera, Inna Platonova, Dave Irvine-Halliday, and Gregor Irvine-Halliday. 2005. "Solid State Lighting for the Developing World—The Only Solution." *Optics and Photonics 2005, Proceedings of SPIE* 5941: 109–23.

Quetchenbach, Thomas, Megan Harper, James Robinson, Kirstin Hervin, Nathan Chase, Chhimi Dorji, and Arne Jacobson. 2013. "The GridShare Solution: A Smart Grid Approach to Improve Service Provision on a Renewable Energy Mini-Grid in Bhutan." *Environmental Research Letters* 8 (1): 014018.

Radecsky, Kristen. 2009. *Understanding the Economics Behind Off-Grid Lighting Products for Small Businesses in Kenya*, Humboldt State University, Arcata, CA. http:// humboldt-dspace.calstate.edu/xmlui/bitstream/handle/2148/508/RadecskyThesis .pdf?sequence=1.

Rekhani, Badri. 2012. Personal communication. March.

Reuters. 2012. "Kenya Regulator Cuts Diesel Price, Petrol, Kerosene Up." Reuters Africa edition, Nairobi, Kenya, May 14. http://af.reuters.com/article/investingNews /idAFJOE84D08Z20120514.

Revolo Acevedo, Miguel. 2011. Personal communication. November.

Smith, Nigel. 1995. *Low Cost Electricity Installation.* Overseas Development Administration, June. http://r4d.dfid.gov.uk/PDF/Outputs/R5685.pdf.

Tan, Rauf. 2012. Personal communication. April.

TANESCO (Tanzania Electric Supply Company). 2010. *TANESCO Tariff Review Application.* Dar es Salaam, Tanzania. http://www.ewura.go.tz/pdf/Notices/Tariff%20 Application%202010%20-%20With%20Covering%20Letter.pdf.

————. 2012. "Electricity Charges." Tanzania Electric Supply Company. http://www .tanesco.co.tz/index.php?option=com_content&view=article&id=63&Itemid=205.

Vaidya, Shankar Lal. n.d. *Cost and Revenue Structures for Micro-Hydro Projects in Nepal.* Alternative Energy Promotion Centre, Kathmandu, Nepal. http://www .microhydropower.net/download/mhpcosts.pdf.

WESCO (Western Electricity Supply Company of Orissa, Ltd.). 2011. *Provision of Revenue Subsidy for Sustainability of Rural Electricity Supply in Villages Being Electrified Under RGGVY.* Bidyut Niyamak Bhawan, Kalyani Complex, Unit VIII, Orissa, India.

When the Big Grid Connects to a Little Grid

Companies may fear that their investment in off-grid solutions may prove worthless if the grid is indeed extended.

—ALLIANCE FOR RURAL ELECTRIFICATION (ARE 2011, 10)

It's our power plant. Villagers work together, build a sense of community, and get electricity that saves money.

—POWERHOUSE OPERATOR FROM HUAI BU VILLAGE, THAILAND, EXTOLLING VIRTUES OF HIS VILLAGE MINI-GRID—WHICH WAS PLANNED TO BE DECOMMISSIONED WITH THE ARRIVAL OF THE NATIONAL GRID LATER THAT YEAR (GREACEN 2004)

Abstract

In chapter 10 we discuss business models and regulatory options that could be used when the big grid connects to a little grid. Five possible business models are described. One involves a conversion from an isolated small power producer (SPP) to a main-grid-connected small power distributor (SPD), a model that is widely used in Asia. We analyze the economic, regulatory, and technical prerequisites for creating viable SPDs and a hybrid model that combines an SPD with an SPP.

From Broad Strategy to Ground-Level Implementation

As discussed in chapter 1, virtually every national electrification strategy in Sub-Saharan Africa contains the recommendation that access to grid-produced electricity is best accomplished by simultaneously pursuing a centralized track that relies on grid extension and a decentralized track that promotes isolated mini-grids. While this two-track strategy (illustrated in figure 10.1) has been widely adopted by African governments, it is often unsuccessful in implementation. If the two-track strategy is going to work, one key requirement is that the government or regulator needs to specify in advance what should happen when the two

Figure 10.1 Base Case: Before the Mini-Grid Connects to the Main Grid

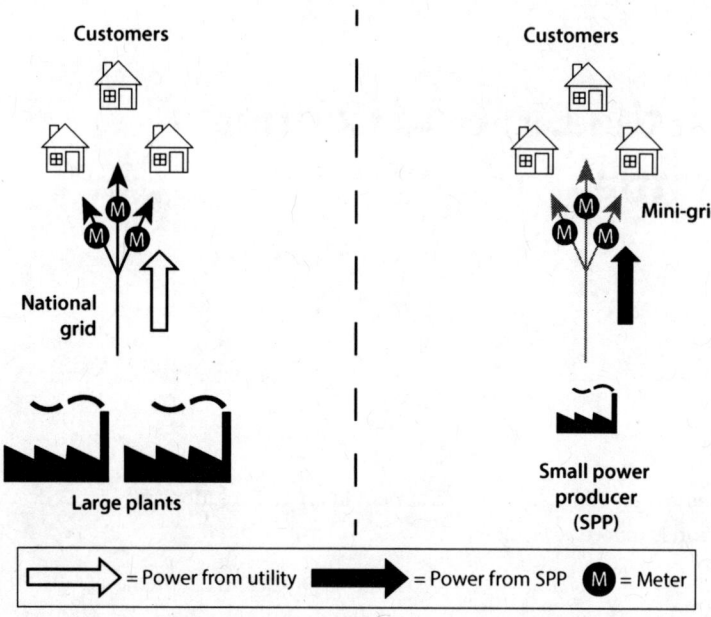

Source: Diagram by Richard Engel and Chris Greacen, 2013. Used with permission.

tracks connect. Making a smooth transition from an isolated mini-grid to a main-grid connection is the subject of this chapter. *If there is no clarity as to what happens when the centralized and decentralized tracks come together, investors will be reluctant to invest in isolated mini-grids.*

This was a significant problem in Cambodia several years ago. The lack of a policy for what to do when the big grid connected to a mini-grid led to underinvestment by hundreds of private mini-grid operators. Many private sector operators of mini-grids limped along with second- and third-hand diesel generators and mini-grid distribution systems using undersized, non-outdoor-rated wiring often tied to trees. Investing in system upgrades made little sense to these entrepreneurs, since they would be out of business and their assets scrapped if the national utility, Electricité du Cambodge, decided to electrify their service area next.

The Cambodian regulator has now addressed this problem by allowing small power producer (SPP) mini-grids that meet sufficient technical standards to connect to the national grid and convert themselves into small power distributors (SPDs). Setting a sufficient margin between the bulk purchase tariff and retail sales tariffs allows the new SPDs to cover their distribution costs and earn a profit (Rekhani 2012). As of 2013, the Cambodian regulator had issued licenses for 82 distribution utilities that were formerly isolated diesel-powered mini-grids. Close to 200 licensed operators of isolated mini-grids still remain (Keosela 2013). Cambodia has been successful in pursuing the centralized and decentralized tracks in parallel, but it appears to be the exception rather than the rule among countries faced with a similar need for rural electrification.

When private and cooperative investors are reluctant to invest in isolated mini-grids, isolated villages suffer because they are denied a chance to receive electricity that they otherwise might have had and instead must wait years or decades for the national grid to arrive, if indeed it ever does. Conversely, policies like Cambodia's that allow mini-grids to connect to the main grid and convert from an SPP to SPD can help to foster win-win-win arrangements for developers, utilities, and the public. If the right policies are in place, both the private sector and community organizations will have economic incentives to build and operate isolated mini-grid systems. They can electrify villages that the national utility is reluctant to serve. This allows the national utility to concentrate on expansion of the high-voltage and medium-voltage grids. The national utility can also potentially benefit from the availability of power generation and end-of-line voltage support when the national grid reaches a previously isolated mini-grid. Rural customers also benefit because they receive electrical service sooner and have the possibility of receiving higher-quality and lower-priced electrical service when the big grid finally arrives.

Recommendations for When the Big Grid Arrives

To encourage investors to invest in isolated mini-grids, regulations and policies should be adopted that give SPPs any of the following options when the big grid arrives:

- *SPD option.* The SPP should have the right—as long as certain conditions are met (discussed below)—to convert from an SPP operating an isolated mini-grid to an SPD that buys electricity at wholesale from the national grid and resells it at retail to its local customers.
- *SPP option.* The SPP sells electricity to the operator of the national grid (or some other designated buyer) but no longer sells electricity to retail customers.
- *Combined SPP and SPD option.* The SPP converts from operating an isolated mini-grid to operating an SPD that buys electricity at wholesale from a national or regional utility and resells it at retail to its local customers. It also maintains an existing or new small generator as a backup generator and/or as a supply source to the main grid and retail customers.
- *Buyout option.* The SPP sells its distribution grid to the national grid operator or some other entity designated by the national government or regulator and receives compensation for the sale of its assets.

Regulators, we believe, should operate under the presumption that any of these options would be approved if the SPP demonstrates that its facilities are built to sufficiently high standards to allow for connection to the national grid and that there are no major and convincing objections from customers.[1] In return for being offered these options, we believe that the SPP that has operated an isolated mini-grid should be required to connect when the national utility's grid reaches the village(s) that the SPP has served.[2]

In the absence of rules that allow these options, or in the event that none of the options above is applicable, the path that remains is:

- *Abandonment.* The distribution grid and generator are abandoned, sold for scrap, or moved. The connecting utility builds and operates a new distribution system to serve customers in the area.

Key Recommendation

Regulations and policies should prespecify the commercial options available to the SPP when the national grid arrives in the SPP's service area. Otherwise, entrepreneurs and investors will not invest in SPP projects in the first place. If this happens, rural households will lose the benefit of access to grid-based electricity until the big grid finally arrives. These postconnection options should include the SPP converting to an SPD, the SPP remaining a stand-alone entity that sells electricity to the main grid, the SPP acting as both an SPP and an SPD, and the SPP selling its assets to the national grid operator or another prespecified entity.

Small Power Distributor Option

An SPD purchases electricity from a national or regional utility (typically at medium voltages such as 33 kilovolt [kV] or 11 kV), and operates a distribution network that delivers this electricity to retail customers. The SPD will usually have a legal right to sell to retail customers in one or more villagers that are specified in its license or permit. A less-ambitious business model involving rural franchisees has been used to India in an attempt to improve revenue collection and service on existing distribution systems that continue to be owned by a state-owned utility (see box 10.1).

SPDs are common in several Asian countries (Nepal, Bangladesh, Vietnam, and Cambodia) that have had major success with scaling up electrification. Sometimes these SPDs started out as SPPs and became SPDs when the grid arrived in the area, as in the case of the 82 Cambodian SPDs discussed in the beginning of this chapter.

Similar considerations apply in the more common cases in which SPDs did not start out as SPPs, but were instead built to function as SPDs right from the beginning. For example, as of July 2010, in Nepal more than 116,000 households received their electricity service from community-owned distribution entities that purchased electricity at wholesale from the national utility and then resold it at retail. These community distribution cooperatives operate under the community electrification bylaws issued by the government in 2003. In accordance with the bylaws, communities must provide 20 percent of the total cost of constructing distribution lines, while the government contributes the remaining 80 percent. The approach has proven successful in electrifying communities more quickly than a conventional national

Box 10.1 Alternatives to Small Power Distributors: Rural Franchisees in India

The central government of India decided to take a new approach to rural electrification when it passed the Electricity Act of 2003. Up to that point, the state electricity boards (SEBs) that provided electricity in each Indian state also had responsibility to serve rural areas. On paper all of the SEBs had a universal service obligation to supply electricity throughout the state, but the reality on the ground was quite different. In 2010 the Central Electricity Authority statistics indicated that peak power deficit was more than 10 percent. When power generation was inadequate, most SEBs gave higher supply priority to urban and commercial customers because tariffs were higher in those areas. Customers in rural areas paid less and they suffered more frequent blackouts (Palit and Chaurey 2011). So even though the rural villages may have had access to electricity infrastructure, this did not translate into actual access to electricity supply on a reliable basis.

The 2003 Electricity Act broke new ground by encouraging the use of rural franchisees (Government of India 2003, Section 5). The law allowed for franchisees to take different forms, the two most common being: revenue franchises (RFs) and input-based franchises (IBFs). By March 2012 it was estimated that 37,000 franchises covering 216,000 villages were in operation (Mukherjee 2013).

Revenue franchises. The RF is the most common form of franchise. Under this business model, the SEB hires an individual or organization to read meters, distribute bills, collect payment, serve as the channel for complaints, and sometimes perform low-level maintenance on distribution facilities. *In essence, the SEB is hiring a local individual to assist it in revenue collection.* This individual is an independent contractor who performs tasks that would otherwise be performed by an employee of the SEB. Hence, it is a form of third-party outsourcing. It is not privatization because the franchisee does not take over ownership of the distribution facilities—ownership remains with the SEB.

As an agent of the SEB, the franchisee does not have its own license, and the retail tariffs do not change. The tariffs are the same that the SEB would charge retail customers if there were no franchisee. The franchisee exists because it has signed a one- to two-year private contract with the SEB. The contract does not require the approval of the regulator or any other state government entity.

In several case studies done in 2007 (TERI 2007), there was evidence that many of these local revenue-based franchisees had achieved significant success in increasing collections and improving service levels. For example, in the Indian state of Karnataka, one study found about a 10 percent increase in billing efficiency, a 20 to 50 percent increase in revenues, and a 10 to 15 percent increase in new customers (TERI 2007). Nevertheless, it appears that there are still several basic problems with the franchisee business model. First, the profit margins are small so the revenue-based franchisee is widely perceived as not a good business. Second, the upstream source of electricity supply remains with the SEB. So if there is an inadequate supply of electricity coming from the SEB, either because it is physically not available or it is not remunerative to supply, the local franchisee gets blamed for the lack of supply even though availability of supply source is clearly beyond its control. Third, while the RF has an incentive to

box continues next page

Box 10.1 Alternatives to Small Power Distributors: Rural Franchisees in India (continued)

collect the amount billed, it is not incentivized to help the utility to reduce losses (Palit and Chaurey 2011).

Input-based franchises. One response to these shortcomings has been the increased use of IBFs. In March 2012 it was estimated that slightly over 1,600 of the 37,614 rural franchisees were served by an IBF. An IBF purchases electricity at a fixed tariff from the distribution utility (typically the government-owned SEB) and then resells it to its retail customers. In essence, it functions as an SPD and assumes responsibility for all commercial activities related to issuing new connections, metering, meter reading, billing, collecting on current bills, collecting debts, disconnecting and reconnecting customers, and addressing customer complaints. If the IBF can reduce distribution losses and improve billing and collection efficiencies, it can earn higher profits. In effect, the IBF functions as an SPD in every respect except for the fact that the SEB is still the official holder of the distribution license.

Sources: ABPS 2011; Dixit 2012; Mukherjee 2013; TERI 2007.
Note: In the case of India, the term *franchisee* does not refer to the licensee. Instead, it refers to a person or entity that has been authorized by a distribution licensee to distribute electricity in a specified area on behalf of the distribution licensee. But the legal obligation to serve still remains with the distribution licensee (that is, a state electricity board).

utility-led expansion. The community-owned SPDs have also substantially reduced electricity theft and improved the timeliness of bill payment by consumers (Mahato 2010).

Similarly, Bangladesh has 70 rural electric cooperatives that provide service to approximately 8.4 million customers. These cooperatives, called Palli Bidyuit Samity (PBS), have been the main mode of electrification through grid extension. PBSs purchase electricity in bulk from the Bangladesh Power Development Board and then resell this electricity at retail to members or nonmember buyers in their service areas (Palit and Chaurey 2011). They serve between 35,000 and 275,000 customers (Chowdhury 2009).

In Vietnam, about 21 percent of the country's 8,000 rural communes are served by private, community, or cooperatively owned local distribution utilities (LDUs) that purchase electricity in bulk from regional power utilities and resell the power to retail customers (Van Tien and Arizu 2011).

Tanzania appears to be the first country in Africa that is seriously considering the SPD option. Under "second-generation" SPP rules currently under consideration by the Tanzanian Energy and Water Utilities Regulatory Authority (EWURA 2013 Section 40 (3)), SPDs are explicitly allowed to apply to EWURA for the right to operate as:

- An SPP selling to a distribution network operator (DNO)[3] that is connected to the main grid
- An SPD that purchases electricity in bulk from a DNO connected to the main grid and then resells that electricity to the SPD's retail customers
- A combination of an SPP and an SPD

If these rules are adopted, previously isolated SPPs (Case 1 from chapter 2: isolated SPP that sells at retail) would have the legal right to convert themselves into connected SPPs selling at wholesale (Case 4: grid-connected SPP that sells at wholesale to a utility) or SPDs (which must be connected to some other supply source) buying at wholesale and reselling at retail or combinations of the two. But the option of converting from an SPP to an SPD, while promising on paper, will be a nonoption unless the SPD has the potential to be commercially viable. Therefore, the proposed rules state that SPDs will be allowed to charge a retail tariff that provides a sufficient margin for an efficient SPD to be commercially viable (EWURA 2012a). But even if this option is legally permitted, it may not be politically feasible because households in the villages now connected to the national grid will argue that they are entitled to the same low tariffs as households served by the national utility in neighboring villages. If the national utility's retail tariffs are a *de facto* cap for an SPD, then the only available option would be to subsidize the SPD's operating or power purchase costs so it can profitably sell electricity at the national utility's retail tariff. Three delivery mechanisms for such subsidies are described in the sections that follow.

The case of an SPP converting to an SPD is shown in figure 10.2. The distribution system that had served the isolated mini-grid continues to sell electricity to the same customers, but now this electricity comes from the national grid network rather than from an isolated SPP generator.

Figure 10.2 Small Power Distributor Option—Mini-Grid System Obtains Bulk Electricity from the National Utility for Local Distribution

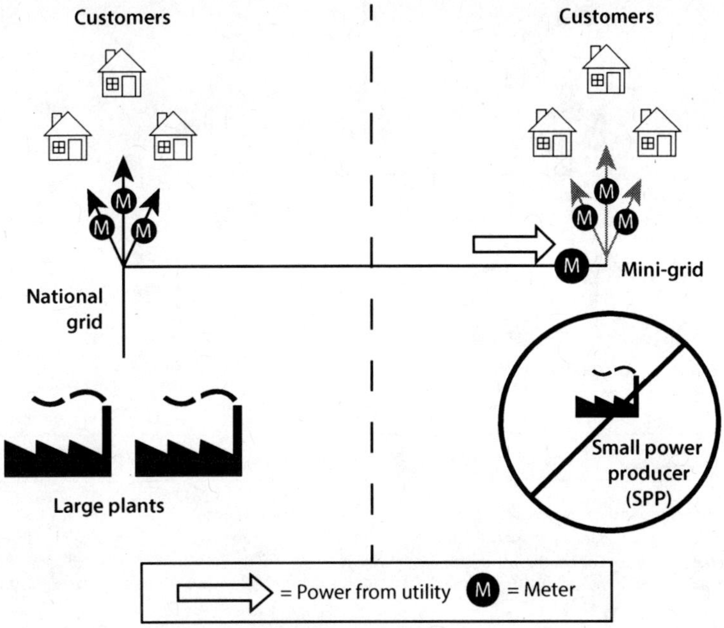

Source: Diagram by Richard Engel and Chris Greacen, 2013. Used with permission.

If regulators allow SPP mini-grids to become SPDs, care must be taken to ensure that the distribution system is built or retrofitted to a standard that can accommodate interconnection with the national grid (Aissa 2011; du Preez 2011). If the SPP developer cuts corners to save money on the cost of installing the initial distribution system, then this system will need to be upgraded when the isolated mini-grid becomes connected to the main grid. Or if upgrading is not feasible, the existing distribution system may even need to be ripped out and totally replaced when the SPP gets connected to the main grid.

SPP Option

When the main grid arrives, some SPPs may prefer to leave the retail sales business and only sell electricity at wholesale to the national grid (see figure 10.3) (Case 4 in table 2.1: grid-connected SPP that sells at wholesale to a utility).

Whether an SPP can make the transition from a mini-grid to the main grid and remain financially viable depends crucially on three factors: the cost of electricity production by the SPP, the feed-in tariff (FIT) that the SPP now connected to the main grid will receive for sales to the national utility, and the capacity factor at which the SPP will be able to operate.

Figure 10.3 SPP Option

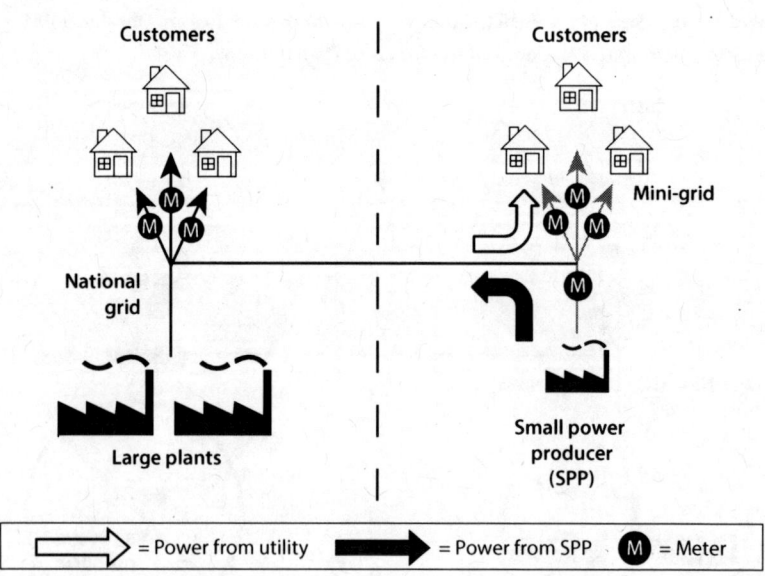

Source: Diagram by Richard Engel and Chris Greacen, 2013. Used with permission.
Note: The SPP generator interconnects with the main grid, becoming another power plant on the grid. The utility takes over distribution of electricity to retail customers. The arrows in this case indicate contracted power flow, not necessarily the flow of electrons. In this case, the SPP is only selling electricity at wholesale to the main grid. It may be the case that some electrons actually flow to the village customers, but this does not really matter—all that matters is that the electricity injected into the grid by the SPP offsets an equal amount that the national grid's other plants would have had to produce.

For some SPP generators such as small hydropower projects, the cost of electricity production can be sufficiently low to compete with conventional generation on the main grid, especially after the bank loans have been paid off. For example, one small hydropower project in southwestern Tanzania is being built to provide electricity to complement the Tanzania Electric Supply Company's (TANESCO's) existing diesel-powered mini-grid. Recall that Tanzania has a dual FIT system: one tariff for sales to TANESCO on one of its existing mini-grids and a second tariff for sales to TANESCO on the main national grid. Until the national grid expands into this area, the SPP generator will receive the mini-grid tariff, which in 2012 was a lucrative 480 T Sh/kilowatt-hour (kWh) ($0.305/kWh) (EWURA 2012c). But once the grid arrives, the project will sell electricity at the much lower national tariff of about 152 T Sh/kWh ($0.097/kWh) (EWURA 2012b).

In countries, such as Tanzania, where SPPs are paid avoided-cost-based tariffs (defined and discussed in chapter 7), only small hydro and some biomass projects with captive agro-industrial waste fuel supplies are likely to be commercially viable if they are connected to the main grid. In countries such as Thailand, with technology-based FITs, more expensive technologies such as solar or wind power may be viable as well, even in an on-grid capacity.

The issue of the capacity factor plays an important role, which may mitigate lower tariffs for on-grid operation. When an SPP is generating electricity for an isolated mini-grid it is only able to sell as much electricity as is being demanded on the mini-grid at any moment. Typically in the middle of the night the demand on a mini-grid system is low, as most residents in the local community are asleep and their appliances are turned off. For a grid-connected SPP, however, the national grid is generally able to absorb full power output from an SPP 24 hours a day. The ability to operate at a greater capacity factor means many more kilowatt-hours are sold, helping to partially or even fully offset the impact of lower tariffs.

Combined SPP and SPD Option

In the combined SPD and SPP option, the SPP simultaneously plays both roles discussed in the two preceding options: it sells electricity to retail customers, as well as generates electricity for sale to the national grid (a combination of cases 3 and 4 in table 2.1, chapter 2). This option should be encouraged in countries that face shortages of generation capacity on their main grids while also facing the challenge of extending rural electrification services to a greater portion of the population; or in areas where the local distribution grid is weak and brownouts or blackouts are common (see figure 10.4).

Electricity sold to retail customers can come from either the SPP generator or as electricity purchased wholesale from the national utility. In this regard, there is a wide spectrum of possibilities, and the position of a given project on the spectrum may shift over time. To take a real-world example, the 4 megawatt (MW) Mwenga hydropower project in Tanzania (commissioned in October 2012) generates most of its electricity for sale to the grid but also supplies

Figure 10.4 Combination SPP and Distributor Option

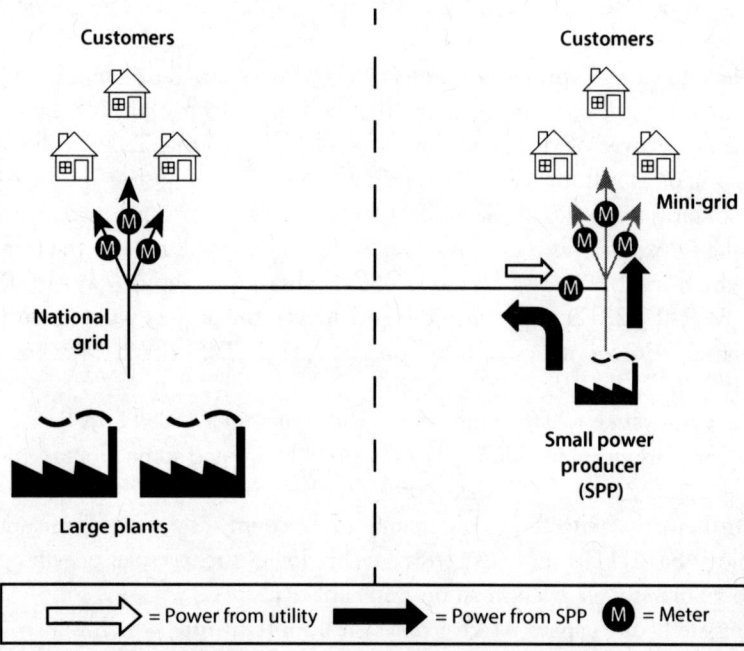

Source: Diagram by Richard Engel and Chris Greacen, 2013. Used with permission.
Note: The arrow indicating power from utility is drawn smaller, representing the fact that electricity purchased from the national grid is reduced (because of the power flowing directly to mini-grid customers from the SPP). In some periods of the day electricity flow from the national grid may cease completely with all local customer load being covered by the SPP.

essentially 100 percent of the electricity used by retail customers. As the project expands from its initial 900 customers to include 4,000 (expected) retail customers spread over 16 villages, the portion of electricity consumed by retail customers will increase and the portion sold to the grid will decrease. This combination SPP/SPD plans to purchase wholesale electricity from the grid only when the hydropower project is shut down for maintenance, or (more likely) for reconnection times of short duration (15 minutes or less) when a disturbance on the national utility grid forces the hydropower project to trip offline. At the other end of the spectrum, diesel generators—inexpensive to own but expensive to operate—provide backup power in villages or for crucial loads (hospitals, mobile-phone repeaters, military bases) that require very reliable electricity or that suffer from frequent grid blackouts.

The combination SPP and SPD has been proposed in India in two recent papers by ABPS (2011) and the World Bank (2011). ABPS refers to this model as "decentralized generation with grid support" and the World Bank report describes it as the "distributed generation and supply model."

Buyout Option

In the buyout option, the utility purchases and operates the existing mini-grid distribution network and possibly the generator (see figure 10.5).

Figure 10.5 Buyout Option

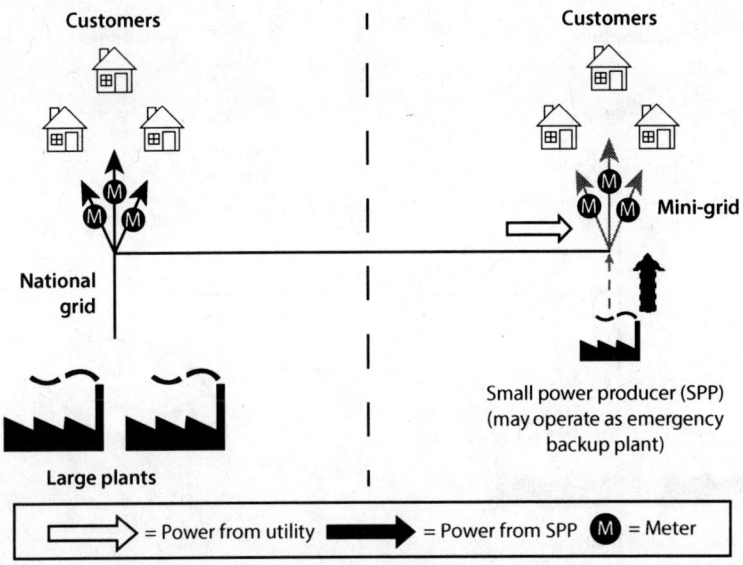

Source: Diagram by Richard Engel and Chris Greacen, 2013. Used with permission.

This option may make sense if the following criteria are met:

- The mini-grid is built to engineering standards comparable to the standards used in the utility's own distribution assets.
- The utility can marshal the human resources necessary to operate the newly acquired mini-grid including bill collection, new hookups, maintenance, and dispute resolution.

In the buyout option, which assets (distribution system assets only or distribution system and generator assets) are to be sold at what cost would need to be worked out on a case-by-case basis. In principle, the sales price would reflect the depreciated value of the assets that remain serviceable. A further consideration in determining a sales price is whether, and to what extent, the mini-grid and/or generator were originally subsidized or built with grant funding.

Assets Abandoned Option

The final option (not likely to be attractive to the SPP) is that the SPP mini-grid assets are scrapped or moved to another location and the national utility treats the area like a greenfield site, building a new distribution system (see figure 10.6). If the quality of the mini-grid is below standard and it is not cost-effective to upgrade it, this may be the only option available. In Thailand, for example, of 59 village-scale micro-hydropower mini-grids installed after 1983, 34 were abandoned by 2004. The vast majority of these communities (31 villages) were connected to the main grid and received completely new distribution systems

From the Bottom Up • http://dx.doi.org/10.1596/978-1-4648-0093-1

Figure 10.6 Abandonment Option

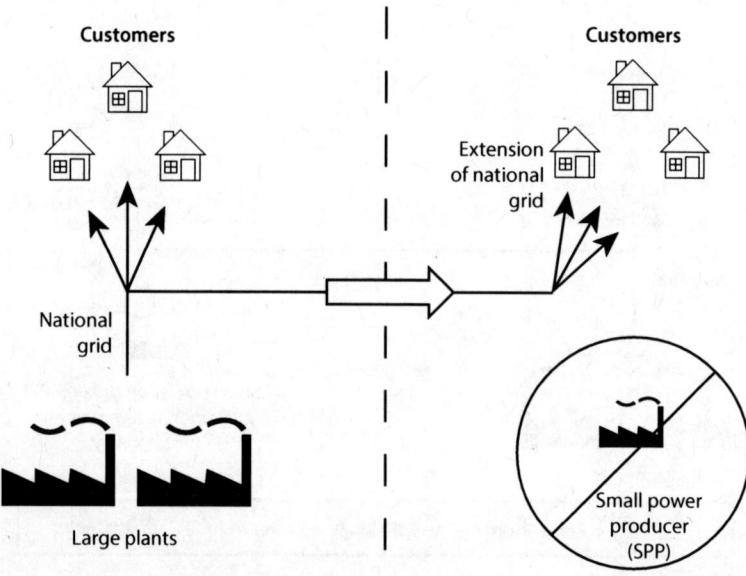

Source: Diagram by Richard Engel and Chris Greacen, 2013. Used with permission.

from the country's rural distribution utility, the Provincial Electricity Authority (PEA) (Greacen 2004, 53).

The Fate of Physical Assets in Each Option

The fate of the generator and mini-grid distribution assets under each option are summarized in table 10.1.

All options except the buyout and abandonment options cannot be unilateral decisions of an existing isolated mini-grid operator. They should require the approval of the regulator and the party who will purchase the electricity (either the local community or the national grid operator). In contrast, the buyout or abandonment options need not require community approval. These last two options can be a unilateral decision of the existing SPP operator because these options imply newly provided service by the national or regional utility—a level of service that is the default option for most other electrified areas in the country. However, if the SPP has operated under a license issued by the regulator, the regulator would normally have to give its approval to both the buyout and abandonment options.

Creating a Viable SPD Option

If a regulator believes that SPDs make sense, there are important details to consider. These include consideration of technical standards for mini-grids, and (if mini-grids are expected to charge uniform national tariffs) the availability of subsidies and how to deliver them.

Table 10.1 Use of Generator and Mini-Grid Distribution Assets in Each Option

Option	Generator	Mini-grid
Small power distributor (SPD)	Scrapped or relocated	Used by SPD to resell electricity purchased at wholesale
Small power producer (SPP)	Used to sell electricity to main grid	Either no longer used, or used by the utility to sell electricity to retail customers
SPP and SPD	Produces electricity for retail sales and to sell to the main grid and/or used as backup supply source	Used to supply electricity to the SPP's/SPD's retail customers
Buyout	No longer used or sold to utility	No longer used or sold to utility
Abandonedment	Scrapped or relocated	Scrapped

Mini-Grid Technical Standards

If a formerly isolated mini-grid is going to interconnect with the national grid (SPD option), or is going to sell its assets to the utility (buyout option), then the distribution system in question will have to meet the national grid's technical standards.

An extremely helpful step that utilities can perform (perhaps requiring encouragement or a formal order from the regulator) is to maintain an updated copy of relevant rural electrification technical standards on the utility's website. An example of this is Vietnam's "Technical Regulations for Rural Electrification/Electric Network" (The Socialist Republic of Vietnam, Ministry of Industry 2006). Making national rural electrification standards accessible to SPP developers makes it easier for developers to elect to build distribution systems that are compliant and are suitable for interconnection or resale to the utility.

Even if interconnection with the national grid is not expected, safety and reliability concerns warrant consideration of minimum technical requirements for mini-grids. Examples of these include Sri Lanka's Village Hydro Specifications (ESD/RERED Government of Sri Lanka 1999) and the International Electrotechnical Commission (IEC)'s *82-62257-9: Recommendations for Small Renewable Energy and Hybrid Systems for Rural Electrification* parts 9-1 through 9-4 (IEC 2008).

Key areas covered in relevant standards—whether for the national grid or for isolated mini-grids include: safety distances and protection corridors; construction of medium- and low-voltage distribution lines including consideration of conductor size and composition; proper insulators and line accessories; lightning protection; switching equipment, poles, hardware, pole stays, cable cross-sections, cable layout, cable joints and terminations, and grounding; meters; boxes; and so on.

Subsidies to Implement the Small Power Distributor Option or the Combined Producer and Distributor Option

If the SPP decides to convert itself from an isolated SPP to a connected SPD or to a grid-connected combined SPD and SPP—*and* the government requires that the new SPD sell electricity to its retail customers at the uniform national tariff—the government must provide the SPD with subsidies, either directly or indirectly, so that it can provide its retail customers with the same tariff levels and structures that are given to the retail customers of the national utility.

As discussed in chapter 9, it is common for political authorities to mandate that a national state-owned utility must charge a uniform national tariff. But a private or community-owned SPD will be able to comply with this mandate only if its business is commercially feasible. Commercial feasibility requires a sufficient inflow of money through tariff revenues and/or subsidies that will cover the SPD's costs. Without this minimum inflow of revenues, the uniform national tariff will be an unfunded mandate that cannot be achieved.

When a national utility serves poor rural customers, it has one major advantage that is not available to SPDs—it is able to subsidize the consumption of its poor rural customers through cross-subsidies from its other well-to-do residential customers or from commercial and industrial customers located elsewhere on its system. An SPD may find it more difficult to cross-subsidize the tariffs of its poor household customers from the tariffs of other customers within its service or concession area, because an SPD will have fewer customers and most of these customers are likely to be other poor households. In addition, SPDs are not likely to have many large commercial and industrial customers among their customers who could potentially subsidize poor households. Therefore, if a government requires that SPDs charge the uniform national tariff, it will have to provide subsidy funding from some outside source.

While it is relatively easy to describe subsidies, it is not easy to implement them in practice. Consider the views of the key players on providing subsidies to SPDs.

Minister of energy

Look, it is a political embarrassment for the government to have two villages near each other with widely different tariffs. You cannot have households paying the uniform national electricity tariff in a village connected to the main grid while you have households in a village just a few kilometers away served by a mini-grid operator who are being forced to pay a tariff to the operator that is two to three times higher than the national tariff. This is unfair and it is simply not sustainable. The only solution is to connect all these small isolated mini-grids to the main grid as soon as possible and then charge them the same retail tariff regardless of whether they are supplied by the national utility or some small private or community distributor.

Managing director of the national utility

I will connect to these isolated villages if the government orders me to do so. But it is going to cost a lot of money to build lines out to these isolated communities and we do not have that money. So if this is the government's policy, then either the government or the donors will have to provide us with grants to pay for the cost of constructing these new lines. And then even after we connect the village, we are going to lose money on every kilowatt-hour that we sell to any households in these villages that are eligible to buy electricity at the lifeline tariff. If the SPP wants to become a distributor of our electricity, that's fine as long as we are not asked to subsidize his tariffs. If the government wants us to subsidize these small distributors, then the government should come up with the money that allows me to do this. The government is happy to make promises, especially before an election, but is very forgetful when it is time for the government to pay for these promises.

Existing mini-grid operator

The minister and the national utility seem to have very short memories. I took the risk of building a generation and distribution system in this community when no one else was interested in providing it with electricity. And I have given good service to this community for many years. Obviously, I do not want to delay the arrival of the main grid because this can lower the cost of electricity to the village and provide the village with more hours of electricity each day. But I should be allowed to continue at least as a distributor after the main grid arrives because I know my customers in this village and I can provide better service and with fewer losses than the national utility. And I can only function as a distributor if there is a workable margin between the price that I pay for the bulk power and the price that I can charge when I resell the power to my retail customers.

Customers in a village currently served by an SPP

It is not fair that we have to pay three times as much for electricity than our friends and relatives in other nearby villages that are served by the national utility. They are able to buy electricity on the social tariff and we pay tariffs that are two to three times higher to the operator of the mini-grid. It is also unfair that the connected villagers get electricity for many more hours than we do. We are lucky if we get electricity for 4 to 6 hours at night while they get electricity for 24 hours a day. And certainly once our village finally gets connected to the national grid, we should pay exactly the same tariffs as everyone else in the country who is connected to the national grid. The ultimate indignity would be to get connected to the national grid but still be forced to pay a private operator more for electricity than those who are served by the national utility.

If the country can afford ongoing operational subsidies for SPDs, decision makers will need to consider how to deliver these subsidies.

How to Deliver Operational Subsidies to Small Power Distributors?

There are three principal ways that a government can provide outside operational subsidies to an SPD to enable it to charge the same tariff as the national utility and also provide social or lifeline tariffs to its low-consumption customers. We consider each of these three methods in turn.

Method 1: Funding from the General Budget

Under this method the government provides subsidy funding from its general budget. The basic problem with this approach is that when most African governments make this promise, they simply do not have the money in their budgets to provide such a subsidy on an ongoing basis. And even if the money were available, it would take time and effort to establish administrative mechanisms for delivering the subsidy funds. Moreover, subsidies agreed to by one government administration might be terminated five years later when another government is in power.

Method 2: A Separate Rural Electrification Subsidy Fund

Here the government or some other entity such as the rural electrification agency (REA) or regulator administers a subsidy fund that receives funding from the government and outside donors. Rural electrification funds (REFs) and REAs now exist in more than 15 Sub-Saharan African countries. Most of them were established, in part, to channel subsidy funding from different sources through a single organization or fund. In almost all instances, the subsidy funds have been used to provide capital cost grants to lower connection costs for households that wish to take service from isolated mini-grids or from the national utility. For example, in Tanzania, the REA is willing to provide a grant of $500 for each new connection made in rural areas. Similarly, AMADER, the REA of Mali, typically provides capital cost grants for about 75 percent of the cost of a new connection—on average about $580 per new connection (Adama and Agalassou 2008).

In Peru the government offers initial capital cost subsidies to both grid and off-grid suppliers in rural areas, which is very similar to ones offered by African REAs/REFs. But the Peruvian government also provides both operational and consumption subsidies in addition to initial capital cost subsidies (see box 5.3). The government has established a fund known as FOSE that provides ongoing consumption subsidies to consumers in rural areas that consume less than 100 kWh/month. The funding for the consumption subsidies comes from all residential, commercial, and industrial customers located anywhere in the country with a monthly consumption greater than 100 kWh/month. Specifically, these consumers are required to pay a 2.5 percent surcharge on their monthly bills to fund a discounted social tariff for any residential customers that consume less than 100 kWh/month.

What is the cost of this cross-subsidy for an average residential electricity consumer in Lima? The average monthly consumption of a typical residential customer is about 200 kWh/month at a price of $0.10/kWh. Hence, the typical monthly urban residential bill without the surcharge would be approximately $20.00, and the FOSE surcharge adds about $0.50 to that bill. The 2.5 percent surcharge is also applied to the monthly bills of commercial and industrial customers. Overall, about 2 million Peruvian electricity consumers pay the 2.5 percent surcharge, which produces a fund of about $36 million per year for subsidizing the monthly bills of 3 million low-consumption customers (Revolo Acevedo 2009). Based on the Peruvian model, similar consumption funds have been established in Brazil, Bolivia, and Guatemala.

Method 3: Discounted Bulk-Supply Tariffs

The government or the regulator requires that the national utility sell wholesale or bulk power to the SPD at a discounted price that will allow the SPD to charge the uniform national tariff (including any social tariff components) to its retail customers and still remain commercially viable. The justification for the discounted bulk-supply tariff is the SPD will be serving many customers at nonremunerative rates under a mandatory social or lifeline tariff and that its distribution costs will be higher because its customers will typically be dispersed over a larger geographic area.

This is the most common subsidy mechanism for three reasons. First, it is administratively simple as its only administrative requirement is that the national utility lower its bulk-supply tariff for some or all wholesale customers. Second, it is a hidden subsidy and does not appear in the government budget nor does it appear as a line item on the monthly bills of other electricity consumers. Third, it does not require any direct contributions from the national government.

It does have drawbacks, however. One is that the subsidy structure does not match the cost structure of SPDs. SPDs have a low share of variable cost and a high share of fixed cost, but the subsidies are tied to sales. Under this structure, the SPD is incentivized to spread its fixed costs over sales of as many kilowatt-hours as possible, which does not work well with energy efficiency or energy conservation.

The one entity that will be most aware of the subsidy is the national utility because it is being forced to sell bulk power at a price below its actual costs. In a sense, the national government is using the national utility as an agent both to provide and deliver a subsidy. The national utility can be made whole if it is allowed to charge its other customers a higher tariff to compensate for the subsidized power that it must provide to SPDs or if it receives some explicit external payment from the government.

A government can always order the national utility to provide such a bulk power supply subsidy, but the utility could find subtle ways to sabotage the implementation of the order. Therefore, it is best to provide the national utility with positive economic incentives to comply with the directive. One way to do this is to require that the national utility periodically and publicly report on the amount of the bulk-supply subsidy that it is providing SPDs and then allow the national utility to include an automatic adjustment component in its general retail tariffs that recovers the cost of the discounts that it has been required to provide. And if the national utility is also selling electricity to retail customers on its own isolated mini-grids at the uniform national tariff, these below-cost subsidies should also be reported and recovered through an automatic adjustment clause (AAC) in its general retail tariffs. This highlights the existence of the subsidies and helps to ensure that the national utility will be "made whole."

In Thailand, where there is a long-standing policy of a uniform national retail tariff, a similar mechanism was used on a national scale. The rural PEA was allowed to purchase electricity from the Electricity Generation Authority of Thailand (EGAT) at a bulk-supply tariff that was 30 percent lower than the tariff paid by the Metropolitan Electricity Authority (MEA), which serves the Bangkok metropolitan area (Barnes and Tuntivate 2009). This helped address the problem of the PEA's distribution costs being much higher per unit of revenue than the MEA's.

The Importance of the Distribution Margin

If this last subsidy method is going to be successful, it requires an adequate differential between the price at which the SPD purchases power using the bulk-supply tariff and the average price at which the SPD resells this power at retail. This differential is usually referred to as the distribution margin. It must be large enough to cover the SPD's distribution costs (annual capital costs of its medium- and low-voltage network and operation and maintenance [O&M] costs to operate these networks) and the tariff discount provided to lifeline tariff customers (if there is no separate subsidy mechanism for these low-consumption customers). If the distribution margin is too small, the SPD will be caught in a price squeeze that may force it into commercial insolvency; if the margin is too large, the SPD will earn unnecessarily high profits. Table 10.2 provides some preliminary estimates of distribution margins that exist in four Asian countries.

Bangladesh

In Bangladesh the bulk-supply price is 3.70 cents, and the average retail price allowed by the national electricity regulator is 3.94 cents for the country's 70 rural electric cooperatives. This leads to a very small distribution margin of 0.24 cents (about one-quarter of 1 cent), which reflects the fact that the regulator has decided to keep the allowed retail tariffs of the cooperatives within a relatively narrow band. But this decision fails to recognize that the cost and load characteristics of the 70 cooperatives are quite different, and the resulting average differential of less than one U.S. cent is simply too small to cover the actual distribution expenses of most of them. As a consequence, many of the 70 cooperatives are commercially insolvent (NRECA 2005; Van Couvering 2011).

Table 10.2 Bulk-Supply and Retail Tariffs of Rural Distribution Entities in Asia

Country	Bulk-supply tariff (U.S. cents/kWh)	Retail sale price (U.S. cents/kWh)	Distribution margin (U.S. cents/kWh)
Bangladesh	3.7 (0–100 kWh)	3.94	0.24
Vietnam	2.4 (0–50 kWh)	3.4	1.0
	6.4 (51–200 kWh)	8.5	0.1
Nepal	4.9	5.5	0.6
Cambodia	13.55	28.0	14.4

Sources: Authors' estimates based on van Couvering 2011; Rekhani 2011, 2012; Van Tien and Arizu 2011; NRECA 2012.
Note: kWh = kilowatt-hour.

Vietnam

The situation is different in Vietnam. Vietnam has proposed implementing a bulk-supply tariff system, where the tariff will vary depending on the consumption levels of retail customers served by the local distribution entity. For example, if the distribution entity is serving a low-consumption customer (that is, 0–50 kWh/month), the distribution entity will be allowed to purchase bulk power at 2.4 cents/kWh and sell it to these low-consumption customers at 3.4 cents/kWh, which allows for a distribution margin of 1 cent. If it is serving retail customers in the 51–200 kWh/month range, it will pay a higher bulk-supply tariff of 6.4 cents/kWh though the allowed retail price will also increase to 8.5 cents/kWh, which provides for a distribution margin of 2 cents. To implement such a system requires having accurate information on the composition of the distribution utility's customers (Van Tien and Arizu 2011), but it is unclear whether this proposed system will ever be implemented. There has been growing customer dissatisfaction with the performance of many existing distribution entities and growing political pressure for a uniform national tariff. As a consequence, the Vietnamese government now seems to favor takeover of distribution entities by the country's provincial power companies.

Nepal

Nepal, like Bangladesh, has a relatively small distribution margin. The 266 community-owned distribution systems buy bulk power from the Nepal Electricity Authority (NEA) at 4.9 cents and are allowed to resell it at 5.5 cents. Hence, the differential of 0.60 cents, although larger than Bangladesh's 0.24 U.S. cent differential, is still relatively small. But the small differential in Nepal may be viable because of one important factual difference. The physical distribution facilities in these villages have been 80 percent funded by the government. This means, in effect, that the community-owned distribution systems are not paying the full capital costs of their distribution system; instead, they are paying a small annual lease payment. This, in turn, may allow them to be financially viable with a small distribution margin (Shrestha 2012).

Cambodia

Cambodia is unique because more than 80 of the SPDs that were functioning at the end of 2012 previously operated as SPPs before their conversion. Three features of the Cambodian tariff system have helped to achieve this successful conversion. The first is that the bulk-supply tariff at which the national utility sells power to the SPDs varies by region of the country and it appears that the bulk-supply tariffs are fully cost-reflective (that is, there is no discounting). For example, the bulk-supply tariff is higher in the northern Highlands and is lower in the rural areas surrounding the capital. A second feature of the Cambodian tariff system, which is not seen in many countries, is that retail tariffs in rural areas are generally higher than the retail tariffs in urban areas. Hence, there are regional tariff differentials and therefore no uniform national tariff. The third is that the allowed distribution margin is generous. Recent statistics from Cambodia show distribution margins of about 14 cents (Rekhani 2011, 2012; Chanthan 2013). This is the highest margin that we have found in Asia. It seems unlikely that these three tariff features could be replicated in most African countries.

Brazil and Peru

Throughout Latin America, distribution margins are routinely calculated by Latin American electricity regulators when retail tariffs are reset every four to five years for distribution companies and these numbers are publicly available. The Latin American numbers are particularly interesting because the allowed distribution margins are calculated based on the density of the distribution enterprise's service area and the composition of customers served. With the exception of Cambodia, Asian distribution margins are considerably lower than those in Brazil. Table 10.3 shows distribution margins for 22 SPDs in Brazil, together with data on number of customers, annual sales, and service territory area. In Brazil

Table 10.3 Distribution Margin of Small Power Developers in Brazil, Sorted by Number of Customers

Company	Area (km²)	Customers	Sales (GWh)	Distribution margin (U.S. cents/kWh)
CAIUÁ-D, Caiuá Distribuição de Energia S/A	9,149	194,000	1,083	3.7
CLFSC, Companhia Luz e Força Santa Cruz	11,850	166,000	767	5.7
EBO, Energisa Borborema—Distribuidora de Energia S.A.	1,984	151,000	551	4.3
EDEVP, Vale de Paranapanema	11,770	147,000	642	5.6
EEB, Empresa Elétrica Bragantina S/A	3,453	110,700	568	5.9
SULGIPE, Companhia Sul Sergipana de Eletricidade	5,946	110,600	251	7.3
CNEE, Companhia Nacional de Energia Elétrica	4,500	90,300	477	5.0
ENF, Energisa Nova Friburgo—Distribuidora de Energia S.A.	933	86,700	287	8.5
DMEPC, Departamento Municipal de Eletricidade de Poços de Caldas	534	60,000	1,654	1.5
CFLO, Companhia Força e Luz do Oeste	1,200	45,000	239	4.2
CLFM, Companhia Luz e Força Mococa	1,844	38,000	183	6.5
COCEL, Companhia Campolarguense de Energia	1,360	34,600	186.7	5.1
CJE, Companhia Jaguari de Energia	252	29,000	505	2.4
COOPERALIANÇA, Cooperativa Aliança	569	29,000	155	4.1
IENERGIA, Iguaçu Distribuidora de Energia Elétrica Ltda	1,252	28,000	198.4	4.9
DEMEI, Departamento Municipal de Energia de Ijuí	45	26,000	96.7	6.7
HIDROPAN, Hidreletrica Panambi	—	14,000	85.4	5.0
UHENPAL, Usina Hidroelétrica Nova Palma Ltda.	—	13,700	58.6	5.8
MUX, Energia, Muxfeldt Marin and Cia. Ltda	—	8,000	50	4.6
FORCEL, Força e Luz Coronel Vivida Ltda	280	5,900	33	7.3
EFLUL, Empresa Força e Luz Urussanga Ltda	237	4,800	73	3.6
EFLJC, Empresa Forca e Luz Joao Cessa	253	2,300	11.1	7.2

Source: Calculations made by Pedro Antmann (World Bank) based on data available on the website of ANEEL (the Brazilian national electricity regulator).

Note: GWh = gigawatt-hours; km² = square kilometers; kWh = kilowatt-hours; — = not available.

distribution margins vary from 1.5 cents/kWh (at an urban utility with 60,000 customers and very high sales of 1,654 gigawatt-hours (GWh)/year) to 8.5 cents/kWh at a more rural utility with only 287 GWh of sales per year.

The last five utilities in table 10.3 are in the size range (if still somewhat larger) than the SPDs that are likely to exist in rural Africa. Distribution margins at these smaller utilities in Brazil vary from 3.6 cents to 7.2 cents, consistent with the rough rule of thumb that in Latin America distribution margins need to be around 4 cents/kWh for SPDs with thousands (but not tens of thousands) of customers. In Peru the necessary distribution margin for a completely rural distribution entity known as a Sector 5 entity was recently calculated as 8.4 cents. For the rural systems in this Sector 5 category, the median number of customers was 10,254 with a median density of 36.1 customers per kilometer of low-voltage lines and a median consumption of 32.3 kWh/month. While further empirical research is required, we think that a comparable SPD operating in most rural areas of Africa will need a distribution margin of at least 4–5 cents. It could be argued that the distribution margin will need to be higher because most distribution entities in Africa will have fewer customers, and probably will not be able to enjoy the economies of scale that are available to the Brazilian and Peruvian rural distributors. But the African utilities may benefit from lower labor costs. *A priori*, it is hard to estimate the overall effect on African distribution margins of these cost-raising and cost-lowering factors. This is an area where more Africa-specific cost analysis is clearly needed.

Key Recommendation

If the government formally or informally requires SPD retail tariffs to be set at uniform national levels, then most SPDs will need subsidies. The subsidies can come from the government's general budget, funding from rural energy agencies, or through mandated discounts on the price paid by SPDs for wholesale power purchases. Minimum distribution margins for SPDs may vary considerably, but based on international experience we estimate that a minimum distribution margin of 4–5 cents/kWh will be required to achieve commercial viability for most rural SPDs in Africa.

Transitioning from an Isolated to a Main Grid SPP: Technical Issues

Whereas the key thorny issues identified above with SPDs relate to the issues of subsidies and tariffs, these issues are more straightforward in the SPP option. With the SPP option, the go versus no-go tariff decision is reduced to a simple commercial calculus: can the SPP generate at a cost sufficiently below the offered FIT (see chapter 7) to make the business worthwhile? While the tariff may be considerably lower when the SPP switches to selling electricity to the national grid (as is true in Tanzania), this lower received price will be somewhat mitigated by the much higher capacity factor that a 24-hour-a-day grid connection affords, compared to the daily load fluctuations and low load factor of most village mini-grid cumulative loads (see figure 8.2). In other words, the SPP may get a lower price per kilowatt-hour sold onto the national grid but it will be able to sell many more kilowatt-hours.

But the technical issues for transitioning from off-grid to the main-grid connection are not trivial. In the remainder of this chapter, we consider the technical transition from stand-alone mini-grid to a main-grid-connected SPP.

How Is Control of an Isolated System Different from Control of a Main-Grid-Connected System?

Electrically, a small generator connected to the national grid is very different than an isolated generator powering a mini-grid. *The biggest difference lies in who is determining the frequency.*[4] Frequency is determined by the rotational speed of the generator shaft; faster rotation generates a higher frequency. Just as a car's motor speed depends on the balance between the fuel going to the motor and whether the car is going uphill or downhill, a hydropower generator's frequency depends on the balance between how much water is flowing past the turbine and the amount of electrical load. With no load, the generator will "freewheel," and run at a very high number of revolutions per minute. If load is excessive, the generator bogs down (spins slower than normal) and the frequency drops below the standard.

In the case of an isolated mini-grid, the generator must maintain frequency control because there is no option of frequency control by any other means.[5] In an isolated micro-hydropower facility, control of frequency is accomplished in one of two ways. One method uses a hydromechanical control that incrementally opens the water supply valve increasing water flow the moment that it detects a drop in frequency, and incrementally closes the valve when it detects that frequency is too high. This feedback loop keeps frequency fairly constant under most conditions.[6] Similarly, renewable energy generators with internal combustion engines modulate the engine throttle in response to slight shifts in frequency, and steam turbines modulate the flow of steam from the boiler to the turbine to keep frequency constant.

The second method used in hydropower mini-grids (where the fuel is "free") is for an electronic controller to manage the load on the generator by varying the amount of power that is wasted in a resistive dump load safely installed to heat air or immersed in water. By adding progressively higher loads, the generator can be slowed until it reaches the exact number of revolutions per minute for alternating current. If the village starts using more power, the increased load causes an incremental reduction in frequency, to which the controller quickly reacts by reducing load. At all times, the controller modulates the dump load to essentially keeping the total load (village + dump load) constant.

Connected to main grid. When a micro-hydropower generator is connected to the main grid, it becomes a very small part of a much larger network of much larger generators all spinning in lockstep. In this case, the micro-hydro does not have to regulate its own frequency. As long as it is connected to the grid, it *will* spin at the grid frequency, which is set by very big generators operating on the grid.[7]

The Technical Requirements of Shifting from Isolated to Grid-Connected Operation

For a formerly isolated mini-grid to operate in a main-grid-connected mode, the SPP must be reconfigured in ways that accomplish the following tasks:

- Remove or disable equipment that modulates fuel supply (for example, water flow in a hydropower project) or that conducts load diversion in response to frequency variations.
- Connect safely to the grid (this is generally an issue of connecting at the correct frequency and phase).
- Inject electricity of sufficient quality (appropriate power factor, low total harmonic distortion).
- Disconnect quickly and safely from the grid in appropriate circumstances (when a disturbance is detected on the grid) and reconnect when it is safe to do so.

With these changes, the grid-connected SPP relies on other (generally much larger) generators in the network to maintain frequency regulation. But if frequency or voltage on the grid network at the site of interconnection deviate sufficiently from agreed-upon standards, the SPP is programmed to disconnect from the grid. The relay devices that measure these conditions and trigger disconnection are carefully chosen and calibrated to match the specific context determined by their location on the network, as well as the electrical characteristics of surrounding generators and loads. The relay's primary function is to prevent islanding, as described below.

Islanding

Islanding refers to the phenomenon in which a portion of the grid becomes temporarily isolated from the main grid, but remains energized by its own distributed generation (DG) resource(s). This is normally considered an undesirable condition, as it can present a hazard to line workers who might assume the lines are not energized during a failure of the central grid and denies the central grid control over power quality.

Intentional Islanding

But there are circumstances in which islanding may be desired. In the case of a mini-grid being integrated with a central grid that has historically shown itself to be prone to reliability problems, the mini-grid interconnection may be designed in a way that permits the mini-grid to intentionally island, that is, to continue operating autonomously and provide uninterrupted service to local customers during outages on the main grid. Policy regarding grid interconnection of previously autonomous mini-grids should allow for maintaining future capability to operate autonomously, provided this can be done safely.

In order to accomplish intentional islanding, the control system needs to be able to quickly disconnect from the main grid, and switch over immediately from the main grid controlling the frequency to a regime in which the frequency control is handled by the SPP generator itself.

Moreover, when the controls sense that stable electricity of proper frequency and voltage has been restored to the national grid, the intentionally islanded SPP should automatically (or with minimal operator effort) resynchronize with the main grid. Islanding, intentional islanding, and relay controls for connecting mini-grids to the main grid are discussed in more technical detail in Greacen, Engel, and Quetchenbach (2013).

Key Recommendation

In the SPP option, regulators and utilities must address technical issues related to the transition from off-grid to grid-connected power generation, especially related to the control of the generator's frequency, and the transition between these two control regimes if the SPP chooses to maintain the ability to intentionally island.

Notes

1. While we do not believe that customers should have unequivocal veto rights on the fate of a mini-grid when the main grid arrives, grievances should be taken into consideration by regulators.

2. However, the obligation to connect should not be imposed on mini- or micro-grids with distribution systems that are technically incompatible with the national utility's grid. For example, in several African countries, developers have built shared solar micro-grids that operate on direct current (DC). The wires and transformers are built to operate on direct current and are technically incompatible with an alternating current (AC) system. Therefore, if the customers of the micro-grid wanted AC power, a new AC distribution system would have to be built. It would be unwise to require that all systems must from Day 1 build to the distribution standards of the national utility, because that would effectively preclude developers from pursuing transitional AC or DC systems that could serve some of the needs of rural households at a low initial capital cost.

3. A DNO is another term for a distribution utility.

4. Most household appliances and motors run on either 50 hertz (Hz) or 60 Hz (depending on what country the grid is located in), as do the major grids that interconnect large generating stations. Frequency on well-functioning national grids seldom deviates more than 0.5 Hz from the country's standard. On isolated mini-grids frequency deviation of several Hz is not uncommon.

5. This assumes that the SPP is the only generator operating on the isolated mini-grid. But in a number of African countries, the SPP may be added to an existing mini-grid that is being supplied with electricity from a diesel generator operated by the national utility (Case 2: isolated SPP that sells at wholesale to a utility). In this situation, there will be a need to coordinate the operations of the two generators to maintain the target frequency on the mini-grid. The technical issues associated with operating two generators, each with a separate owner, on a single mini-grid are discussed in chapter 8.

6. Problems occur when large loads (for example, welding equipment or large motors) are suddenly turned on or off. In this case, electromechanical frequency regulation devices cannot open or shut valves quickly enough and sometimes overcorrect, creating oscillations in frequency and voltage.

7. To use a metaphorical example, an isolated mini-grid is like someone walking down a train track by himself. He can walk fast or slow, run, or stop. A system connected to the main grid is like someone walking down a railroad track—while tied to a slowly moving freight train. He can walk the same speed as the train, or he can push against the train to try to make it go faster (putting energy into the system), or he can drag his feet to try to slow down the train (extracting energy from the train). But no matter what he does, he is not going to affect the speed of the freight train (the frequency of the main grid) very much because his capacity to inject power into the system is small compared to that of the freight train.

References

ABPS Infrastructure Advisory Pvt. Ltd. 2011. "Policy and Regulatory Interventions to Support Community-Level Off-Grid Projects." New Delhi and Mumbai, November. http://www.forumofregulators.gov.in/Data/Reports/CWF%20Off-grid%20final%20 report%20nov%202011_Latest_feb2012.pdf.

Adama, Sissoko, and Alassane Agalassou. 2008. "Mali's Rural Electrification Fund." Presentation at Sustainable Development Week, World Bank, Washington, DC, February. http://siteresources.worldbank.org/INTENERGY2/Resources/presentation8 .pdf.

Aissa, Moncef. 2011. "Techniques to Reduce Costs of Rural Distribution Networks in Tunisia." Presentation at Africa Electrification Initiative Practitioners Workshop, Dakar, Senegal, November.

ARE (Alliance for Rural Electrification). 2011. *Hybrid Mini-Grids for Rural Electrification: Lessons Learned.* Brussels. http://www.ruralelec.org/fileadmin/DATA/Documents/06 _Publications/Position_papers/ARE_Mini-grids_-_Full_version.pdf.

Barnes, Douglas F., and Voravate Tuntivate. 2009. "The Challenge of Grid Rural Electrification: Experience of Successful Programs." Presentation at the AEI Practitioners Workshop, Maputo, Mozambique, June 9. http://siteresources.worldbank .org/EXTAFRREGTOPENERGY/Resources/717305-1264695610003/6743444 -1268073442212/1.1.Overview_bestpractice_institutional_issues.pdf.

Chanthan, Ky. 2013. Personal communication. February.

Chowdhury, Nazmul Hossain. 2009. "Rural Electrification: Bangladesh Experience." Presentation at the AEI Practitioners Workshop, Maputo, Mozambique, June 9. http:// siteresources.worldbank.org/EXTAFRREGTOPENERGY/Resources/717305

-1264695610003/6743444-1268073476416/3.2.Rural_Electrification_Cooperatives _Bangladesh2.pdf.

Dixit, Shantanu. 2012. "Powering 1.2 Billion People: The Case of India's Access Efforts." Presentation at World Bank Energy Days 2012, Washington, DC, February 23.

du Preez, Jaap. 2011. "Design of Low-Cost Options for Distribution Networks in South Africa." Presentation at Africa Electrification Initiative (AEI) Practitioners Workshop, Dakar, Senegal, November.

ESD/RERED Government of Sri Lanka. 1999. "Village Hydro Specifications Sri Lanka: Line Distribution." http://www.energyservices.lk/pdf/techspecs/vh_w_b/line.pdf.

EWURA (Tanzanian Energy and Water Utilities Regulatory Authority). 2012a. "The Electricity (Development of Small Power Projects) Rules." Proposed for Public Consultation, Dar es Salaam, Tanzania.

———. 2012b. *Detailed Tariff Calculations for Year 2012 for the Sale of Electricity to the Main Grid in Tanzania under Standardized Small Power Purchase Agreements in Tanzania.* Dar es Salaam, Tanzania. http://www.ewura.go.tz/pdf/SPPT/2012/2012%20 SPPT%20Calculation%20for%20Main%20Grid.pdf.

———. 2012c. "Detailed Tariff Calculations for Year 2012 for the Sale of Electricity to the Mini-Grids in Tanzania under Standardized Small Power Purchase Agreements in Tanzania." Dar es Salaam, Tanzania. http://www.ewura.go.tz/pdf/SPPT/2012/2012%20 SPPT%20Calculation%20for%20Mini-Grid.pdf.

———. 2013. The Electricity (Development of Small Power Projects) Rules, 2013. Proposed for Public Consultation, Dar es Salaam, Tanzania, June.

Government of India. 2003. *The Electricity Act, 2003.* http://guj-epd.gov.in/extra_no_46 .pdf.

Greacen, Chris. 2004. "The Marginalization of 'Small Is Beautiful': Micro-hydroelectricity, Common Property, and the Politics of Rural Electricity Provision in Thailand." PhD thesis, University of California, Berkeley. http://palangthai.org/docs /GreacenDissertation.pdf.

Greacen, Chris, Richard Engel, and Thomas Quetchenbach. 2013. "A Guide on Grid Interconnection and Island Operation of Mini-Grid Power Systems Up to 200 kW." Document 6224E, Schatz Energy Research Center, Humboldt State University, Arcata, CA, and Lawrence Berkeley National Laboratory, Berkeley, CA.

IEC (International Electrotechnical Commission). 2008. *Recommendations for Small Renewable Energy and Hybrid Systems for Rural Electrification, 82-62257-9.* Geneva, Switzerland: IEC.

Keosela, Loeung. 2013. "Status of the Power Sector in Cambodia." Presentation at the Renewable Energy Workshop, Chiang Mai, Thailand, January 24.

Mahato, Rubeena. 2010. "Power Sharing, Nepali Style." *Nepali Times*, July 23.

Mukherjee, Mohua. 2013. *Lessons Learned from Two Decades of Experience with Private Sector Participation in the Indian Power Sector.* India Power Sector Diagnostic Review, World Bank, Washington, DC.

NRECA (National Rural Electric Cooperative Association). 2005. *Bangladesh Rural Electrification Program at the Crossroads.* Report Submitted to U.S. Agency for International Development, USAID, Washington, DC.

———. 2012. *Affordability Analysis and Options for a Program to Make the Cost of Rural Household Grid Connections Affordable.* Unpublished draft report, June.

Palit, Debajit, and Akanksha Chaurey. 2011. "Off-Grid Rural Electrification Experiences from South Asia: Status and Best Practices." *Energy for Sustainable Development* 15 (3): 266–76.

Rekhani, Badri. 2011. Personal communication. November.

———. 2012. Personal communication. March.

Revolo Acevedo, Miguel. 2009. "Mechanism of Subsidies Applied in Peru." Presentation at the AEI Practitioners Workshop, Maputo, Mozambique, June 9. http://siteresources .worldbank.org/EXTAFRREGTOPENERGY/Resources/717305-1264695610003 /6743444-1268073611861/11.3Mechanism_subsidies_applied_in_Peru.pdf.

Shrestha, Binod. 2012. Personal communication. September.

Socialist Republic of Vietnam, Ministry of Industry. 2006. *Technical Regulations for Rural Electrification/Electric Network.* http://ppp.worldbank.org/public-private-partnership /sites/ppp.worldbank.org/files/documents/Vietnam11Technical1standards10Part01 .pdf.

TERI (The Energy and Resources Institute). 2007. *Evaluation of Franchise System in Selected Districts of Assam, Karnataka and Madhya Pradesh.* Draft final report, Energy and Resources Institute, New Delhi, India, April. http://recindia.nic.in/download /Franchisee_Eval_TERI.pdf.

Van Couvering, Jim. 2011. Personal communication. March.

Van Tien, Hung, and Beatriz Arizu. 2011. Personal communication. January.

World Bank. 2011. *One Goal, Two Paths: Achieving Universal Access to Modern Energy in East Asia and the Pacific.* Washington, DC: World Bank. https://openknowledge .worldbank.org/handle/10986/2354.

CHAPTER 11

Final Thoughts

Abstract

In this final chapter of the guide, we offer closing thoughts on the factors necessary for successful small power producer (SPP) programs. We also consider whether or not small SPPs serving isolated mini- or micro-grids should have their prices regulated during an initial period of operation. Finally, we present some recommendations on what countries can do to launch or improve such programs, and what the international development community can do to support the development of SPPs.

What Else Is Required for a Successful SPP Program?

If the only tool you have is a hammer, you tend to see every problem as a nail.

—Abraham Maslow

When one has expertise or experience in a particular area—whether it is engineering, economics, marketing, law, regulation, or another field—there is a natural tendency to define key problems and solutions in terms of one's expertise.[1] It is important to resist this temptation. Though the focus of this guide has been on the ground-level implementation of regulations and policies for small power producers, we recognize, as was stressed in chapter 1, that this constitutes only one component of what is needed to create an enabling environment for successful SPP development. In this final chapter we briefly examine the other key components.

Access to Finance

Historically, the bulk of SPP financing has come from donors or government grants. In Africa in recent years it has not been unusual to see 70–80 percent of the initial capital cost of SPP projects being paid for by donor or government grants. While such grants are critical in a project's early stage, grant capital is often unreliable and unsustainable. Moreover, it is unrealistic to expect that there will ever be enough donor or government funding to support a large-scale ramping up of SPP projects throughout Africa.

For SPPs to bridge the gap from blueprint to scale, they will need to access both debt and equity capital (Koh, Karamchandani, and Katz 2012). And to tap into these funding streams, they will need to demonstrate commercial viability. A recent, hopeful sign in the area of debt capital is that local banks in some African countries have begun to recognize the bankability of well-executed SPP projects, though still with some guarantee support from international donors. In the future, it is conceivable that loans from larger, international banks and institutions could also provide a new source of debt financing for these projects, perhaps at lower interest rates, if there is more certainty about SPPs' revenue flows. But financing from external sources, whether from private sources or international financial institutions, may trigger the need for tighter, more-detailed documents (for example, more comprehensive power-purchase agreements, PPAs) than would be required if the financing is mostly from domestic sources.

On the equity side of finance for SPPs, in addition to traditional equity investors, newer forms of investment are emerging as options for SPPs. Impact-investing firms—venture capital or other investment houses that seek financial returns as well as social or environmental impact returns on their investments[2]—are starting to take an equity stake in SPPs (for example, Bamboo Finance investing in Husk Power in India). The financial returns are often lower than conventional venture capital investments, but impact investment firms also measure returns in quantifiable social and/or environmental returns like households illuminated or tons of carbon dioxide (CO_2) emissions avoided by investing in a mini-grid or household solar lighting enterprise. The impact investment industry is predicted to grow rapidly—from an estimated $4 billion of investments in 2012 to $1 trillion in 10 years (Dichter and others 2013). In parallel, there is growing interest by philanthropies and even government aid agencies to enter this investment space as impact investors seeking financial and social returns: for example, the Shell Foundation with Husk Power, the Acumen Fund with Avani Bio Energy, and FMO (the Dutch development bank) with Clean Energy/Newcom LLC. These organizations often take a partnership role, providing working capital funds, loan guarantees, and management and strategy advice to help develop sustainable businesses. But for SPPs to access these nondonor sources of equity financing at a significant level, they will need to prove their commercial viability.

One problem that equity financing for SPPs faces is the high transaction cost of investing in these companies. Generating deal flow, conducting due diligence, and providing business support to companies in which they invest cost investment firms and philanthropies significant time and money. Another problem is that SPPs require large capital infusions to scale up their activities from a single project to multiple projects throughout a region or country in order to serve a much larger customer base. These investments for expansion are sometimes too expensive for any single investor. To alleviate these two problems, and to bring big money to small projects, the international investment community is working on strategies to aggregate from the bottom up and from the top down.

Working from the bottom up, bundling similar SPP projects into a single asset class or even a single investment, for example, could decrease risk uncertainty and attract larger investors. Strategies for aggregation from the top down include creating funds of funds, where large public and private investors invest in a portfolio of smaller, more specialized funds that already have expertise in a specific industry or geographical area.

A good example of an aggregation strategy was developed by Dr. David Jhirad and his colleagues at Johns Hopkins University, Prayas Energy Group, TARA, and elsewhere. These energy and finance experts propose to aggregate "big money" to reach "small projects" in the following way. A large pool of capital, in the form of grants, loans, and equity would be pooled by a financial institution acting as custodian for the capital, which would be used to establish a sustainable-development "B corporation" (a for-profit company that has been certified as having met certain standards for social and environmental performance, transparency, and accountability).[3] Asset managers at the financial institution would work with the B corporation to establish portfolios of investment-ready sustainable-development projects. The B corporation would then serve as the funding authority and delivery mechanism for these projects. For a diagram of this aggregation strategy, see Jhirad (2013).

Human Capital and Technical Capacity

In chapter 1 we noted that in addition to financial capital (the "seeds"), successful SPP projects also require human capital (the "fertilizer"). Human capital is required from all those involved in the development of an SPP project—including the developer, those financing the project, and regulators and policy makers. Just as different forms of financing will help an SPP grow from blueprint to scale, different forms of human capital are necessary at different stages of a project. Here, traditional concepts of capacity building, which might focus more on general business skills and engineering or technical knowledge, are certainly important, but what is particularly valuable to an SPP is project-specific and problem-targeted assistance at a particular phase of an SPP's development. For example, the Rural Energy Agency in Tanzania has played an important role in capacity building in the launch phase of SPPs— helping developers create bankable business plans and offering site-specific advice. To develop the overall SPP sector in a particular country, donors and national governments have also assisted in capacity building for banks, regulators, and policy makers by helping these entities and individuals understand the economic and regulatory characteristics of the emerging SPP sector in their country.

In the growth phase from a single project to multiple projects, there will be a need for internally driven human capital development to complement the earlier technical assistance received from donors, governments, and rural electrification agencies (REAs). For this scale-up phase, SPP developers need three kinds of knowledge. First, they need detailed knowledge of the local communities and environments into which they plan to expand. Second, they need business and technical knowledge. They must be able to scale up the right technology in the right places at the right time to create a commercially viable energy company. Third, once the equipment is chosen and put into place, the SPP developer and his employees must have the technical capacity to operate, maintain, and repair the equipment. If there is going to be significant scale-up of SPPs, we think that the requisite technical and business expertise is more likely to be acquired if it comes from within, that is, from the SPP's own leadership team. Business models that include franchising options or just-in-time training for employees are likely to be good models for scale-up success, as already witnessed by Omnigrid Micropower Company (OMC) and Husk Power. We think similar models will work in Africa.

Market Data

As with any new business, access to market data will be a key requirement for success in SPP projects. It is also a requirement of donors that are considering grants and banks that are contemplating loans. How many customers an SPP can expect to serve, how much customers are willing to pay for electricity, what the comparable national or regional utility rates are for electricity, and how quickly the main grid is expanding are among the questions that ideally should be answered before investments are made. When data and information are not available, an SPP will need to absorb the costs of collecting its own market data through surveys and observations. This can be a significant cost, both in terms of time and money, so gathering market information and providing access to it are two areas where the development community can intervene to support SPPs. In Tanzania the Rural Energy Agency is providing grants of up to $100,000 per project to gather these data and prepare business plans.

To help collect and provide electricity demand and supply data on a national and regional scale, the International Finance Corporation's (IFC's) Investment Climate Group has been partnering with local and national agencies to gather statistics on the size of and segments within the customer base, customer willingness to pay for electricity services, and current electrification rates and sources (Arias 2013). But past consumption patterns may not always be a good predictor of future consumption. For example, past consumption expenditures on phone communications provided few, if any clues, as to how rapidly mobile phone usage would spread in rural Africa.

The Financial and Operational Viability of the Buying and Backup Utility

For nearly all grid-connected SPP projects, the national or regional utility is a key partner—as a customer or supplier or both. The utility's ability to supply good-quality electricity when needed and its tariffs are key factors for SPPs who rely on the utility for backup power. In addition, a utility's ability to make good on its promises to purchase and pay for electricity from an SPP, either as the SPP's anchor customer or as its only customer, are critical for an SPP to be commercially viable. Finally, in the case in which the main grid finally connects to the SPP's mini-grid, the utility needs to be willing to work with the

SPP to strike a deal that makes business sense for the SPP and its investors and creditors. Chapter 10 highlights some business models for when the two grids connect.

Policy and Regulation

An SPP developer will also need to look at current or expected policies and regulations that will impact the business before deciding to go through with a project. The focus of our guide has been on this component—how policy makers and regulators can help, not hinder, the development of SPP projects. Good policies and regulations can help SPPs to become successful, sustainable enterprises. As a general rule, we think that it is beneficial to use light-handed regulation for SPPs, or in some cases, not to regulate them at all. The next section offers some additional thoughts on the question of whether or not the retail prices of electricity from certain types of SPPs should be regulated.

To Regulate or Not to Regulate?

Do no harm.

—Paraphrased excerpt from the Hippocratic Oath

The Hippocratic Oath taken by medical doctors should also be a fundamental principle for regulators. In this guide, our focus has been on how to create regulatory systems that do good and not harm. In the process, it was necessary to delve into the details of many first-level policy and second-level regulatory decisions that policy makers and regulators must make if they are going to achieve fair and efficient outcomes for customers and project developers, whether private or community owned. But we also recognize that it is very easy to get lost in the details of regulatory rules and processes and end up ignoring what should always be the threshold question: is it necessary to regulate?

This is an especially important question in the case of one type of SPP: a small, privately owned SPP that wishes to supply a previously unserved rural community. *Our starting assumption for this guide is that regulation is not an end in itself.* Instead, regulation is a means to an end. And that end is the goal of bringing reliable, grid-based electricity at the lowest possible cost to unserved rural villages as soon as possible. So the basic decision whether to regulate or not, especially with respect to tariffs, must be judged against this goal.

In making that evaluation, one must always be conscious of the fact that regulation is not without cost. Regulation costs time and money for private and community-based developers who have to comply with the regulator's rules and processes. And when these costs become too high, there is a considerable risk, given the fragile economics of many mini-grid systems, that a developer may give up on a project because it is no longer commercially viable. If this happens, then the villagers, whom the regulator with good intentions was trying to protect from excessively high prices, may lose an opportunity to get grid-based electricity until sometime in the distant future when the national grid finally arrives. Or as one participant at the 2009 AEI Maputo workshop observed, what regulators and other government officials sometimes forget is that "the most expensive electricity is no electricity at all."

Five Reasons to Not Regulate the Retail Prices of Small, Isolated, Rural Mini-Grids

If we start regulating private providers along the lines of a large centralized grid ... we are dooming the off-grid populations to never having anyone interested in serving them.

—Mohua Mukherjee, Senior Energy Specialist, World Bank, 2013

Those who support price deregulation for small, rural providers serving isolated mini- and micro-grids usually present five reasons why price regulation is neither necessary nor desirable. Let us consider each of these in turn.

The first reason is that most successful decentralized rural electrification schemes for isolated mini-grids have involved little or no price regulation. For example, as discussed earlier (box 2.1 in chapter 2), in Cambodia, close to 300 diesel-fired mini-grids have come into existence without any government grants or approvals. Instead, these rural systems developed spontaneously because there were willing buyers and willing sellers in rural villages and hamlets throughout the country. When these isolated mini-grids were created by village entrepreneurs, there was no functioning national regulator in the country. Hence, the deregulation experiment was accidental rather than intentional.

Cambodia is not the only country that has experienced unplanned tariff deregulation. In many African countries, neighbors or businesses sell electricity to other neighbors or businesses without seeking formal government approval. These small, informal electricity suppliers have operated successfully because, in part, they do not show up on the radar screens of the electricity regulator or any other government entity. Similarly, in Sri Lanka more than 250 village electricity consumer societies (small community-owned, hydro-based distribution systems, each serving typically 20 to 80 households) have operated for many years without any price regulation. Unlike Cambodia and Africa, the *de facto* price deregulation in Sri Lanka had a clear legal basis and resulted from the fact that the monthly usage payments of each household were legally interpreted as membership dues rather than as tariff payments. Therefore, there was no legal requirement for regulatory review. The common feature in all three cases is that electricity has been supplied in rural villages without a regulator setting maximum or minimum prices.

The second reason is that substitutes already exist for the electricity that the new mini- or micro-grid operator proposes to supply, and that a new operator knows it will have to offer a better deal to get customers. Households in the village are not forced to purchase from these new electricity suppliers. They can either continue to use the energy sources that they currently have, or they can switch to the new supplier of electricity. Presumably, a household would purchase from the new supplier only if he or she offers lower prices and better-quality energy services than what is currently available to the household from other energy sources such as kerosene lanterns, solar lanterns, batteries, and small diesel generators. The same would be true for small businesses that choose to become customers of the new supplier. Given the presence of these currently existing substitutes, there is no compelling reason why the regulator should second-guess these consumer decisions.

Early anecdotal evidence is that these new suppliers of electricity are offering, as a matter of business strategy, better deals than the options currently available to villagers from other energy sources. For example, in India, OMC is offering a monthly rental fee of Rs 100 ($1.70) for rechargeable lanterns using light-emitting diodes to replace polluting kerosene lanterns that provide inferior light at a current monthly cost of about Rs 180/month ($3.06). In Tanzania Devergy (see below) charges a connection fee of T Sh 20,000 (about $15) to hook up a new household versus the T Sh 178,000 ($111) that is charged by the national utility for a basic connection. Presumably, these lower charges for both connections and energy simply reflect recognition by the new suppliers that they need to increase the number of customers and the average consumption per customer if they want to become commercially viable.

The third reason is based on the assertion that these operators are serving rural electricity markets that are "contestable."[4] The essence of contestability is that there are no significant legal or economic barriers to prevent other suppliers from coming into the market, even when there is currently a single supplier. In other words, if there is low-cost entry and exit, new competitors can easily come into the market to provide fresh competition. The presumption is that the threat of competition may be just as powerful a limiting influence on the prices charged by the existing supplier as the actual entry of new suppliers. In fact, this was the justification explicitly given by the Government of India when it decided to eliminate tariff regulation and licensing for small, new, rural electricity providers in the regulations issued to implement the 2003 Electricity Act (Government of India 2003, Section 14, Proviso 8).

The fourth reason is that a country is likely to benefit if small private operators of mini- and micro-grids are given the chance to experiment with different business models. Around the world, private operators are currently experimenting with or proposing a variety of business models for supplying electricity to rural villages. These include:

- OMC (India; http://www.omcpower.com), a hybrid generating system selling electricity to a telecommunications tower operator and to households and businesses using rechargeable battery boxes delivered daily
- *Gram Power* (India; http://www.grampower.com), a connected or isolated micro-grid selling power to a local village entrepreneur who resells electricity to villagers using prepaid smart meters that can detect theft
- *Devergy* (Tanzania; http://www.devergy.com), a village-size solar-powered, direct-current (DC) micro-grid using prepaid cards for household and businesses combined with sales of DC appliances

- *Sincronicity* (Tanzania proposed; http://sincronicitypower.com), a hybrid generating system selling electricity to a telecommunications tower operator and to households and businesses using a wired distribution network, combined with a kiosk that sells other energy products

These are clearly not the rural electrification business models of a large, vertically integrated utility that uses the traditional model of extending the existing high- and medium-voltage grid. Some of these new business models may succeed and others may fail. Developers may revise their business models after they start operating to achieve commercial viability. And a business model that works in one country may not work in another. A shared characteristic of the first three business models listed above is that they have operated, at least initially, with little or no price regulation. And despite the fact that no regulator was protecting households in these communities, the early anecdotal evidence shows that these new suppliers have charged prices that reduced monthly lighting and other household energy expenditures by 30–50 percent. These developers have argued that too much regulation at an early stage could cut off experimentation and make it impossible for them to provide rural households with more-reliable, less-polluting, and less-expensive energy options. The basic argument, then, is that these experiments in electricity delivery will be "stillborn" unless the experimenters are given an explicit incubation period that allows them to test different delivery and pricing approaches without having to get the preapproval of the regulator or some other government entity.

The fifth reason is that the regulator will have neither the time nor the resources to regulate many different small electricity providers. If the regulator had to regulate every small rural electricity provider using a traditional enterprise-by-enterprise, cost-of-service tariff-setting approach, the regulator's administrative capacity could be quickly overwhelmed by hundreds of applications. (See the discussion in chapter 9 of the Philippine electricity regulator's early attempts at regulating more than 100 isolated rural electricity cooperatives.) And if this happens, the regulator could find that it is unable to devote the needed resources to regulating the national utility with a large monopoly franchise whose tariffs and performance will affect many thousands of customers.

What Should a Regulator Do?

... there are no riskless options.

—Professor Richard Schmalensee, MIT, 1993

The arguments of the previous section are compelling. Hence, we see considerable merit in explicitly allowing an initial grace period of five years or so, during which private operators of small mini- and micro-grids in rural areas would not need to obtain the national regulator's approval for their retail tariffs or a full license to operate. (But they would still be subject to safety regulation.)

But, admittedly, there are risks in creating a grace period free of tariff regulation. If the regulator provides this freedom from price regulation, it is conceivable that some private operators could and would take advantage of this pricing freedom. For example, an operator might initially offer low prices and then increase prices in the expectation that it will have a *de facto* monopoly because the regulator has promised to leave it alone for five years. If this happens, it is highly likely that the president or prime minister will receive a call from a member of parliament who represents the village served by this operator. One would expect that the member of parliament would say something like the following:

> Sir, do you know that the electricity regulator you appointed is totally ignoring his legal responsibilities? He is allowing [name of the operator] to charge my constituents obscenely high prices for electricity in [name of village]. This is not fair for these poor villagers. Please also remember that these villagers voted for us in the last election. You must order this so-called regulator to do his job or fire him and replace him with someone who will follow the law.

After this initial call, it is likely that the regulator will receive a follow-up call from the president or prime minister's office with orders to end the deregulation experiment immediately and impose tough tariff ceilings on this and other SPP operators. Hence, the price deregulation experiment for small rural providers could backfire.

The universal reality is that regulators, while legally independent on paper, always operate in a political environment. Therefore, we think that the only workable solution is to combine the proposed pricing grace period with pre-specified backstop measures to protect village consumers. The backstop would include the following features:

- *Annual reporting requirement.* In return for an exemption from the need to obtain the regulator's approval for retail tariffs, the operator would be required to file annual reports specifying annual sales, hours of service, number of customers by category, the average consumption by customer type, and the tariffs charged by customer category. To reduce the reporting burden on the operator, provision should be made to have the same reporting form used by the regulator, the REA, or whatever other government body is charged with promoting rural electrification.

- *Review of operations in response to customer complaints.* If 25 or 30 percent of the operator's customers sign a petition complaining about the prices or services provided by the operator, the regulator will initiate a review of the project's operations. But if any remedy is imposed by the regulator, it should be prospective rather than retroactive. The possibility of a retroactive rate being set would likely create too great a risk for most investors. Also, the regulator must specify in advance the standards that it will apply to grace-period SPPs (see below) in judging whether their performance is acceptable. Investors need to know in advance the performance standards by which they will be judged.

- *Registration rather than licensing.* The regulator would register the project rather than issue a license. This means that the operator's registration would be for informational purposes. Registration would *not* give the operator an exclusive right to serve this territory and other competitors could also register to serve the same area, ensuring that the market will remain contestable. But if the mini- or micro-grid operator seeks a license with an exclusive monopoly for a defined period of time, then the regulator would have the option (though not the obligation) of imposing stricter pricing and service standards on the operator.

- *Review after five years.* If the operator seeks extension of its registration after five years, the regulator will have the option of conducting a review to determine whether the pricing grace period should be extended. The decision will depend, in part, on whether the operator's service area can still be characterized as "contestable." For example, if the operator has built a distribution network and does not give nondiscriminatory access to the network to other potential electricity suppliers, then it would be hard to conclude that the market is still contestable.

Where to Go from Here

We hope *From the Bottom Up* can serve as a useful reference guide for policy makers and regulators in developing countries around the world, recognizing that it is but one contribution to an ongoing conversation among regulators, policy makers, developers, and investors (and those who advise them). Our broader goal is to inspire and enable action to support SPPs as one element of the decentralized track for scaling up access to grid-based electricity in Africa and elsewhere. We therefore want to share some of our thoughts on actionable, immediate next steps for three groups of readers: regulators and policy makers in countries without an SPP program who would like to start one; regulators and policy makers who would like to improve their existing SPP program; and donors in the international development community who would like to support SPP programs in developing countries.

Starting from a Blank Slate

Regulators and policy makers working from a blank slate who are looking to put in place SPP regulations or implement an SPP program can begin by taking the following initial actions:

- Review existing electricity laws and regulations to see if they would allow the development of SPPs and private sale of electricity. If restrictive or unclear laws or regulations exist, consider ways to adapt or update them so SPPs are granted the legal right to exist. Altering an existing regulation or law can be easier than repealing one and starting entirely from scratch.
- Document the *de jure* and *de facto* practices and conditions in the electricity sector, especially in areas related to customer tariffs, electricity quality

and reliability, and utility company solvency. This will help potential SPPs understand the existing technical and economic factors that they will need to navigate.

- Arrange for a study tour to countries that have successful SPP programs. The study participants should include personnel from the national and other large utilities, relevant ministries, the regulatory entity, the SPP community, and local banks. As a condition for going on the trip, the participants should be required, upon their return, to debrief other sector stakeholders on what they learned that might be applicable to their own country.
- Decide how SPP power sold to the national or other large utility will be priced: will it be cost-based technology-specific tariffs or avoided-cost tariffs or some hybrid pricing arrangement?
- Decide how backup tariffs will be structured for SPPs that need to purchase backup power from the national or other large utility.
- Issue draft rules and regulations for public comment.
- Once the policy rules and regulations have been drafted or approved, develop training for local banks that might be in a position to offer loans to SPPs.
- If investors are concerned about nonpayment by off-taker utilities, explore the possibility of obtaining guarantees from the World Bank and other international or regional organizations on late payment or nonpayment for SPP purchases made by the national utility.
- Consider offering a grace period in which new SPPs that propose to serve isolated mini- and micro-grids in rural areas would be exempted from tariff regulation.

Improving an Existing Program

Regulators and policy makers who are looking to improve or evaluate their current SPP program can begin by doing the following:

- Commission case studies of successful and failed SPP projects. These studies should, at a minimum, cover: ownership structure, business model, interconnection (where relevant) and operating requirements, regulatory rules and procedures, wholesale and retail tariffs, sources of financing (equity and debt), capital structure, proportion of capital costs paid for by grants, backup tariffs (where applicable), and connection charges for customers.
- Document the principal nonelectricity rules, regulations, and taxes that affect different types of SPPs. Make recommendations on how to reduce barriers created by nonelectricity rules, procedures, and taxes.
- Understand differences between *de jure* and *de facto* practices to make regulations more effective and in tune with reality, and to eliminate unnecessary and burdensome regulations.
- Review the operations and policies of the regulatory agency and rural energy agency/fund. Are there ways through which the actions of the two entities can be better harmonized to eliminate duplication and delays? Ideally, the evaluations should be performed by regulators, REA officials, and SPP

developers from other African countries, assisted by independent consultants from Africa and elsewhere. The African Electricity Regulator Peer Review and Learning Network provides a good model for how this could be done (Kapika and Eberhard 2013).

Donor Support for SPP Programs in Developing Countries

Members of the international development community who are looking to support SPP programs in developing countries can begin by funding:

- Reviews of electricity laws and regulations in the targeted countries to see if they allow for SPPs, and if not, encouraging the country's policy makers and regulators to adopt changes that explicitly allow and promote SPPs.
- Studies of issues involved in operating a "top-up" payments program (discussed in chapter 7) for increasing payments to grid-connected SPPs.
- Preparation of regional regulatory and policy guidelines for promoting grid and off-grid SPPs in partnership with regional associations of African regulators.
- Mapping of the renewable energy resources in the country. These might take the form of wind energy maps, solar insolation maps, maps of distribution of biomass of different types, and spatial/temporal assessment of small hydropower sites in the country or region.
- Studies of substations, feeders, or utility mini-grids where distributed generation is particularly valuable to utilities to defer investments in new transmission capacity, reduce diesel consumption, or improve power quality.
- Collection and dissemination of information on national and regional electricity markets, including market size, market segments and customer classes, willingness to pay, and electrification rates and sources of electricity.
- The establishment of partnerships with SPPs and government agencies to provide business advisory services, technical assistance and grants for capital costs, connection costs, or feed-in tariff (FIT) "top-ups" (when appropriate; see chapters 7 and 9).
- Research on some of the outstanding questions surrounding SPPs, including: gathering new electrification data based on the improved definition presented in chapter 2; cataloging SPP business models to identify what works well, where, and why; documenting the terms and conditions of different donor grant programs for buying down the initial capital costs of SPPs and creating a searchable database of existing SPP regulations, programs, rules, and policies.
- Development of a system for cross-country benchmarking of the electricity sector approval processes required for grid-connected and isolated SPPs to be allowed to operate. (This would be similar to the cross-country rankings of general business regulatory processes that appear in the World Bank's annual *Doing Business* publication.)

At the very beginning of this guide, we described how governments in Sub-Saharan Africa often propose a two-track approach to grid-based electrification.

The centralized or top-down approach relies on a large, usually state-owned utility to expand its grid into more and more rural areas. In contrast, the decentralized track is a bottom-up approach. It encourages some SPPs (renewable and hybrid) to serve rural communities not yet connected to a main grid and other SPPs to sell renewable energy to the main grid. We think that early evidence shows that privately owned and operated SPPs can succeed under the decentralized approach. In doing so, they can play a role in promoting both renewable energy and electrification in rural Africa and elsewhere. But if this is to happen, one key element (though clearly not the only element) is a rational and supportive system of economic regulation. In this guide, we have tried to address in detail what that system should look like. At the outset, we recognized that "[o]ne has a choice between making trivial statements with absolute certainty, and making important statements with uncertainty" (Galal and others 1994). We clearly opted for the second approach. Our readers in Africa and elsewhere will be the final judges as to whether we succeeded.

Notes

1. This section draws from two presentations at a 2013 World Bank conference in London: Desai and Moreno (2013) and Ranade (2013).

2. For a good description of the emerging impact investment industry, see the Global Impact Investing Network's resources page at http://www.thegiin.org/cgi-bin/iowa /resources/about/index.html.

3. See http://www.bcorporation.net/.

4. See Baumol (1982) for a more in-depth description of contestable markets.

References

Arias, Ricardo. 2013. "Demand Assessment: Is There a Market for Modern Energy Services?" Presentation at Incubating Innovation for Off-Grid Rural Electrification: London Investors' Conference, London, United Kingdom, March 21.

Baumol, William J. 1982. "Contestable Markets: An Uprising in the Theory of Industry Structure." *American Economic Review* (March): 40–54.

Desai, Vyjayanti, and Alejandro Moreno. 2013. "Enabling Environment for Off-Grid and Small Power." Presentation at Incubating Innovation for Off-Grid Rural Electrification: London Investors' Conference, London, United Kingdom, March 21.

Dichter, Sasha, Robert Katz, Harvey Koh, and Ashish Karamchandani. 2013. "Closing the Pioneer Gap." *Stanford Social Innovation Review* (Winter). http://www.ssireview.org /articles/entry/closing_the_pioneer_gap.

Galal, Ahmed, Leroy Jones, Pankay Tandon, and Ingo Vogelsang. 1994. *Welfare Consequences of Selling Public Enterprises: An Empirical Analysis.* Oxford, U.K.: Oxford University Press.

Government of India. 2003. "The Electricity Act, 2003." http://guj-epd.gov.in/extra _no_46.pdf.

Jhirad, David J. 2013. "SPEED: Smart Power for Environmentally-Sound Economic Development." Presentation at Incubating Innovation for Off-Grid Rural Electrification: London Investors' Conference, London, United Kingdom, March 21.

Kapika, Joseph, and Anton Eberhard. 2013. *Power Sector Reform and Regulation in Africa: Lessons from Ghana, Kenya, Namibia, Tanzania, Uganda and Zambia.* Cape Town, South Africa: HSRC Press.

Koh, Harvey, Ashish Karamchandani, and Robert Katz. 2012. "From Blueprint to Scale: The Case for Philanthropy in Impact Investing." Produced by Monitor Group in Collaboration with Acumen Fund, April.

Ranade, Monali. 2013. "Unbundling Business Models." Presentation at Incubating Innovation for Off-Grid Rural Electrification: London Investors' Conference, London, United Kingdom, March 21.

Hybrid Small Power Producers

What Are Hybrid Power Systems?

The term *hybrid power system* refers to the use of two or more fuel sources (generally one of which is a renewable fuel, and the other a fossil fuel) to generate electricity.[1] Throughout the world, there are thousands of mini-hybrid systems in operation. For hybrid mini-grid systems, the power system will typically combine a diesel generator with a solar, wind, or a mini-hydro generator. The diesel generator is generally used as a backup source of supply during periods of high loads or when there is little renewable power available. Most hybrid power systems also include a battery, which works as a buffer to smooth out the hour-by-hour or even second-by-second variations in the difference between what the renewable energy source is supplying and what mini-grid customers are demanding. The battery also helps optimize dispatch of the diesel generator so that the diesel can operate at close to full capacity when turned on,[2] usually during the nighttime peak in many rural villages. Excess generation above what is needed to serve the village's peak load can be stored in batteries to be used at other times. Finally, a hybrid system with solar photovoltaic (PV) panels will include a multifunction inverter. The inverter converts the direct-current (DC) power of the PV panels and batteries into alternating-current (AC) power. It also serves as a battery charger, storing excess generation in batteries for future use.

In some hybrid systems, the renewable generator will provide up to 80 percent or 90 percent of the total generated electricity. While the economics of hybrid systems are very site specific, there are four general justifications for installing new or retrofitted hybrid systems. They can:

- Supply electricity at lower cost than either a pure renewable system or a diesel-only. (figure A.1)
- Provide electricity for more hours of the day than a pure diesel system.[3]
- Improve the operational efficiency of diesel generators by allowing them to operate at higher capacities, which in turn causes them to use less diesel fuel per kilowatt-hours (kWh) generated.

Figure A.1 Levelized Cost of Electricity versus Renewable Energy Penetration for an Island Mini-Grid

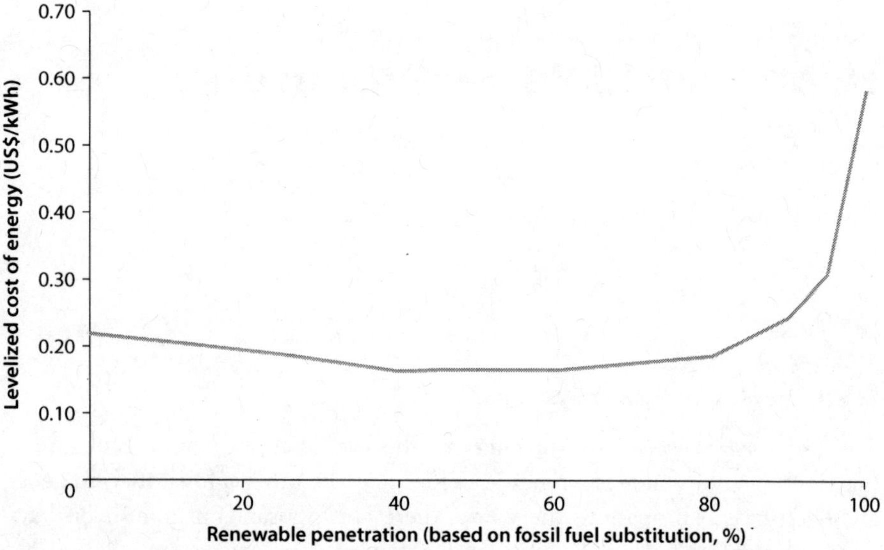

Source: Lilienthal 2013. Graph courtesy of HOMER Energy, used with permission. http://www.homerenergy.com.

- Extend the lives of diesel generators by reducing the number of hours of operation.

Therefore, as discussed in chapter 2, we think that it makes sense for a regulator to allow hybrid small power producers (SPPs) for both isolated mini-grid SPPs and grid-connected SPPs.[4]

Should Hybrid SPPs on Isolated Mini-Grids Be Allowed?

For isolated mini-grids, common hybrid combinations include solar/diesel, wind/diesel, and hydropower/diesel or solar/wind/diesel. Hybrid systems range in size from tens of watts (individual household scale) to tens or hundreds of kilowatts (kW) (village scale) or even megawatts (MW) (town scale). Besides village power, hybrid systems are widely used in communications (cell phone, TV) repeater applications.

Because the sun does not always shine and the wind doesn't always blow, diesel backup is an important option to mitigate the intermittency of renewable energy sources. At the same time, solar or wind power can lower costs and increase reliability compared to a diesel-only system by reducing expensive diesel consumption and also reducing wear and tear on diesel engines.[5] This is especially true in remote areas where it is expensive to run a fossil-fueled generator and to supply spare parts and provide repairs to diesel generators.

By the same token, allowing mini-grid SPP electricity developers the flexibility to use fossil-fuel generators can help lower the cost and/or improve reliability in providing electricity to rural areas. *If the developer had to use solar or wind energy only to power a mini-grid, the system would require prohibitively costly amounts of battery storage to provide sufficient backup for cloudy or windless periods.* In addition, the amount of renewable energy investment would have to be over-sized compared to the load for significant portions of the year. Diesel oil, while expensive, has the advantage of being storable in tanks over weeks or months and provides a less-expensive backup supply than the installation of a large bank of batteries. It allows the SPP to generate electricity when there are shortfalls in renewable resources.

Hybrid systems often have lower life-cycle costs compared to diesel-only or renewable-energy-only systems. For example, the levelized cost of electricity from a diesel-only system used in a remote island in Thailand was calculated to be $0.84/kilowatt-hour (kWh). An optimized hybrid system with batteries, inverter, and solar panels powering the same load lowers the levelized cost to $0.57/kWh (Greacen and others 2007). Similarly, in Senegal, ASER (Agence Sénégalaise d'Électrification Rurale, the rural electrification agency) with assistance from GIZ (Gesellschaft für Internationale Zusammenarbeit, the German aid agency) has installed, under the PERACOD[6] program, 35 hybrid mini-grid systems using a standard configuration (5 kilowatts peak PV/48 kWh battery/15 kilovolt-ampere [kVA] diesel generator) with plans under way to install 41 more by the end of 2014. It has been calculated that the lifetime levelized cost of a pure diesel system serving an isolated mini-grid will be $0.98/kWh, whereas the hybrid generating system that uses PVs, batteries, and a diesel generator will have a lower lifetime levelized cost of $0.69/kWh. This is true despite the fact that the estimated capital costs of the pure diesel system are much lower than the capital costs of a hybrid system ($33,000 for the diesel system versus $55,000 for a hybrid system). Even though the diesel system's capital costs are lower, the pure diesel system's running costs will be much higher than the running costs of a hybrid system. When both the capital and operating costs are calculated over the life of the project (to generate life-cycle costs), the hybrid system is the lower-cost option.

As noted earlier, the economics of hybrid systems versus conventional generation are very site specific, and depend on many factors including the discount rate; hourly and weekly load curve; the hourly and seasonal variations in renewable energy resources; the relative costs of equipment including batteries, inverters, generators; as well as the cost of diesel fuel. Fortunately, excellent free software is available to simulate as well as find the optimal investment in solar panels, batteries, and generators for a given situation. The HOMER software model allows users to enter both technical and financial parameters and compares the life-cycle costs of many different configurations to select the configuration with lowest levelized cost of energy.[7]

The ownership arrangements for isolated hybrid systems will affect ease of operations. Ownership arrangements are especially important for hybrid systems because hybrid systems are more complicated to operate than a pure diesel system. Unlike a purely diesel system, a hybrid system requires daily and seasonal decisions about when each of the potential sources of electricity supply—the diesel generator, the PV panels, and the batteries—should be used to minimize the overall cost of supply. The simplest arrangement for ownership of a hybrid system is for a single entity to own all components: the renewable generator, the diesel generator, the inverter, and the batteries. The single owner then decides how to optimize electricity production using the four principal components that it owns and controls. This will be the norm in Mali, where Agence Malienne pour le Développement de l'Energie Domestique et de l'Electrification Rurale (AMADER) and the World Bank intend to provide grants to support "hybridization" of existing diesel-fired mini-grid systems. A more complicated ownership arrangement seems to be emerging in Tanzania. In Tanzania, several private developers have proposed to add renewable generators to existing diesel-fired mini-grids currently owned by the national utility. Presumably, they are motivated by the high feed-in tariffs (FITs) offered on Tanzania Electric Supply Company's (TANESCO) isolated mini-grids (more than 24 U.S. cents in 2012). In this situation, the diesel generators will continue to be owned by the national utility, whereas the renewable generator and probably the batteries (if any) will be owned by the private developer. Under this dual ownership arrangement, the two owners will need to develop contracts and protocols to ensure efficient and reliable moment-to-moment operations of the overall hybrid system. The technical requirements to ensure reliable operation are discussed in chapter 8. *In our view, the best solution would be to encourage one of the two parties to take over ownership and operation of both the renewable and diesel generator on a separate isolated mini-grid.*

Should Grid-Connected Hybrid SPPs Be Allowed?

SPPs connected to the utility's main grid have a big advantage over SPPs that operate their own isolated/remote mini-grid. The advantage of getting a grid connection is that if the sun stops shining, the wind stops blowing, the water stops flowing, or the biomass resource is burned up, then the renewable energy SPP may simply cease generating and let the grid continue to serve customers in the area. By the same token, the grid can also absorb excess electricity beyond what is required to meet local loads. In practical terms, this means that unlike operators of most mini-grids, SPP generators connected to the main grid do not need to invest in diesel generators that stand idle much of the time, waiting to be dispatched when necessary to keep the lights on, or in an expensive battery storage system. The operator of the main grid provides the backup supply.

Despite this, there remain cases where hybridization of renewable and conventional energy sources makes sense for grid-connected renewable energy SPPs. Consider a biomass-powered SPP plant selling electricity to the grid. If there are strong seasonal variations in fuel supply, or if the renewable energy source (rice husk, sugarcane) has a bad production year, then the decision to use a limited amount of fossil fuels allows for better capital asset utilization. The SPP's boiler and turbine generator, which are generally biomass powered, may be fired using coal or other conventional fuel, or a mix of biomass and coal, thus producing revenue for the company and producing needed electricity for the country.

To remain consistent with the clean energy goals of an SPP power program, and to keep the country from overinvestment in small merchant coal generators, it makes sense for the regulator to limit the allowable quantity of conventional fuels burned in an SPP. For example, in Thailand grid-connected renewable energy SPPs may use up to 25 percent fossil fuel on an annual basis.[8] The fossil fuel is typically coal, used to cofire with biomass. In Tanzania the regulator has recently proposed that grid-connected SPPs can use up to 25 percent fossil fuel and still be eligible to sell power under the feed-in tariff established for renewable systems (EWURA 2013, Cap 141).

Notes

1. A good discussion of hybrid mini-grids can be found in ARE (2011) and IEA PVPS (2013).

2. A diesel that operates at 20 percent capacity produces electricity at about half the efficiency of a diesel that operates at 90 percent capacity. With no battery, a diesel must modulate its power production to match load. With the addition of a battery, the diesel can operate instead at high output (for example, 90 percent capacity) with electricity produced in excess of what the load is demanding going toward charging the battery. When the battery is sufficiently charged, the diesel can then be shut off completely and the load served by the battery (converted to AC [alternating current] electricity by use of an inverter).

3. Hybrid systems can also offer a higher degree of reliability. In India, Omnigrid Micropower Company (OMC) has a contractual obligation with its mobile-phone tower customers to supply electricity with 99.95 percent reliability. If it fails, it must pay significant penalties.

4. In Mali, the World Bank is initiating a project with AMADER (the rural electrification agency) to convert existing mini-grids that are purely diesel fired to hybrid mini-grids that will combine diesel generator, solar photovoltaic cells, and batteries. A similar project has been proposed in Kenya under the Scaling-Up Renewable Energy Program (SREP) which will be supported by a number of bilateral and multi-lateral donors.

5. A general rule of thumb is that a diesel generator will have to be replaced after about 25,000 hours of use.

6. Promotion de l'Electrification Rurale et de l'Approvisionnement Durable en Combustibles Domestiques (Programme to Promote Rural Electrification and a Sustainable Supply of Domestic Fuel).

7. http://www.homerenergy.com.

8. See section 4.3 of the Provincial Electricity Authority's very small power producer model power purchase agreement at http://tinyurl.com/ThaiPPA or http://www.eppo.go.th/power/vspp-eng/PPA%20Model%20-VSPP%20Renew%20-10%20MW-eng.pdf.

References

ARE (Alliance for Rural Electrification). 2011. *Hybrid Mini-Grids for Rural Electrification: Lessons Learned.* Brussels. http://www.ruralelec.org/fileadmin/DATA/Documents/06 _Publications/Position_papers/ARE_Mini-grids_-_Full_version.pdf.

EWURA. 2013. *The Electricity (Development of Small Power Projects) Rules, 2013. Proposed for Public Consultation.* Dar es Salaam, Tanzania. http://www.ewura.go.tz/sppselectricity .html

Greacen, Chris, Sirikul Prasitpianchai, Tawatchai Suwannakum, and Christoph Menke. 2007. *Renewable Energy Options on Islands in the Andaman Sea: Hybrid Solar/Wind/ Diesel Systems.* Study for the Tsunami Aid Watch Programme of the Heinrich Böll Foundation Southeast Asia Regional Office. http://www.palangthai.org/docs /KohPoKohPuEng.pdf (associated HOMER file: http://www.palangthai.org/docs /KohPo.hmr).

IEA PVPS (International Energy Agency, Photovoltaic Power Systems Programme). 2013. *Rural Electrification with PV Hybrid Systems: Overview and Recommendations for Further Deployment.* Report IEA-PVPS T9-13. http://www.iea-pvps.org/index .php?id=1&eID=dam_frontend_push&docID=1590.

Lilienthal, Peter. "The Problem with 100% Renewable Energy." *HOMER Energy* (blog), (accessed June 30, 2013). http://blog.homerenergy.com/the-problem-with-100 -renewable-energy/?utm_source=Microgrid+News+by+HOMER+Energy&utm _campaign=18f8238c1b-Microgrid_News_June_20136_13_2013&utm _medium=email&utm_term=0_0f7f799f46-18f8238c1b-164784662.

Conversion of Flat Monthly Charges to per-kWh Charges

Table B.1 shows a sample output from a simple spreadsheet (available at http://tinyurl.com/tariffconversion) that converts flat appliance-based tariffs (used in some mini-grid systems) to kWh-based tariffs.

Table B.1 Conversion of Flat Monthly Charges to per-kWh Charges

Device(s)	Wattage	Quantity of devices	Estimated daily usage (in hours)	Number of kWh consumed daily	Number of kWh consumed monthly[a]
Interior lightbulbs	60	5	5	1.5	45
Lightbulbs (CFL)	30	0	0	0	0
Fan(s)	20	0	0	0	0
Fluorescent tube lights	30	0	0	0	0
Television	120	0	0	0	0
Radio	100	1	8.5	0.85	25.5
Exterior security lightbulb	60	1	12	0.72	21.6
Total(s)	420	7	25.5	3.07	92.1

Estimated monthly cost

Item	Number	Monthly flat charge (in T Sh)[a]	Monthly charges (in T Sh)
Sockets for lightbulbs	5	1,500	7,500
Outlets for other devices	5	1,500	7,500
Total flat monthly charges (in T Sh)			15,000

Estimated cost/kWh

In Tanzanian shillings (T Sh)[b]	162.9
In U.S. dollars	0.109

Note: CFL = compact fluorescent lamp; kWh = kilowatt-hour.
a. Monthly flat charge per device used numbers from one project in Tanzania. It is assumed that the user pays a flat monthly charge per electrical device or per outlet or socket. This nonmetered system of charging is often combined with a load-limiting device to limit the user's maximum consumption.
b. Exchange rate between US$ and T Sh taken as $1 = 1,500 T Sh.

Technical and Commercial Quality-of-Service Standards in Rural and Urban Areas of Peru

In Peru technical and commercial quality-of-service standards set by the regulator are different in urban and rural areas. Quality-of-service standards are lower for isolated electricity service providers in rural areas, the rationale being that it is more difficult and costly to provide comparable service in rural areas at a price that would be affordable for generally poorer rural customers.

Quality of Supply and Quality of Product

For example, table C.1 shows the different required service levels for SAIFI (System Average Interruption Frequency Index) and SAIDI (System Average Interruption Duration Index), two universal measures of quality of supply in rural and urban areas of Peru. SAIFI is a standard measure of the number of outages during a specified calendar period; SAIDI refers to the total duration of these outages, measured in hours per year.

In setting retail tariffs, OSINERGMIN (El Organismo Supervisor de la Inversión en Energía y Minería, the Peruvian electricity regulator), has categorized service areas of all retail electricity providers based mostly on the density of their service areas. There are five categories of service area, and a single operator may supply several different categories of service areas within its overall franchise or concession area. In general, isolated mini-grids would be serving rural, dispersed areas, which are the least dense service areas. As shown in table C.1, under Peruvian regulatory rules, a mini-grid operator is allowed 40 interruptions per year without triggering a penalty versus the 12 interruptions permitted for a provider of service in an urban high-density area. Isolated rural service providers are also allowed to provide electricity with a higher measured drop in voltage (7.5 percent) than providers in urban areas (5 percent).

Table C.1 Targeted SAIFI and SAIDI Standards in Peru

Types of service areas	SAIFI (number of interruptions per year)	SAIDI (hours/year)
Urban high density	12	7
Urban medium density	16	9
Rural concentrated	25	10
Rural dispersed	40	10

Source: Revolo Acevedo 2011.
Note: SAIDI = System Average Interruption Duration Index; SAIFI = System Average Interruption Frequency Index.

Table C.2 Maximum Time for Making a New Connection
Days

		Without network adaptation	With network adaptation	With installation of new network equipment
Urban	Up to 50 kW	7	21	360
	Above 50 kW	21	56	360
Rural	Up to 50 kW	15	30	360
	Above 50 kW	30	90	360

Source: Revolo Acevedo 2011.
Note: kW = kilowatt.

Quality of Commercial Service

There are also wide differences between urban and rural areas in the required commercial levels of services. Table C.2 shows the maximum allowed time for making a new connection. The standard specifies the maximum number of days for an operator to connect a new customer after receiving the customer's application. In less-dense rural areas, the number of allowed days is 15, if the customer has a maximum demand below 50 kilowatts (kW) and can simply be connected to existing distribution equipment. But if the connection requires the installation of new equipment, such as a larger step-down transformer, then the operator has 30 days in which to make the connection. In both cases, the number of allowed days is higher in rural areas than the number of days allowed in urban areas for a customer with comparable demand. Table C.3 shows the required standards of performance for other dimensions of commercial service. Once again, the required service standards are significantly lower in less densely populated rural areas.

One paradox in the Peruvian system is that the approximately 350 isolated mini-grids owned and operated by small municipalities are not required to satisfy these standards. These standards apply only to entities that receive concessions, and municipal systems do not need a concession. The concession law that exempts municipalities states that the "service standards will be established by mutual agreement between the municipality and its customers." As a general phenomenon, the service provided by isolated municipal small power producers (SPPs) is not good—they charge very low tariffs, and do not maintain their

Table C.3 Standards for Various Commercial Services

	Urban	Rural concentrated	Rural dispersed
Reconnection of service (maximum hours after payment of outstanding bill)	24	24	48
Opening hours of commercial office (hours/day)	8	8	4
Resolution of billing errors (maximum number of days)	30	30	30
Hours of service by call center (minimum hours a day)	24	24	12

Sources: Quality of Service Standard for Electricity Providers in Peru (seventh chapter of Supreme Decree No. 020-1977-EM) and Quality of Service Standard for Rural Electricity Providers in Peru (sixth chapter of Decree No. 016-2008-EM/DGE).

equipment presumably because they expect that the national government will provide a new capital grant if their equipment is no longer able to provide service. In other words, the small Peruvian municipal systems generally do not pursue commercial viability because they expect to be bailed out by the national government when their system fails.

Reference

Revolo Acevedo, Miguel. 2011. Personal communication.

Calculating the Effect of Cost-Reflective Technology-Specific Feed-In Tariffs on Retail Tariffs

Sri Lanka

Table D.1, taken from the tariff order for the January–June 2011 period issued by the Sri Lankan electricity regulator (PUCSL 2011), shows the estimated additional costs to Sri Lankan electricity consumers that result from the 2007 requirement that the national utility (Ceylon Electricity Board [CEB]) purchase electricity from small power producers (SPPs) based on feed-in tariff (FIT) prices using technology-specific, cost-reflective calculations. As described in appendix F, prior to 2007 all SPPs in Sri Lanka were paid a FIT set at an estimate of the avoided cost of the national utility (CEB). Under this avoided-cost FIT, the only SPPs that were economically viable were small hydro facilities and some biomass generators.

In 2007 Sri Lanka decided to offer technology-specific, cost-reflective FITs to expand the pool of renewable technologies. Under the new system, developers were given the option of a single levelized FIT for the life of the power-purchase agreement or a tiered FIT with higher prices in the early years and lower prices in the later years. The new FIT values were higher than CEB's avoided costs starting in 2008. For example, under the new rules, wind generators that use locally manufactured turbines were eligible to be paid a levelized price of 20.81 SL Rs/kWh (US 19.92 cents/kWh) compared to CEB's estimated weighted average avoided cost of 11.25 SL Rs/kWh (US 10.23 cents/kWh). Consequently, on every kilowatt-hour purchased from wind SPPs in 2011, CEB will be required to pay 9.69 cents/kWh more than its avoided costs.

The overall estimated cost of the new higher FITs in the January–June 2011 period is displayed in table D.1 in the column "Additional burden on customers." As shown in the row for wind-powered SPPs, it is estimated that CEB will pay wind-generating SPPs SL Rs 1,140 million ($10.4 million) rather than SL Rs 556 million ($5.1 million) if the FIT had continued to be based on CEB's avoided cost. On a per-kWh basis, the estimated effect in 2011 of paying

Table D.1 Estimate of the Retail Tariff Impact in Sri Lanka of Technology-Specific FITs for Six-Month Period of January–June 2011

Type	Pricing basis of agreement	Forecast energy purchased (GWh)		Forecast price (SL Rs/kWh)		Allowed payments (SL Rs million)	Payment on avoided costs (SL Rs million)	Additional burden on customers	
		Filed	Adjusted	Filed	Adjusted	Adjusted		SL Rs million	SL Rs/kWh of end-use sales
Mini-hydro	Avoided cost	188.0	197.7	11.98	11.50	2,273	2,273	0.0	0.00
	Three-tier	9.6	10.1	14.27	11.77	119	116	2.7	0.00
Biomass	Avoided cost	1.2	1.3	14.00	11.50	15	15	0.0	0.00
	Three-tier	12.0	12.6	22.00	9.90	125	145	−20.2	0.00
Wind	Three-tier	46.0	48.4	24.73	23.58	1,140	556	584.2	0.12
Total		256.8	270.0			3,672	3,105	566.8	0.12

Source: PUCSL 2010.
Note: "Filed" means the numbers submitted by the Ceylon Electricity Board to the regulator; "adjusted" means the corrections and adjustments made by the regulator; and "allowed payments" are the costs that the regulator will allow to be recovered in tariffs based on the adjusted payments. FIT = feed-in tariff; GWh = gigawatt-hour; kWh = kilowatt-hour; SL Rs = Sri Lankan rupees.

the higher FIT prices on all renewable SPP technologies is to raise the average retail tariff by about 1 percent. This relatively low impact reflects the fact that at present more than 95 percent of SPPs continue to be paid using the "avoided-cost" FIT, which was designed to have no incremental impact on retail tariffs. But the tariff impact will grow over time because all new SPPs will be paid a technology-specific, cost-reflective FIT. There will be some downward adjustments when the SPPs begin to receive lower, tier 2 and tier 3 FIT values over time.

In this same order, the Sri Lankan regulator mandated that the higher costs be passed on to all retail customers through a surcharge on the allowed bulk supply costs that would be used in calculating retail tariffs. Previously, CEB was supposed to recover this premium (the difference between an avoided-cost FIT and a technology-specific, cost-reflective FIT) through a separate payment from Sri Lanka's Sustainable Energy Authority (SEA) to CEB (which meant that it would be paid for by Sri Lankan taxpayers). This proved to be unworkable because the SEA did not receive funds from the government to pay for the premium. But even though CEB did not receive the promised payments, it continued to honor its power-purchase agreements (PPAs) with the SPPs. In a later order, the Sri Lankan electricity regulator eliminated the likely growing financial burden on CEB when it decided that in the January–June 2011 time period the premium would be paid for directly by Sri Lankan retail electricity customers in a surcharge that would be added to their electricity bills. It is likely that this regulatory policy of passing through all FIT costs to retail customers will be continued in the future.

Thailand

Similar calculations have been performed in Thailand, where renewable energy SPPs and very small power producers (VSPPs) receive premiums above the avoided cost of 2.58 baht/kWh ($0.086/kWh). The Thai Ministry of Energy has estimated that the premiums will increase the average retail tariff by about 2.8 percent in 2010–11. As in Sri Lanka, the customer impact will become greater over time. It is projected that the retail tariff effect will increase to 10.7 percent in the years 2012–16, as thousands of megawatts of SPPs and VSPPs come online. Over 60 percent of this tariff impact increase is expected to come from hundreds of new solar farms, with a total installed capacity of 2,890 MW coming online. The solar farms have a dominant impact because many photovoltaics (PVs) receive a premium of 8 baht ($0.24)/kWh supplied which is added to the avoided-cost "bulk-supply" tariff of 2.58 baht/kWh that is the estimate of the buying utility's avoided cost. Hence, the premium increases the FIT received by renewable energy SPPs and VSPPs by about 300 percent. Like Sri Lanka, the additional costs of the FIT are paid through a surcharge on all consumers' bills.

The Thai government soon realized that this premium could lead to significant increases in retail electricity tariffs. Hence, on July 28, 2010, the Thai National Energy Policy Council (NEPC) passed a resolution (http://www.eppo .go.th/nepc/kpc/kpc-131.htm#5) that lowered the 10-year FIT adder from 8 baht/kWh to 6.5 baht/kWh for all solar projects that were still in the pipeline but not yet approved as of the date the resolution was issued. The resolution also stated that after July 28, 2010, no new solar applications would be accepted.

There is no simple rule for calculating the impact of paying technology-specific, cost-reflective FITs on average retail tariffs. The calculation has to be done on a country-by-country basis. The effect on retail tariffs will vary depending on the size of the premiums, the number of kilowatt-hours generated by SPPs that are eligible for the premiums, the number of kilowatt-hours consumed by retail customers, and decisions concerning the allocation of tariff increases among customer classes.

References

PUCSL (Public Utilities Commission of Sri Lanka). 2010. *Cost Reflective Methodology for Tariffs and Charges: Methodology for Charges.* Colombo, September 30. http://www .pucsl.gov.lk/english/wp-content/themes/pucsl/pdfs/methodology_for_charges.pdf.

———. 2011. *Decision on Electricity Tariffs, Table 22.* July 4. http://www.pucsl.gov.lk /english/wp-content/themes/pucsl/pdfs/decision_on_electricity_tariffs-2011.pdf.

Evaluation of Risk Allocation in a Power-Purchase Agreement for a Mini-Hydro Project in Rwanda

Table E.1 Evaluation of Risk Allocation in a Power-Purchase Agreement for a Mini-Hydro Project in Rwanda

Description of risk	Party allocated the risk
Government, policy, and regulatory risks	
Permits and approvals. Risk that required approvals (for example, environmental permits, water-use rights, generation license) may not be obtained or obtained subject to conditions that increase costs.	The seller is responsible for acquiring all permits for the duration of the agreement.
Government policy. Risk that a change in law, policy, or other government action increases the estimated cost of the power supply.	Not clearly allocated.
Environmental liabilities. Risk that power production over the contract term results in significant environmental liabilities (greater than anticipated at contract signing) other than a change in law.	Not clearly allocated.
Financing risks	
Availability of finance. Risk that debt and/or equity is not available when required by the seller to develop the project.	Not clearly allocated.
Sponsor insolvency. Risk that seller is unable to provide required services due to insolvency.	Should the seller become insolvent, it would result in the seller defaulting on their contract. The buyer is then permitted to terminate the power-purchase agreement (PPA) and pursue remedies as permitted by the PPA or law.
Interest rate. Risk that interest rates move adversely after the contract is signed.	Not clearly allocated.
Exchange rate. Risk that exchange rates may move adversely after the contract is signed, affecting the seller's ability to service foreign-denominated debt and obtain its expected return on investment.	Not applicable.
Refinancing impact. Risk that seller can/cannot refinance as expected after project commissioning.	Not clearly allocated.
Tax changes. Risk that before or after completion, tax rates change.	Not clearly allocated.

table continues next page

Table E.1 Evaluation of Risk Allocation in a Power-Purchase Agreement for a Mini-Hydro Project in Rwanda (continued)

Description of risk	Party allocated the risk
Completion risks	
Site. Risk that unanticipated conditions at the site are discovered during construction, adding cost or delay.	Seller has 17 months total to complete the construction of the power plant. Site risk lies with the seller.
Design. Risk that design of the facility is not able to deliver supply at expected cost and specified level of service.	Buyer enforces their service standards, and those enforced by best practice guidelines. In pursuing these standards, all design risk lies with the seller.
Construction. Risk that events occur during construction that prevent the facility being delivered on time and on cost.	Seller has 17 months total to complete the construction of the power plant. Construction risk lies with the seller.
Industrial relations. Risk that industrial action (for example, strikes, lockouts, work bans, work-to-rules, blockades, go-slow action, and so on) negatively affects the viability of the project.	Not clearly allocated.
Commissioning. Risk that commissioning tests required for supply to commence cannot be successfully completed on time, or have higher-than-anticipated costs.	Seller has one month to begin commercial operations once construction is complete. Commissioning risk lies with the seller.
Operating risks	
Inputs and fuel supply. Risk that required inputs (such as fuel) cost more than anticipated, are of inadequate quality, or are unavailable in required quantities.	Not clearly allocated.
Maintenance and refurbishments. Risk that the facility incurs higher-than-anticipated maintenance and refurbishment costs.	Buyer enforces their service standards, and those enforced by best practice guidelines. In pursuing these standards, all maintenance and cost risk lies with the seller.
Plant performance. Risk that plant does not provide contracted capacity, energy, or other services (reserves, black-start capability), or experiences higher-than-expected outage rates.	Seller is responsible for providing the buyer with a monthly forecast of likely generation capacity. There is no penalty on the seller for failing to meet these forecasts, meaning the risk of insufficient electricity generation remains with the buyer.
Output requirements. Risk that output requirements (capacity or energy) are changed after contract signing, whether before or after project commissioning.	Not clearly allocated.
Operator failure. Risk that operator (including an operating subcontractor) fails financially or fails to provide contracted services to specification.	Should the seller fail, it would result in the seller defaulting on their contract. The buyer is then permitted to terminate the PPA and pursue remedies as permitted by the PPA or law.
Transmission and distribution risks	
Transmission access. Risk that seller is not provided access to networks required to deliver power as per contract conditions.	Not clearly allocated.
Transmission investment. Risk that cost of connecting facility or transporting power to buyer's facility requires further investment in the transmission network.	It appears that if the buyer commits to constructing the transmission line so it is ready by the commissioning date of the power plant, then the risk for transmission access lies with the buyer. If the buyer cannot commit to this deadline, or both parties otherwise agree, then the seller can construct the transmission line. The transmission cost would therefore lie with the seller. This should be clarified as the relevant clause (4.7) is not clear.

table continues next page

Table E.1 Evaluation of Risk Allocation in a Power-Purchase Agreement for a Mini-Hydro Project in Rwanda
(continued)

Description of risk	Party allocated the risk
Transmission constraints. Risk that transmission constraints impose costs on power deliveries under contract terms, for example, if voltage instabilities or constraints prevent dispatch.	The seller is responsible for these costs (both in terms of additional wear and tear on assets and foregone revenue from not being dispatched).
Commercial and market risks	
Demand risk. Risk that the demand for a service or the use of a facility will vary from forecast levels, generating less revenue from users than expected.	The buyer assumes all retail demand risk and must accept all electricity at the termination point providing it is within certain quality assurance parameters.
Nontechnical losses. Risk that end users of the service will fail to pay for electricity (due to either theft, nonbilling, or nonpayment of bills).	Not clearly allocated.
Nonpayment. Risk that buyer is unable or unwilling to pay purchase price for contracted services.	The buyer is obligated to purchase all electricity at the termination point. The risk of lost revenue however lies with the seller in the event that the buyer refuses or is unable to purchase this electricity.
Economic obsolescence. Risk that contracted services are able to be provided at lower cost from alternative suppliers.	Not clearly allocated.
Other risks	
Security of supply risks. Risk that plant outages will negatively affect security of supply on the buyer's electricity system.	Risk lies with the buyer as the seller is not obligated or liable for any amount of electricity at the termination point, providing the lack of electricity is not the result of electricity being sold to a third party.
Force majeure. Risk that inability to supply power (pre- or postcompletion) is caused by reason of *force majeure* (no-fault) events.	In the event of a *force majeure* an affected party is exempt from its obligations under this PPA, to the extent necessary from whatever performance is affected by the *force majeure*. Therefore, revenue risk lies with the seller and electricity supply risk with the buyer.

Source: Unpublished analysis of Ben Gerritsen, Managing Director, Castalia Strategic Advisors, April 2011.

Feed-In Tariff Case Studies: Tanzania, Sri Lanka, and South Africa

This appendix describes feed-in tariff (FIT) policies and implementation issues in three countries with which the authors have personal experience. Tanzania currently sets FITs based on avoided cost (FIT method 1), while Sri Lanka began with an avoided-cost methodology in 1996 and then shifted to a standardized, technology-specific, cost-reflective methodology (FIT method 2) in 2007. South Africa started with fixed FITs, and shifted to competitive procurement (bidding).

Tanzania's Feed-In Tariffs

Background

During most of the last five years, Tanzania has faced serious power generation crises caused by periods of drought combined with unplanned generator outages. The most recent crisis occurred in 2011 when Tanzania experienced up to 16 hours of load shedding per day in certain parts of the country. The national electrification rate is 14 percent (U.S. Foreign Commercial Service 2010) and only 3 percent of the population in rural areas has access to electricity (Gaddis 2012). The inadequate provision of electricity has been recognized as one of Tanzania's major bottlenecks to economic growth.

To help meet Tanzania's need for power, improve electricity access, and foster domestic private sector investment in small clean power sources, the Ministry of Energy and Minerals developed the small power producer (SPP) program in mid-2009. The SPP regulations enable development of renewable and cogenerated electricity through standardized power-purchase agreements (SPPAs) and FIT payments, and streamlined interconnection and licensing requirements. The regulations provide the legal basis for customers to interconnect renewable energy generators to both the national grid and Tanzania Electric Supply

Company's (TANESCO's) existing isolated mini-grids, and to export excess power (up to 10 megawatts) to TANESCO (EWURA 2010, 4).[1] For 16 existing isolated TANESCO mini-grids, SPPs have the potential to replace, in whole or in part, TANESCO's existing diesel generation. The SPP policy also allows for SPPs to construct new isolated mini-grids to service communities without electricity and sell directly to new customers.

Feed-In Tariff Calculations and Values in Tanzania

Tanzania currently has two different FIT levels for wholesale sales of electricity by SPPs (table F.1). The first is for SPPs selling electricity to TANESCO on its main grid; the second and higher FIT is for SPPs that sell electricity to TANESCO on any of its existing isolated mini-grids. Both of these FIT prices are based on estimates of TANESCO's average avoided costs, either on the main grid for the first FIT or as an average of the avoided costs of the main grid and isolated grid for the second FIT. At the time of this writing, it appears that Tanzania is the only country in Africa to use an avoided-cost methodology in setting FITs.

Why Was the Avoided-Cost Method Selected in Tanzania?

There were several reasons for choosing to set FITs keyed to TANESCO's avoided financial costs and to create different tariffs for main-grid and isolated mini-grid SPPs. A primary consideration was that TANESCO stated it would only accept a FIT arrangement that did not require it to incur costs above its avoided costs for electricity. When different FIT methodologies were being discussed, there were no indications that the Tanzanian government or outside donors would be able to provide additional funding for the additional cost of setting FIT values above TANESCO's avoided costs. Setting tariffs at TANESCO's avoided financial costs was a way to get started. The SPP policy program was able to move forward without having to resolve the potentially deal-breaking issue of where subsidies might come from to cover incremental FIT costs (from technology-specific, cost-reflective tariffs).

Table F.1 Feed-In Tariff Levels in Tanzania for SPPs Selling to TANESCO's Mini-Grids and TANESCO's Main Grid

	2011		2012	
	T Sh/kWh	$/kWh	T Sh/kWh	$/kWh
Mini-grid				
Annual average	380.22	0.243	380.22	0.243
Main grid				
Annual average	112.43	0.072	152.54	0.096
Dry season	134.92	0.086	183.05	0.116
Wet season	101.99	0.065	137.29	0.087

Sources: EWURA 2011a, b; 2012a, b.
Note: kWh = kilowatt-hour; SPP = small power producer; TANESCO = Tanzania Electric Supply Company; T Sh = Tanzanian shillings.

Another primary consideration was that rural electrification is a priority for the country—and higher tariffs to support SPP investments to displace diesel generation on existing isolated mini-grids might encourage more hours of operation on these existing grids. TANESCO's mini-grid avoided costs are very high because the mini-grids use expensive diesel generation. Purchasing electricity from SPPs, even at the relatively high price of 380.22 T Sh/kilowatt-hour (kWh) ($0.243/kWh) in 2011 (table F.1) would save TANESCO money over generating the electricity itself from diesel generators, which was estimated to cost about 629.55 T Sh/kWh ($0.402/kWh) in 2011, and slightly less in earlier years. Moreover, TANESCO was selling electricity to its customers on the isolated mini-grids at the national uniform tariff—an average price of 118.8 T Sh/kWh (approximately 8.8 cents/kWh). Hence, TANESCO found itself losing on average about 31.4 cents for every kWh that it sold on isolated mini-grids. If, instead, TANESCO were able to displace its own diesel-generated electricity with purchases from an SPP, then TANESCO would be able to reduce its per kWh losses from 31.4 cents to 15.5 cents.

The Buyer's Marginal Cost: Which One?

The value for the main-grid FIT was calculated as an average of TANESCO's long-run marginal cost (LRMC) and short-run marginal cost (SRMC). The LRMC was based on TANESCO's least-cost generation supply plan and the SRMC on TANESCO's projected energy costs for thermal generation in the next year. TANESCO's LRMC is dominated by large hydro, gas-fired, and coal-fired power plants using indigenous fuels, while the near-term estimates of SRMC are dominated by higher-cost oil-fired generation. Similarly, the value of the FIT for an SPP connected to an existing TANESCO mini-grid was calculated as an average of short- and long-run marginal costs. In this latter case, the LRMC continued to be the cost of electricity on the main grid (TANESCO's LRMC),[2] but with a different SRMC. The SRMC used for the FIT on an existing isolated mini-grid was calculated as the incremental cost of fuel for new diesel-fired generation on SPP mini-grids. The rationale for choosing the same LRMC is that in the (very) long run, currently isolated rural centers will become grid connected. The rationale for choosing a different SRMC is that the near-term avoided cost for TANESCO on its isolated mini-grids is based on the running costs of its diesel-fired generating units.

Another benefit of using a higher SRMC is that it helps to ensure the commercial viability of SPPs connected to an existing isolated mini-grid. An SPP connected to the main grid has the potential to sell its entire electrical output to TANESCO because TANESCO has a mandated "must-take" obligation. But if the SPP is connected to one of TANESCO's existing mini-grids, TANESCO may not be able to purchase all of the SPP's potential output for technical reasons. On isolated mini-grids, it is likely that the SPP generator will have to curtail its level of generation at certain times every day to satisfy the engineering requirement that generation and load must be balanced. From a technical perspective, the SPP cannot be allowed to produce electricity if there is insufficient demand to absorb that electricity. The SPP's output may also need to be constrained to meet minimum operating requirements of the diesel generator.[3] Considering the fact that the SPP will be constrained to produce for fewer hours and at less than its maximum level of production, the tariffs for SPPs selling to an existing isolated mini-grid will need to be significantly higher than the tariffs for main-grid-connected SPPs (see table F.1) if the SPP selling to an isolated mini-grid is going to be commercially viable.

What Happens When an Isolated Grid Connects to the Main Grid?

When the main grid expands to interconnect with a mini-grid to which an SPP is selling electricity, the power-purchase agreement (PPA) and tariff will be converted to those applicable to other main-grid-connected SPPs.[4] Six months prior to the expected date for interconnection with the main grid, TANESCO is required to notify the SPP of its intention to terminate the mini-grid PPA and to enter into a new 15-year main-grid PPA with the SPP. Some SPP developers have strongly objected to the provision. They have said that the size of the price reduction (from about $0.24/kWh to $0.07/kWh using the 2011 FIT) and the uncertainty as to when it will occur will make their projects unviable.

What Are the Allowed Floors, Ceilings, and Adjustments?

Tanzania's FITs are adjusted annually, based on the best estimates of the projected avoided cost for electricity. In other words, the tariffs are not locked in or fixed for the life of the PPA. A FIT paid to an existing SPP can go up or down depending on the annual update of avoided-cost calculations made by the regulator. For example, as shown in table F.1, the average FIT for SPPs connected to TANESCO's main grid went up by 33 percent between 2011 and 2012. Though there is some protection from annual adjustments for both the SPP and TANESCO through the use of price floors and price caps, the protection is asymmetric. The floor is equal to the tariff in the year in which the PPA between the SPP and TANESCO enters into force. That floor price is locked in for the duration of the PPA to protect the SPP against possible reduction in the standardized tariff in future years. For example, if a main-grid-connected SPP signed a PPA with TANESCO in 2010, it is assured that the price that it will receive during TANESCO's dry season will never go below T Sh 134.92 (8.6 cents)/kWh for the entire 15 years of the PPA. But if the same SPP signed its PPA in 2012, it would be guaranteed that its sale price to TANESCO would never go below T Sh 183.05/kWh (11.6 cents/kWh), an increase in the guaranteed floor price of almost 35 percent representing a significant revenue bonus to the SPP assuming that TANESCO is able and willing to pay these tariffs.

In contrast, TANESCO as the buyer has only partial protection from upward price adjustments. The FIT price can go up subject to an initial price cap, equal to 1.5 times the standardized tariff for the year the PPA enters into force. But the price cap itself is not fixed and will be adjusted on an annual basis to reflect changes in the consumer price index (CPI).[5] Hence, there are two possible upward adjustments: one adjustment is to the price cap, which will depend on changes in the CPI, and the other adjustment is to the FIT value under the price cap, which will depend on the annual avoided-cost calculation.

What Is the Annual Process for Setting FIT Values in Tanzania?

Annual calculations of FITs are currently performed by Tanzania's Energy and Water Utilities Regulatory Authority (EWURA) based on information supplied by TANESCO. EWURA receives comments on these calculations from an SPP working group comprising representatives from the Ministry of Energy and Minerals and other sector stakeholders, and also allows a 21-day public consultation period on the proposed calculations. After addressing comments, the EWURA board issues a final approval.

Is There an Automatic Retail Tariff Pass-Through Mechanism?

At present, there is no automatic pass-through of SPP purchase costs in the TANESCO retail tariffs. But even if this were put in place by expanding TANESCO's current fuel adjustment clause to become a fuel adjustment *and* power purchase clause, this, in itself, is no guarantee that an individual SPP will actually get paid. In other words, TANESCO may have an automatic mechanism to recover SPP purchase costs without waiting for the next tariff case, but this, by itself, provides no certainty that any individual SPP will be paid if TANESCO finds that it does not have enough money to pay all of its suppliers, creditors, and employees. Consequently, SPPs have asked for the additional protection of a payment guarantee to ensure that they will be at the front of the payment queue in addition to any automatic adjustment mechanism that would automatically allow TANESCO to adjust its retail tariffs for any increases in its SPP purchase costs.

Sri Lanka's Feed-In Tariffs

Since FITs were first established in Sri Lanka, the country has used two different approaches. In 1996 the country initiated a FIT program based on the value of the SPP purchases to the national, government-owned utility, the Ceylon Electricity Board (CEB) (that is, the avoided-cost or FIT method 1, as in Tanzania). Sri Lanka made key modifications in 1999, and then in 2007 switched to a technology-specific, cost-reflective (FIT method 2) tariff methodology for all new SPPs. When the switchover occurred in 2007, SPP developers who had signed PPAs under the avoided-cost methodology were given the option of staying with that methodology or switching over to the new technology-specific, cost-reflective methodology, and a transition package was offered.

What Were the Key Elements of the 1996 Avoided-Cost Feed-In Tariff Methodology?

The FIT announced in 1996 was equal to the avoided cost of energy of the Ceylon Electricity Board. On November 1 of each year, CEB would calculate the avoided cost for the subsequent year, based on the following key parameters: (a) generation forecast from each non-SPP power plant in the forthcoming year, (b) forecasted operating costs per kilowatt-hour of each thermal power plant based on the fuel prices that prevailed on November 1, and (c) estimated avoided maintenance costs of thermal power plants and avoided transmission network losses. Since all SPPs are embedded in the distribution network, it was assumed that there would be no use of the transmission grid (and, hence, a savings on transmission losses) since the SPPs' entire output would be physically consumed by local CEB customers served by the substation to which the SPPs were connected, even though the contractual buyer would be CEB.

For all standardized power-purchase agreements (SPPAs) signed within a year, a floor price was fixed at 90 percent of the FIT announced for the year of signing. *No adjustments for inflation were allowed in the floor price.* Surprisingly, there was no price cap included in the FIT/SPPA, which was later seen as a major negative factor by CEB. But in spite of many requests from CEB and recommendations in various studies, a price cap was never introduced, and remained so until the last SPPA on the basis of avoided-cost tariffs was signed in 2007.

The CEB thought that the combined FIT-SPPA was quite favorable to SPP developers because (a) the SPPA was a "lenient" SPPA with a buy-back guarantee of 15 years and no penalties for nondelivery of energy, and (b) it provided healthy (or even "high") profits to the developers, whose costs were lower than the avoided-cost FIT price. The SPP developers, on the other hand, argued that: (a) the FIT was not adequate to make commercially viable investments, (b) the FIT was based on forecast avoided costs, and for any increases in oil prices within the year, the avoided costs, were not immediately adjusted in the FIT price and hence CEB made additional savings,[6] (c) the methodology, input data, and step-by-step calculation of avoided costs and the FIT were not transparent and were not published with adequate details by CEB.

Three-Year Rolling Average: A Revision to the Avoided-Cost-Based Feed-In Tariff Methodology

As Sri Lanka's economy grew throughout the 1990s, the nation's electricity generation mix moved from being hydropower dominant to becoming mostly reliant on imported oil. In 1999 declining oil prices led to a drop in the forecast avoided cost (and FIT) of about 28 percent. Developers of operational power plants as well as those aspiring to sign PPAs petitioned CEB for relief from these new significantly lower tariffs. To prevent such a large drop in the FIT, CEB agreed to make the FIT equal to the moving average of forecast avoided costs for three years. The averaging period was defined as the two previous years and the current year. After averaging, the reduction in FIT in 1999 was limited to 12 percent. The practice of fixing the FIT to be equal to the three-year moving average continues to date. Thus, the SPP developer has two lines of defense against falling avoided costs: the first line of defense is the three-year averaging, and the second line of defense is the floor price (90 percent of the tariff in the year that the PPA was signed).

What Were the Key Elements of the Cost-Reflective, Time-Tiered, Technology-Specific Feed-In Tariff Methodology Introduced in 2007?

By the end of 2006, 61 SPPs were in operation (57 of them mini-hydro). There were no significant biomass or wind-power plants because the FIT was not high enough to make such SPPs financially viable. In the same year the government announced its goal of generating 10 percent of Sri Lanka's grid electricity from nonconventional[7] renewable energy by 2015. It was clear, however, that this goal could not be met with existing FIT pricing policies for two reasons: most of the country's low-cost mini-hydro sites had already been developed, and the avoided-cost FIT prices were not high enough for nonhydro renewable-energy projects. (Sri Lanka has a resale market in mini-hydro permits that appears to have created a bias to oversizing of mini-hydro plants. See box F.1.)

A new FIT pricing policy was announced in 2007 to encourage nonhydro renewable-energy projects and to base the FIT prices on factors that were less volatile than oil prices. This new policy adopted a pricing methodology that was cost reflective and technology specific (FIT method 2) and also gave developers the choice of two cost-based options: a levelized (that is, flat) FIT and a time-tiered (that is, front-end-loaded) FIT. Under both options, the FIT would be paid in Sri Lankan rupees. Under the levelized option, the FIT remains fixed for the full 20 years of the PPA and there would be no allowed adjustments to the FIT price. (The levelized FIT values by technology for PPAs signed in 2011 are shown in table F.2.) Under the time-tiered system, the highest tariff was set for years 1–8 (tier 1), followed by lower tariffs in years 9–15 (tier 2), and still lower tariffs for years 16–20. (See box F.2 for a discussion of the rationale for the time-tiered option.) The time-tiered option also included a prespecified adjustment mechanism for operation and maintenance (O&M) costs—and for both O&M costs and fuel costs for biomass projects. When presented with these two cost-based options, virtually all SPP developers chose time-tiered FITs. And unlike the earlier avoided-cost FIT approach, neither of the new technology-specific, cost-reflective options (levelized and time-tiered) included floors or ceilings.

Where Will the Money Come from to Pay for the Amount by Which Feed-In Tariffs Exceed the Ceylon Electricity Board's Estimated Financial Avoided Cost?

Sri Lanka's national energy policy states: "The New and Renewable Energy (NRE) strategy shall not cause any additional burden on the end use customer tariffs. If justified, the Government may subsidize the energy utilities for this purpose" (Government of Sri Lanka 2006, 15). The FIT policy revision in 2007 caused the total payment for all SPPs (regardless of technology and fuel source) to be higher than the avoided cost of CEB's thermal generation. The first year of significant impacts occurred in 2011 when PPAs for wind plants that were signed in 2008 became fully operational. The regulator estimated the incremental impact on the cost of supply to customers, after discounting for the fuel saved at the avoided cost of fuel, was about $0.0012/kWh sold to customers or about a 1 percent increase in the average retail tariff. To address this funding shortfall,

Box F.1 Sri Lanka: The "Sizing Decision" and the Permit Resale Market for Mini-Hydro SPPs

The sizing decision without a resale market. An important initial decision for any small power producer (SPP) developer is the sizing decision: how large a generating facility should be constructed. Normally, a private developer, if confronted with an externally established feed-in tariff (FIT) and an externally established demand for the facility's output, would be expected to size the generator to maximize his operating profits (that is, the difference between total costs and total revenues). But this prediction assumes that the initial developer will both build and operate the plant. In fact, in some countries such as Sri Lanka, it is quite common for hydro-plant developers (sometimes referred to as promoters) to sell their operating permit or approval to some other party before a plant is built or becomes operational (see chapter 4). The existence of a resale market for permits, especially one in which the buyer and seller have unequal access to information about likely operating conditions, may create a bias for oversizing of plants that are constructed, especially for mini-hydro plants.

Hydropower output is highly site specific, and is subject to seasonal and annual variations. The amount of money that will need to be invested on a mini-hydro power plant will depend on the (a) plant capacity (MW), (b) designed maximum water-flow rate (cubic meters per second, m^3/s), (c) distance between the water intake and the power plant (the "head"), (d) terrain, (e) length of the transmission line that needs to be built to connect to the buyer's grid, and (f) other investment costs such as the length of the access road that needs to be built.

Let us take a closer look at the first two design parameters—plant capacity (MW) and designed maximum water flow. The height difference between the water intake point and the turbine (the water head), and the designed water-flow rate jointly determine the plant's potential energy output (kWh). A power plant design engineer would always try to maximize the water head, by lowering the turbine location to the safest lowest level. Usually, the river's high flood level is the lowest safest level to locate the turbine. The second parameter, the water-flow rate, depends on the maximum amount of water that can be captured to go through the turbines. Usually, the design flow rate of the plant will be significantly lower than the maximum flow rate of the river. This is because high flow rates in the river exist only for a short period every year or season, and it is not economical to invest a lot of capital on wide channels, bigger intake structures, and additional turbines, just to capture a high flow rate that occurs only for a few hours every year.

Of the six listed factors affecting the size of investment ([a] through [f] above), the highest share of investment is usually for electromechanical equipment (for capacity) and water-flow rate (for example, channels to carry water). Taken together, these investments typically account for 60–80 percent of the total project cost. Capacity (MW) and water-flow rates create energy output (kWh), which, in turn, allows for FIT sales that produce revenue. A mini-hydro developer would normally attempt to maximize both the capacity and the water-flow rate if the additional capital costs to do so will also increase profits.

The sizing decision with a resale market. The sizing incentives change when there is a resale market with asymmetric information, as appears to be the case in Sri Lanka. Some observers believe that project promoters in Sri Lanka have an incentive to oversize the project

box continues next page

Box F.1 Sri Lanka: The "Sizing Decision" and the Permit Resale Market for Mini-Hydro SPPs
(continued)

and to inflate hydrological flow estimates. The incentive to oversize the plant and inflate projected water flows is most likely to occur if the promoter of a plant intends to sell his rights to the project before the project is actually built and becomes operational.

In this preoperational resale market, a rational promoter will naturally try to maximize the payment that he receives when he resells his mini-hydro permit. As the Sri Lankan market is currently structured, the sale price for a permit is keyed to the capacity of the project, measured in megawatts. In other words, this secondary market for mini-hydro permits has project price tags denominated in Sri Lankan rupees per megawatt. A purchaser of a permit, if he is not able to independently validate estimated flow data (which would be difficult to do because most smaller rivers or streams do not have historical flow data) and critically review the electromechanical equipment capacity before crucial investment decisions are made, may discover that he has paid for a hydro capacity that is not able to produce sufficient electricity to be profitable under the current allowed FITs. This means that when the project becomes operational, the new owner will find that the plant operates at a lower capacity factor than was assumed by the regulator in calculating cost-reflective FITs for hydro projects. As a consequence, the new owner will receive lower revenues and a lower rate of return than he had anticipated.

Pressure on the regulator to lower the targeted capacity factor. It should not be a surprise that mini-hydro plant operators, when faced with the unpleasant reality of low-capacity factors, will lobby the government and the regulator to use lower-capacity factors in calculating FITs for mini-hydro plants. The typical complaint from operators is that their sites have lower capacity factors and therefore the FIT should be increased. By and large, the Sri Lankan government and the regulator have resisted this request, the one exception being in 2010 when the regulator reduced the target capacity factor from 42 percent to 38 percent for mini-hydro plants. This had the effect of raising the mini-hydro FIT by about 10 percent. Over time, once the better sites are developed, the newer, undeveloped sites will have lower potential capacity factors. So if hydro is still cheaper than other forms of renewable energy, the regulator will have to lower the capacity factor used in calculating future FITs for mini-hydro plants. But the need for this change over time does not imply that the regulator has an obligation to "bail out" developers who were "snookered" into paying too much money for capacity acquired on the permit resale market.

---

the government decided that in initial years, CEB would continue to pay its own avoided cost for electricity—and the Sri Lanka Energy Fund (SLEF), a fund established to promote green energy, would pay the balance. *But invoices sent by CEB to the administrator of the SLEF for amounts paid above CEB's avoided cost have not been honored.* The basic problem is that the SLEF has no funds for this purpose.

A resolution to the problem may be on the horizon. The Public Utilities Commission of Sri Lanka (PUCSL) has recently made a decision that would allow CEB to pass all FIT costs on to consumers as an interim solution to the problem. In the tariff applications that CEB makes every six months to adjust retail tariffs

Box F.2 Rationale for Introducing Tiered Feed-In Tariffs in Sri Lanka

The rationale for introducing time-tiered feed-in tariffs (FITs) (also sometimes referred to as front-end-loaded tariffs) is that the costs of a typical small power producer (SPP) will vary over time and are likely to be higher in the early years, as an SPP developer will need to make interest and principal payments. Once these loans are paid off (usually after 7–10 years), the project's out-of-pocket costs will drop. Tiered FIT values are designed to track how an SPP's costs will vary over time rather than assuming that the project's costs will remain constant over the entire life of the project. The key feature of a time-tiered or front-end-loaded FIT system is that tariffs are higher in initial periods and lower in later periods. For example, a mini-hydro project that signs a power-purchase agreement (PPA) in 2012 would have the option of taking a levelized FIT of SL Rs 13.04/kWh over the entire life of the project or a tier 1 FIT of SL Rs 14.25/kWh in years 1–8, with significantly lower FITs in the years of tiers 2 and 3.

In Sri Lanka SPPs are usually able to receive bank loans with durations of 6–8 years after commissioning. Hence, in fixing tier 1 of the FIT at 8 years, the FIT methodology is implicitly assuming an 8-year loan repayment period. Once the loan is paid off, the investor's costs will drop because it will no longer have to pay interest and principal on its loan. But if it is going to remain a commercially sustainable operation, it will still need to receive revenues that will allow it to earn a return on its equity and cover its ongoing operation and maintenance (O&M) costs. Since principal and interest payments are no longer being made, the tariff in the tier 2 time period can be set lower. At the end of tier 2 (at the end of year 15 in Sri Lanka), the period that the Sri Lankan government allows investors to earn profits from a natural resource (for example, a hydro plant) comes to an end. But it would not make sense to shut down the plant at the end of year 15, because the plant is still able to produce electricity and at relatively low cost. Hence, from a national perspective, there needs to be a pricing scheme that incentivizes the operator to continue producing electricity. From year 16 onward, the tariff is a fee plus the O&M costs. The fee included in the tariff is about $0.01/kWh. Therefore, the FIT price in tier 3 is even lower than the FIT price paid in tier 2. An overview of the three tiers with the tariffs for hydrogenerators using domestic turbines is given below:

• *Tier 1:* Period of loan repayment while providing a return on equity (14.25 SL Rs/kWh that is, $0.13/kWh for a mini-hydro standardized power-purchase agreement [SPPA] signed in 2011).
• *Tier 2:* Loan-free period. FIT continues to provide a return on equity (6.67 SL Rs/kWh that is, $0.07/kWh for a mini-hydro SPPA signed in 2011).
• *Tier 3:* Renewable energy benefits are returned to society, investor is paid a fee, but not a return on equity. Tier 3 is extendable beyond the twentieth year by mutual consent, at the same tariff (3.29 SL Rs/kWh that is, $0.03/kWh for a mini-hydro SPP signed in 2011).

The O&M cost component that exists in all three tiers is inflation adjusted. The fee in tier 3 is also indexed to inflation, with the inflation adjustment tracked from the first year of operation, but paid only from year 16 to 20. Resource costs and royalties for the use of hydro and wind resources are not changed at any time over the 20-year SPPA or during an extended

box continues next page

Box F.2 Rationale for Introducing Tiered Feed-In Tariffs in Sri Lanka *(continued)*

period after 20 years. Hence the need to return the renewable energy benefits to society is perceived to be achieved through a very low fee in tier 3.

SPP developers also have the alternative of selecting a levelized tariff for the entire life of the plant rather than a three-tier tariff. The levelized tariff assumes indices in the year of the FIT announcement to prevail for 20 years, and calculates the expected payments to the SPP in each year under tiers 1 through 3, and then levelizes the payment using a discount rate that reflects the estimated weighted average cost of capital (WACC) to the investor. The WACC, too, is based on the indices in the year of the FIT announcement.

The concern about "walk-aways." There was a concern that developers who opted for tiered FITs might walk away from the PPA and the project after enjoying the higher tariffs in the tier 1 period. If this happened, developers would get the benefit of higher prices in years 1 to 8 but without fulfilling their obligation to deliver power in later years. It was thought that "walk-aways" were more likely to occur in biomass power plants because of their shorter lifetimes and their continuing need to acquire fuel. Two options to deal with this possible problem were considered. The first was to add a clause in the PPA that would impose a legal obligation on the SPP to pay back the difference between the tier 1 and tier 2 prices and the levelized price if the SPP fails to deliver power from years 16 to 20. The second was to add a clause that required developers to provide CEB with a bank guarantee for the amount of annual excess payments (that is, the difference between the front-end-loaded and levelized FITs) received during tier 1. Developers strongly opposed this second option on the grounds that the cost of acquiring the guarantees would significantly increase their operational costs. The Sri Lankan government adopted the first option. To date, no power plants built under Sri Lanka's tiered tariffs have reached the tier 2 period.

Table F.2 Levelized/Fixed Feed-In Tariff Values for SPPs Signing Power-Purchase Agreements in 2011 in Sri Lanka

Technology/source	Fixed tariff (SL Rs/kWh)	Fixed tariff ($/kWh)
Mini-hydro	13.04	0.118
Mini-hydro—local	13.32	0.120
Wind	19.43	0.175
Wind—local	19.97	0.180
Biomass (dendro)	20.7	0.187
Biomass (agro and industrial waste)	14.53	0.131
Municipal waste	22.02	0.199
Waste heat	6.64	0.060

Source: PUCSL 2010.
Note: Only flat tariff values for the flat tariff option are shown. A more complex table showing the full variety of tiered and scalable tariffs is available at: http://www.pucsl.gov.lk/images/stories/pdf/PUCSL_NCRE_Advertisement.pdf.
SL Rs = Sri Lankan rupees.

for expected changes in generation costs, CEB now includes the full projected cost of purchasing renewable energy at the prices applicable to each SPP contract for that particular period. At the end of the six-month period, there is a "true-up": the projected costs are adjusted upward or downward based on the energy actually purchased by CEB from SPPs. Previously, CEB was only allowed to pass on the avoided-cost component of its FIT to its retail customers. (See appendix D for calculations of the estimated increase in retail tariffs that result from paying above avoided-cost FITs in Sri Lanka and Thailand.) Since CEB's retail customers will now pay for the premium in their tariffs, the premium, while relatively small at present, does constitute an "additional burden on the end use customer tariffs." If this provision is strictly implemented, the target of serving 10 percent of grid electricity from nonconventional renewable energy (NCRE) by 2015 may not be achieved. The most likely alternative would be for the government to pay the premium, but that simply means that most Sri Lankans will end up paying for the premium through their taxes rather than through their electricity bills.

South Africa's Feed-In Tariffs and Competitive Procurements

South Africa provides a striking example of major shifts in government policies for promoting renewable generation. Initially, the National Energy Regulator of South Africa (NERSA) proposed to use FITs to set prices for purchases of energy from renewable generators, both large and small. In 2009 NERSA announced specific FIT prices for a variety of technologies using estimates of levelized cost by technology. Then, in April 2011, citing lower-than-expected inflation and debt costs, NERSA declared its intent to lower FIT prices by up to 40 percent, as shown in table F.3 (Sguazzin 2011). In fact, no contracts were ever signed at either the higher proposed 2009 FIT prices or the lower proposed 2011 FIT prices because in June 2011 South Africa's renewable energy policy took yet another unexpected turn. The South African government announced that the regulator's planned administrative determination of FITs was not legal. Instead, the government decided that all independent renewable energy generation would be acquired through competitive procurements.

The stated reason for this policy "U-turn" was that the treasury department had determined that FITs established through administrative determination by NERSA were inconsistent with the South African constitution and existing laws. One South African newspaper quoted a treasury department spokesperson as saying that "different pieces of legislation prohibit South Africa from embarking on a FIT procurement process where the tariff is set upfront by the regulator, ... according to the legal opinions we have received, the only possible procurement process option the Department of Energy currently has, is a competitive bidding process" (Gosling 2011). It was reported that the legal analysis had determined that Eskom, the designated state-owned buyer of the renewable energy, was required to always seek competitive bids on price under South Africa's public procurement law, which applies to all state-owned enterprises. Hence, fixed prices set in advance by the regulator based on the regulator's estimates of the

Table F.3 Feed-In Tariff Values in 2009 and Proposed Values for 2011

	REFIT 2009	REFIT 2011	REFIT 2009	REFIT 2011	% change
	R/kWh		USD/kWh		2011/2009
Wind ≥1 MW	1.25	0.938	0.184	0.138	−25.0
Landfill gas ≥1 MW	0.9	0.538	0.133	0.079	−40.2
Small hydro ≥1 MW	0.94	0.671	0.138	0.099	−28.6
CSP trough ≥1 MW with 6 hours storage	2.1	1.836	0.309	0.270	−12.6
CSP trough ≥1 MW without storage	3.14	1.938	0.463	0.285	−38.3
CSP central receiver ≥1 MW with 6 hours storage	2.31	1.399	0.340	0.206	−39.4
Photovoltaic ≥1 MW	3.94	2.311	0.580	0.340	−41.3
Biomass solid ≥1 MW (direct combustion)	1.18	1.06	0.174	0.156	−10.2
Biogas ≥1 MW	0.96	0.837	0.141	0.123	−12.8

Source: Long 2011.
Note: CSP = concentrating solar power; kWh = kilowatt-hour; MW = megawatts; R = South African rand; REFIT = renewable energy feed-in tariff.

production costs of specific renewable technologies would be inconsistent with this law. The treasury department decided that FITs in South Africa must be set through competitive bidding with a heavy weight on the prices bid rather than through either administrative determination of avoided cost (FIT method 1) or administrative determination of levelized technology-specific costs (FIT method 2), which was NERSA's proposed approach in 2009 and 2011.

The South African government also stated that its overall goal was to acquire 3,725 MW of operating renewable-energy-generating capacity by 2016 through five separate rounds of competitive bidding. As of November 2012, two separate rounds of competitive procurements had been completed. In the first round, 53 bids were received and 28 preferred bidders (that is, winners) were selected whose accepted bids totaled 1,415 MW. In the second round, 79 bids were received and 19 preferred bidders were selected whose accepted bids totaled 1,043 MW of capacity. In both rounds, the bidding was conducted on a technology-by-technology basis with separate maximum-quantity ceilings set by the Department of Energy for each of four technologies (solar photovoltaic [PV], wind, small hydro, and concentrated solar power). The bids were evaluated on a multi-attribute basis with 70 percent weight given to the bidder's levelized price in rand and 30 percent weight to 17 other factors such as job creation, socioeconomic impact, and local content. After selection, preferred bidders were given 6 months to achieve financial closure to maintain their status as preferred bidders (Republic of South Africa, Department of Energy 2012).

These first two rounds of bidding were open to both large and small renewable power producers. Five of the winning bidders proposed projects that were 10 MW or less in size. Three were PV and two were hydro. Table F.4 shows the levelized prices of all winning bidders, small and large, in rounds 1 and 2. In all

Table F.4 Average of Winning Bids for Rounds 1 and 2 in South Africa's Bidding Program

	Round 1 *Average of winning bids (U.S. cents)*	*Round 2* *Average of winning bids (U.S. cents)*
Solar photovoltaic	33.6	20.0
Wind	13.9	10.9
Small hydro	n.a.	12.5
Concentrated solar power	32.8	30.6

Source: Republic of South Africa, Department of Energy, May 2012.
Note: n.a. = not applicable.

instances, the average bid prices showed significant declines between the first two rounds, even though the bid rounds were only four months apart (November 2011 versus March 2012). To ensure that these were serious bids, bid bonds of $12,500/MW of proposed nameplate capacity were required of all bidders and this amount was doubled if the bidder was given preferred bidder status. In November 2012 implementation, direct, and power-purchase agreements were signed between the government, Eskom, and the 28 successful bidders. If these projects are built, it is estimated that they will result in a total investment of close to $6 billion (Eberhard 2013).

To date, the South African competitive bidding program has been quite successful. But it is still in its early stages so it is premature to reach any firm conclusions as to its ultimate effectiveness. A key question is how many of the selected projects will achieve financial closure and actually become operational. In other countries that have used competitive bidding, it has become clear that some winning bidders bid too low and find themselves unable to build and operate the projects at the prices that they bid. This phenomenon is known as winner's curse and has been observed in India. But it seems less likely that this will happen in South Africa because of the bid bond requirements. It is also important to note that the South African bidding program has significant administrative costs. The Department of Energy reported using at least 11 separate consulting firms to help it in designing and conducting the bidding process and then in evaluating the bids received. Presumably, the 132 entities that submitted bids also expended considerable money to develop their bids.

The South African government has recognized that the full-fledged bidding program used to date might not be best for promoting SPPs. At the time of this writing, the Department of Energy had announced that it will conduct a separate Small Projects Procurement Programme for renewable energy generators between 1 and 5 MW. One hundred MW of the total 3,725 MW of installed renewable capacity has been specifically allocated for acquisition from small renewable producers. In creating this alternate track, the Department of Energy stated that its goal was to produce "a more simplistic procurement programme with less requirements and cheaper for bidders to participate" (Government of South Africa 2011). It remains to be seen how the bidding program for smaller power producers will be streamlined relative to the bidding process for larger projects.

The South African Department of Energy also runs a separate rural-electrification program that does not overlap in any obvious way with its renewable-energy bidding program. To date, almost all rural electrification has taken place through extension of the main grid and the installation of solar home systems (Barnard 2011). This bifurcated approach also seems to be used in most other Sub-Saharan African countries. Presumably, this reflects the fact that it is more complicated to design a competitive bidding program that is simultaneously supposed to promote two different outcomes. For example, in Tanzania, there are two separate programs for promoting rural electrification and renewable energy and both operate on a first-come, first-served basis. And since the programs operate separately, there has been no need to decide between projects that might offer differing proportions of electrification and renewable energy.

Notes

1. See EWURA (2010). The SPP rules impose a mandatory purchase obligation on any distribution network operator (DNO). At present, TANESCO is the only functioning DNO in Tanzania.

2. The long-run marginal cost (LRMC) is based on TANESCO's least-cost generation supply plan, adjusted to reflect reduction in transmission losses. The LRMC was the same for both the main grid and isolated mini-grid FIT calculations.

3. For further discussion of these operating issues, see chapter 8.

4. The economic, institutional, and technical implications of connecting a previously isolated mini-grid to the main grid are discussed more fully in chapter 10.

5. In the Tanzania case, a CPI adjustment is provided for the ceiling but not for the floor. Mathematically it makes sense that both the ceiling and the floor would be CPI adjusted, but in Tanzania the decision to not adjust the floor was political, to broker TANESCO buy-in to the concept of SPPs.

6. Oil prices, and hence the avoided costs, dropped from 1998 to 1999 to 2000. Otherwise, throughout the period from 1996 to 2011, the avoided costs continued to increase, owing to the increasing share of thermal generation in the Sri Lankan grid, increases in oil prices, and depreciation of the Sri Lankan rupee against stronger currencies.

7. In Sri Lanka "nonconventional" refers to all renewable energy excluding large hydropower.

References

Barnard, Wolsey. 2011. *Renewable Energy and Rural Electrification: South Africa Experience.* Department of Energy, Republic of South Africa, Pretoria. http://www.energy.gov.za /files/media/presentations/2011/20111206_WolseyBarnard_RE_electrification Presentationv2.pdf.

Eberhard, Anton. 2013. *Feed-in Tariffs or Auctions: The Renewable Energy IPP Procurement Process in South Africa.* Viewpoint, Financial and Private Sector Vice Presidency, World Bank Group, Washington, DC.

EWURA (Energy and Water Utilities Regulatory Authority). 2010. *The Electricity Act (CAP 131): The Electricity (Development of Small Power Projects) Rules, 2010.* Dar es Salaam, Tanzania. http://www.ewura.com/pdf/SPPT/PROPOSED%20RULES/The%20Electricity%20(Development%20of%20Small%20Power%20Project)%20Rules-2010.pdf.

————. 2011a. *Detailed Tariff Calculations for Year 2011 for the Sale of Electricity to the Main Grid in Tanzania under Standardized Small Power Purchase Agreements in Tanzania.* Dar es Salaam, Tanzania. http://www.ewura.go.tz/pdf/SPPT/PROPOSED%20GUIDELINES/PROCESS%20GUIDELINES/2011%20SPPT%20Calculation%20for%20Main%20Grid.pdf.

————. 2011b. *Detailed Tariff Calculations for Year 2011 for the Sale of Electricity to the Mini-Grids in Tanzania under Standardized Small Power Purchase Agreements in Tanzania.* Dar es Salaam, Tanzania. http://www.ewura.go.tz/pdf/SPPT/2011%20SPPT%20Calculation%20for%20Mini-Grid.pdf.

————. 2012a. *Detailed Tariff Calculations for Year 2012 for the Sale of Electricity to the Main Grid in Tanzania under Standardized Small Power Purchase Agreements in Tanzania.* Dar es Salaam, Tanzania. http://www.ewura.go.tz/pdf/SPPT/2012/2012%20SPPT%20Calculation%20for%20Main%20Grid.pdf.

————. 2012b. *Detailed Tariff Calculations for Year 2012 for the Sale of Electricity to the Mini-Grids in Tanzania under Standardized Small Power Purchase Agreements in Tanzania.* Dar es Salaam, Tanzania. http://www.ewura.go.tz/pdf/SPPT/2012/2012%20SPPT%20Calculation%20for%20Mini-Grid.pdf.

Gaddis, Isis. 2012. "Only 14% of Tanzanians Have Electricity: What Can Be Done?" *End Poverty* (blog), October 31. http://blogs.worldbank.org/africacan/node/2187.

Gosling, Melanie. 2011. "Government's U-turn on Wind Energy Rates." *Cape Times,* June 20.

Government of South Africa, Department of Energy. 2011. "Fact Sheet for the Media Briefing Session on 31 August 2011 re the Renewable Energy Independent Power Producer (IPP) Programme." Government of South Africa, August 31. http://www.info.gov.za/speeches/docs/2011/reippp.pdf.

Government of Sri Lanka, Ministry of Power and Energy. 2006. *National Energy Policy and Strategies of Sri Lanka.* http://www.futurepolicy.org/fileadmin/user_upload/PACT/Laws/Sri_Lanka_Energy_Policy_2006.pdf.

Long, Bridget. 2011. "Presentation to SAIPPA: Recent NERSA COFIT and REFIT Public Hearings." IES Energy, San Diego, CA, May 24. http://0101.nccdn.net/1_5/1d3/1a7/263/Bridget-Long-Presentation.pdf.

PUCSL (Public Utilities Commission of Sri Lanka). 2010. *Cost Reflective Methodology for Tariffs and Charges: Methodology for Charges.* Colombo, September 30. http://www.pucsl.gov.lk/english/wp-content/themes/pucsl/pdfs/methodology_for_charges.pdf.

Republic of South Africa Department of Energy. 2012. "Preferred Bidders—Window 2." *Energy,* May 22. http://www.ipprenewables.co.za/#blog/post/view/isAjaxRequest/true/id/182.

Sguazzin, Antony. 2011. *South Africa Cuts Proposed Renewable Energy Prices, Business Day Says.* Bloomberg, South Africa, March 23. http://www.bloomberg.com/news/2011-03-23/south-africa-cuts-proposed-renewable-energy-prices-business-day-says.html.

U.S. Foreign Commercial Service. 2010. *Doing Business in Tanzania: 2011 Country Commercial Guide for U.S. Companies.* http://photos.state.gov/libraries/tanzania/231771/PDFs/Country_Commercial_Guide_2011_Tanzania.pdf.

Topping Up Feed-In Tariffs by Donors: Key Implementation Issues

The topping up of feed-in tariffs (FITs) by donors, as pioneered by Uganda, has considerable potential for accelerating the development of renewable generators in Africa. Some of the key issues that would need to be resolved by ministries and regulators include the following:

Eligible Renewable Technologies

Should the top-up be available to all renewable technologies or just the ones that require the least top-up on a per-kWh generated basis?

Comment

Presumably, donors will want to get the most renewable energy production possible in return for their top-up grants. This then implies that donors should subsidize those renewable technologies that require the smallest top-up on a per-kWh basis. In East Africa small hydro and some forms of biomass are the renewable technologies that are likely to require the least amount of subsidy per kilowatt-hour.

But there may be other risks in choosing renewable technologies solely on the basis of minimizing top-up charges. Consider the case of small hydro generators— if these are built at locations that are equally vulnerable to the multiyear drought that has affected various East African countries, then the country has not really diversified its portfolio of supply sources. If the small hydro facilities are located in the same river basin as a large hydro plant that is experiencing low electricity production because of drought conditions, small hydro may not be very helpful in improving a country's security of supply.

Uniform versus Particularized Top-Ups

Should the top-up amount depend on project-generating capacity or be size neutral (uniform across all projects using that the same technology)?

Comment

All renewable energy technologies have some degree of economy of scale: big projects generally have lower costs per megawatt or per megawatt-hour (MWh) than smaller projects of the same technology. For this reason it is not uncommon for projects that are below 500 kilowatts (kW) to receive a higher tariff, with a somewhat lower tariff for projects between 500 kW and 1 MW, and yet a still lower tariff for those above 1 MW. Because economies of scale vary by technology, the difference between tiered FITs will also vary by technology.

Disbursement

Should the top-up be disbursed over the entire life of the project or should disbursement be accelerated so that the subsidy is completely disbursed over some shorter period of time?

Comment

In Uganda it has been recommended that the subsidy be disbursed in two tranches: the first tranche of 50 percent of the total subsidy amount would be disbursed on the date of the project's commercial operation and the second tranche for the remaining 50 percent of the net present value of the top-up would be disbursed over the first five years of commercial operation. The justification for this approach is that it will be difficult for donors to provide a credible up-front guarantee of a future top-up payment in every year of the 15–20 years of a typical power-purchase agreement (PPA). The reality is that donor priorities change. A bilateral donor may have the money now, but it cannot provide a credible commitment that it will have the money 15 or 20 years in the future. This argues for early and full disbursement of the top-up payment even though it is intended as a payment for 15–20 years of expected renewable energy production. But early and full disbursement also raises at least two concerns. The first is whether the developer will have an incentive to continue operations after the last subsidy payment is received in year five. The second is that it would require estimating the expected production of electricity from the project over its full life.

Selection

Should the selection of the projects that will receive the subsidy be performed on a first-come, first-served or on a competitive basis?

Comment

Considerable time and resources are required to set up a fair and efficient competitive selection. Many potential renewable producers are small, so they do not have the time or resources to participate in a competitive procurement. In most African countries, there are usually a handful of projects that are at a reasonably advanced stage of development. Hence, the better approach would be to provide the subsidies and technical assistance to these ready-to-go projects on a first-come, first-served basis and then, if these projects become operational, consider developing a system for competitively awarding top-up payments at a later stage.

To date, there has been little experience with establishing FITs through competitive bidding by renewable energy projects under 10 MW. The major problem is that smaller project developers do not have the deep pockets necessary to develop a project when it is uncertain whether their bid for a tariff will be successful or not. Similarly, the government's costs in running such bidding programs can be high. Successful renewable-energy-subsidy-bidding programs, such as Thailand's bidding program for small power producers (SPPs), have generally focused on large (tens of megawatts) projects. As discussed in appendix F, South Africa has established a competitive bidding program for both large and small renewable projects because the government concluded that administratively set FITs would violate the South African constitution. The bidding for the larger projects appears to have been quite successful. The selected bidders bid prices that were significantly below the regulator's previously proposed FITs (Eberhard 2013). The South African government plans to conduct a separate multi-attribute bidding process for renewable energy projects under 10 MW. But it is too early to know if the bidding program for smaller renewable projects will be equally successful.

Relationship to Carbon Credits

Should expected carbon credits be ignored in calculating the size of top-up payments?

Comment

If it is prespecified that the size of top-up payments will be reduced for every dollar or euro that the developer receives through carbon credits, this policy is likely to reduce or eliminate any incentive for developers to apply for carbon credits. This may not be a big concern now because the level of revenues expected to be received through carbon credits has declined with the recent worldwide decline in the price of certified emission reductions (CERs). But this might be an important concern in the future if the price of CERs increases.

Top-Up, Buy-Down, or Both?

Should the program also include a buy-down to lower the cost of purchases by the buying entity or should it be limited to just a top-up for the selling entity?

Comment

The Global Energy Transfer Feed-in Tariff (GET FiT) program described earlier (chapter 5) is designed to provide extra top-up payments to developers of potential renewable energy projects. The top-up payments are in addition to any base payment that the project developer will receive from the buying utility. The top-up or premium payments are intended to ensure the commercial viability for some renewable technologies. To date, we are not aware of any program that would subsidize the base FIT payments made by buying utilities. The underlying assumption seems to be that the buying utility will make these purchases if the government imposes a legal obligation on it, but legal mandates may not be effective if they are inconsistent with the buyer's economic incentives.

Consider the case when a top-up of 2 cents is added to a previously established FIT of 9 cents so that the total price received by the renewable generator is 11 cents. But if the buying utility's avoided cost is 6 cents, then it would find itself being forced to pay 3 additional cents per kilowatt-hour above the cost that it would have incurred if it had self-supplied or purchased electricity from some other source (that is, its avoided cost without the purchase from the SPP). So even if the buying utility has a legal obligation to make these purchases, it will be reluctant to do so (and may find subtle ways to avoid making some or all of the purchases) because it views itself as being forced to pay 3 cents too much for every kilowatt-hour that it purchases under the program. The fact that the renewable generator is getting a top-up payment of 2 cents does not help the buyer who would rationally prefer to supply its power needs from some less-expensive source. This implies that a premium payment program like the proposed GET FiT program would be more successful if the buying entity is also given an explicit financial incentive above and beyond the top-up incentive given to the renewable generator.

Such incentives could take the form of "buy-downs" so that the buyer's net purchase costs would be equal to or less than their avoided costs. While this would raise the overall cost of the program because it would require giving both "top-ups" and "buy-downs," it might lead to active support rather than passive resistance from an otherwise reluctant buyer.

Extra Top-Up for Electrification?

Should a project receive extra payments if it can also commit to electrifying households and businesses?

Comment

African government officials often state that electrification is much more important to their country than renewable energy. This suggests that higher subsidies should be granted to those projects that can offer both renewable energy and electrification. But the counterargument is that it is more complex to design and implement a single program that tries to simultaneously achieve two different outcomes. For example, in selecting a renewable energy project from among competing projects, it would be necessary to decide how much weight should be given to the electrification versus the renewable energy outcomes. Another consideration is that subsidies for connecting rural households and businesses are already provided through donor grants to rural electrification agencies or national utilities in a number of African countries. Therefore, in an initial phase, we think that it would be best to allow the two subsidy programs to operate separately rather than trying to merge them into a single program.

Guarantees of Payment

Does a top-up program also need to be accompanied by a payment guarantee mechanism?

Comment

As noted earlier, many African utilities are commercially insolvent (Eberhard and others 2008). This means that they may be unable to make the basic FIT payment to renewable generators to which the top-up payment is added. So the top-up payment will be of little value to renewable generators if the basic payment is not made or is made with considerable delay. Therefore, it has been proposed that there should be some additional mechanism to guarantee that the basic payment will be made. In Kenya, the World Bank has given payment guarantees (known as partial risk guarantees [PRGs]) for purchases by Kenya Power from larger independent power producers (IPPs). It is proposed that similar guarantees backed by a government counterguarantee be given to small renewable generators who would be eligible to receive top-up payments. But the counterargument is that a payment guarantee is simply a band-aid because it does not solve the basic underlying problem: the financial insolvency of the buyer. While it would give SPPs a more favorable position in the buying utility's accounts payable queue, presumably it would mean that some other supplier of goods and services to the buying utility will not get paid or will get paid with a longer delay. It remains to be seen whether the World Bank or any other financial institution would be willing to give such a payment guarantee unless tariff and operating reforms are first made to ensure that the buying enterprise has some minimum level of financial viability.

Concessional versus Market Financing for Equity and Debt

Should the core equity and debt financing for SPPs be reserved for commercial sources of financing?

Comment

Top-up programs will work only if there is core financing of equity and debt for the SPP developer. Without the provision of this initial capital, the projects will not be in a position to receive top-up FIT payments from donors. Commercial banks, who have done a lot of the initial groundwork and due diligence for GET FiT programs, argue that it is unfair for them to be crowded out by donor organizations who piggyback off their groundwork and who are able to provide equity and debt on concessional (that is, below-market) terms. The banks contend that if concessional sources of financing are allowed to come in at the last minute, it effectively eliminates any incentive for them to do the costly developmental work. Moreover, donor funding comes and goes. The long-term sustainability of renewable generation projects will require reliance on regular commercial financing channels. Developers, on the other hand, argue that they should be able to seek financing from whomever is willing to offer them the lowest financing costs and there should be no requirement that certain commercial banks be given exclusive rights to finance their projects.

References

Eberhard, Anton. 2013. *Feed-in Tariffs or Auctions: The Renewable Energy IPP Procurement Process in South Africa.* Viewpoint, Financial and Private Sector Vice Presidency, World Bank Group, Washington, DC.

Eberhard, Anton, Vivien Foster, Cecilia Briceño-Garmendia, Fatimata Ouedraogo, Daniel Camos, and Maria Shkaratan. 2008. "Underpowered: The State of the Power Sector in Sub-Saharan Africa." Background Paper, Africa Infrastructure Country Diagnostic, World Bank, Washington, DC. https://openknowledge.worldbank.org/handle /10986/7833.

Glossary

Avoided costs: Avoided costs are the incremental costs that can be avoided by acquiring electricity from a small power producer (SPP). Avoided costs are used in some countries to calculate feed-in tariffs. The three categories of avoided costs are financial, economic, and social. Financial avoided costs are based on how much it would cost the utility to generate the electricity provided by the SPP or another supplier. Economic avoided costs are based on how much it would cost the national economy to replace the electricity generated by the SPP. Economic avoided costs do not include subsidies or taxes because these are internal transfers within the national economy. Social avoided costs are calculated as the economic avoided costs plus the environmental and health costs that would be incurred locally and globally if the electricity had to be generated from a source other than the SPP.

Backup tariff: A tariff that compensates the national utility or other supplier for providing electricity to an SPP when the SPP is not generating enough electricity to meet its loads. The SPP may need to buy backup power for one or more reasons: the SPP's generator may be too small to meet its own or its retail customers' demand; the SPP's generator may need an external source of power to restart after it was shut down because of a planned or unplanned outage on its system or the system to which it is selling power; or the SPP may need supplementary supply to service its own retail customers when the SPP is not generating (for whatever reason).

Bulk supply tariff: The tariff applied to sales of electric power in bulk to a reseller, usually a distribution entity, that resells the electric power to retail customers.

Centralized electrification track: The centralized electrification track is a top-down approach to electrification that typically occurs through expansion of medium- and high-voltage power grids built and operated through the separate or joint actions of a national or regional power company, a government ministry, or a rural electrification agency.

Certified emission reduction (CER) credits: CER credits are payments offered by the United Nations Clean Development Mechanism or other emissions abatement programs to entities that are able to offer a specified and audited reduction in carbon emissions against an estimated business-as-usual benchmark.

Cogeneration: *See* Combined heat and power.

Combined heat and power (CHP): CHP is a power generation source that simultaneously generates both electricity and useful heat.

Commercially sustainable: A commercially sustainable outcome is one in which an entity is able to recover its operating costs and depreciation on all capital assets (whether supplied by the operator or others), a return on invested equity, and debt payments (if any), while also setting aside reserves to deal with emergency repairs and replacements.

Connection charge: A connection charge is the payment required from new customers for their initial physical connection to an electricity supplier.

Cross-subsidy: A cross-subsidy is a tariff structure in which some customers (such as businesses) pay a higher tariff to subsidize the tariffs of other customers (such as poor households).

Debt service coverage ratio (DSCR): The DSCR is equal to net operating income divided by the sum of interest, principal, and lease payments.

Decentralized electrification track: The decentralized electrification track is a bottom-up approach to expanding access to electricity in which electrification is achieved through the creation of isolated or grid-connected mini-grids operated by private, cooperative, or community-based organizations.

Distribution margin: The distribution margin is the difference between the average retail tariff of an entity that is providing distribution service, and the average bulk supply tariff that the entity pays to purchase wholesale electricity.

Distribution network operator (DNO): As defined in Tanzania, a DNO is an entity responsible for the operation of a distribution network serving 10,000 customers or more.

Embedded generator: An embedded generator is a single generator or a group of generating plants connected to a medium-voltage distribution network (typically 33 kV or less).

Feed-in tariff (FIT): A FIT is a tariff-support mechanism for renewable energy generators or cogenerators in which the generator is guaranteed a payment, usually over a long-term period, for every kWh generated and fed into the grid.

Grid-connected SPP: An SPP that is connected to the main or a regional grid.

Hybrid system: A system of two or more energy sources used together to provide increased system efficiency and lower costs is referred to as a hybrid system. Hybrid systems often involve one or more renewable energy generators, together with a fossil fuel–powered source such as a diesel generator.

Interconnection: Interconnection refers to all of the physical facilities needed to connect an SPP to an existing grid.

Interconnection costs: Interconnection costs are those costs paid by an SPP to connect to a purchaser, whether that purchaser is the national utility or some other entity. They include any costs required to upgrade a grid operator's system to receive electricity produced by the SPP.

Interconnection point: The interconnection point is the point at which a power seller's electric output line (or electric system) feeds into the electric system to which it delivers power, whether owned by the buyer or another entity.

Isolated mini-grid: An isolated mini-grid is an electricity generation and distribution network that is physically isolated from the main grid or a regional grid.

Letter of intent (LOI): A letter issued by a power buyer to an SPP stating its intent to connect to the SPP and purchase power from it.

License: Regulatory authorities authorize operators to generate, transmit, distribute and sell power by issuing consolidated or separate licenses to them.

Liquidated damages: When one party to a power-purchase agreement (PPA) fails to perform as specified by the agreement, it may have to compensate the other party through a payment known as liquidated damages.

Load factor: The load factor is the ratio of the average electric load (measured across one billing interval, typically one month) to the peak load (measured in intervals consistent with those specified in the grid code, typically 15 minutes) averaged over a period of time corresponding to the billing interval.

Main grid: This term refers to the interconnected electricity transmission network of a country or region. Typically most sizeable electricity generating facilities in a given country or region are connected to the main grid.

Micro-grid: A micro-grid is like a mini-grid, but smaller. Some micro-grids operate on DC current.

Mini-grid: A mini-grid is a small electricity generation and distribution network, typically with a generation capacity of less than 10 MW. It may be physically separate (isolated) from the main grid in the area. Alternatively, it may be connected to the main grid but have a separate owner and operator that performs commercial functions (metering, billing, and collections) and technical functions (repairs, maintenance, and replacement of distribution facilities) that would otherwise be performed by the main grid operator.

Must-take: A power-purchase agreement (PPA) with a "must-take" clause obligates the buyer to take all of the electricity produced by the supplier. SPPs usually have a "must-take" clause in their PPAs.

Partial risk guarantee (PRG): A PRG protects private lenders against the risk of a national electric utility or other public entity failing to perform its obligations under an agreement with an SPP or private power producer. It can cover a range of risks, including the governmental or quasi-governmental entity's failure to meet contractual payment obligations, changes in law, obstruction of an arbitration process, expropriation and nationalization, foreign currency availability and convertibility, nonpayment of a termination amount or an arbitration award following a covered default, and failure to issue licenses, approvals, and consents in a timely manner.

Point of common coupling (PCC): When an SPP connects to a grid, the PCC is the point on the DNO-owned grid beyond which other lines, customers, and other SPPs are connected.

Point of interconnection (POI): The POI is the point in a connection between a distribution network and a small power producer beyond which all technical matters are the responsibility of the utility.

Point of supply (POS): The POS is the location of the metering point at which the SPP sells power to the utility that owns and operates the main grid.

Power-purchase agreement (PPA): A PPA is a multiyear contract between a generator and a buyer of power. The agreement details the rights and obligations of the two parties. The PPAs applied to SPPs usually take one of several forms that have become standardized over the years.

Provisional license: Regulatory authorities issue provisional licenses (valid only for a limited period) to allow operators to conduct preparatory activities (such as assessments, studies, and acquisition of land and resource rights and other non-sector-specific government approvals) necessary to apply for a full license.

Regional grid: A regional grid is an electric power system that serves one or more regions of a country. It may or may not be connected to the country's main grid.

Renewable energy: Renewable energy comes from natural resources (such as sunlight, wind, water, tides, biomass, and geothermal heat) that are continually replenished over a short time period—not millions of years, as with fossil fuels.

Reseller: A reseller purchases power at wholesale for the purpose of reselling it at retail to end-use customers (households and businesses).

Retailer: A power retailer sells electricity to end-use customers. It sells electricity that it generated from its own plants or that was purchased from one or more wholesale suppliers.

Small power distributor (SPD): An SPD is an entity that purchases electricity at wholesale prices from a bulk supplier (such as a distribution network operator) and resells it at retail prices to end-use customers.

Small power producer (SPP): An SPP is an independently operated, small-scale electricity generator. An SPP can operate on isolated mini-grids or mini-grids that are connected to a larger national or regional grid or on a direct connection to a national or regional grid. SPPs may sell the electricity they generate at wholesale to a distribution network operator, at retail directly to end-use customers, or both. SPPs are typically defined as having a power export capacity smaller than some threshold (for example, 10 MW).

Standby tariff: *See* Backup tariff.

Tariff: A tariff is any charge, fee, price, or rate that must be paid to purchase electricity.

Uniform national tariff: Under a uniform national tariff system, all customers in the country in a given tariff category are charged the same price regardless of geographic location or differences in the cost of supply.

Wholesale: The wholesale sale of electricity refers to the sale of electricity for resale.

Bibliography

Regulatory and Policy Documents from Countries with Small Power Producer Programs

Cambodia

EAC (Electricity Authority of Cambodia). 2012. "On Determination of Electricity Tariff Based on Fuel Adjustment Mechanism for Consumers in the Distribution Area of Mr. Chet Layhim." Decision 098-SR-12-EAC. Phnom Penh, Cambodia.

Mahé, Jean Pierre, and Ky Chanthan. 2005. *Rehabilitation of a Rural Electricity System.* GRET and Kosan Engineering, Phnom Penh. http://www.gret.org/wp-content /uploads/07409.pdf.

India

ABPS Infrastructure Advisory Pvt. Ltd. 2011. *Policy and Regulatory Interventions to Support Community-Level Off-Grid Projects.* Final Report. http://www.forumofregulators.gov .in/Data/Reports/CWF%20Off-grid%20final%20report%20nov%202011_Latest _feb2012.pdf.

———. 2012. *Model Draft Regulations for Off-Grid RECs for Community-Level Off-Grid Projects.* Prepared for the Forum of Electricity Regulators. New Delhi, India.

Government of India. 2003. *The Electricity Act, 2003.* http://guj-epd.gov.in/extra_no_46 .pdf.

Government of India, Ministry of Power. n.d. *Definition of Electrified Villages.* http://rggvy .gov.in/rggvy/rggvyportal/def_elect_vill.htm.

———. 2005. *National Electricity Policy.* http://218.248.11.68/energy/NationalElectPolicy6 .asp?lnk=26.

———. 2006a. *Rural Electrification Policy, 2006.* http://www.powermin.nic.in/whats_new /pdf/RE%20Policy.pdf.

———. 2006b. *Tariff Policy.* http://www.powermin.nic.in/whats_new/pdf/Tariff_Policy.pdf.

———. 2009. *Guidelines for Village Electrification through Decentralized Distributed Generation (DDG) under Rajiv Gandhi Grameen Vidyutikaran Yojana in the XI Plan— Scheme of Rural Electricity Infrastructure and Household Electrification.* http://www .indiaenvironmentportal.org.in/files/Guidelines_for_Village_Electrification_DDG _under_RGGVY.pdf.

WESCO (Western Electricity Supply Company). 2011. *Provision of Revenue Subsidy for Sustainability of Rural Electricity Supply in Villages Being Electrified under RGGVY.* Bidyut Niyamak Bhawan, Kalyani Complex, Unit VIII.

Kenya

Castalia Strategic Advisors. 2009. *Kenya Electrification Investment and Policy Prospectus*. Africa Renewable Energy Access Program, World Bank, Washington, DC, October.

ECA (Economic Consulting Associates), and Ramboll Management Consulting. 2012a. *Connection Guidelines for Small-Scale Renewable Generating Plant*. Government of Kenya, Ministry of Energy, August.

———. 2012b. *Feed-In Tariff Policy: Application and Implementation Guidelines*. Government of Kenya, Ministry of Energy, August.

———. 2012c. *Feed-In-Tariff Policy on Wind, Biomass, Small-Hydro, Geothermal, Biogas and Solar Resource Generated Electricity*. 2nd rev. Government of Kenya, Ministry of Energy, August.

———. 2012d. *Standardized Non-negotiable Power Purchase Agreement for Renewable Generators of Less than 10 MW*. Government of Kenya, Ministry of Energy, August.

———. 2012e. *Technical and Economic Study for Development of Small Scale Grid Connected Renewable Energy in Kenya*. London, United Kingdom: Economic Consulting Associates.

Kenya Ministry of Energy. 2010. *Feed-in-Tariffs Policy for Wind, Biomass, Small Hydros, Geothermal, Biogas and Solar Resource Generated Electricity*. http://www.erc.go.ke/erc /fitpolicy.pdf.

Kenya Power. 2008. *Rates and Tariffs*. http://www.kplc.co.ke/index.php?id=45.

Mali

AMADER (Malian Agency for the Development of Household Energy and Rural Electrification). n.d. *Concession Contract*. World Bank. http://ppp.worldbank.org /public-private-partnership/sites/ppp.worldbank.org/files/documents /Mali11CONCESSION0CONTRACT0YK.pdf.

———. n.d. *Specifications Annexed to Concession Order*. World Bank. http://ppp .worldbank.org/public-private-partnership/sites/ppp.worldbank.org/files/documents /Mali0Specifications.pdf.

Nepal

Government of Nepal. 1992. *Electricity Act, 2049 (1992): An Act Made for the Management and Development of Electricity*. http://www.propublic.org/tai/download/Electricity%20 Act%201992.pdf.

Government of Nepal, Ministry of Energy. 2012. *Directive on Licensing of Hydropower Projects, 2068 (2011)*. Government of Nepal, January 29.

NEA (Nepal Electricity Authority). 2003. *Nepal Electricity Authority Community Electricity Distribution Bye Laws, 2060*. http://www.nea.org.np/images/supportive_docs /Community%20Electricity%20Distribution%20Bylaw.pdf.

Vaidya, Dr. n.d. *Cost and Revenue Structures for Micro-Hydro Projects in Nepal*. AEPC. http://www.microhydropower.net/download/mhpcosts.pdf.

Peru

Government of Peru. 2007. "General Rural Electrification Law." *Article 25*. Lima, Peru.

The Philippines

Republic of the Philippines, Energy Regulatory Commission. 2009a. *A Resolution Adopting the Rules for Setting the Electric Cooperatives' Wheeling Rates.* http://www.erc.gov.ph.

———. 2009b. *Rules for Setting the Electric Cooperatives' Wheeling Rates.* http://www.erc.gov.ph.

———. 2011. *A Resolution Adopting the Rules Governing the Tariff Glide Path Pursuant to Article VII of the Rules for Setting the Electric Cooperatives' Wheeling Rates.* Resolution No. 08, Series of 2011. http://www.erc.gov.ph.

Portugal

Government of Portugal, Ministério das Actividades Económicas e do Trabalho. 2005. *Decreto-Lei N. 33-A/2005.* http://dre.pt/pdf1s/2005/02/033A01/00020009.pdf.

Rwanda

Castalia Strategic Advisors. 2009. *Rwanda Electricity Sector Access Programme.* Vol. 1: *Investment Prospectus.* Washington, DC: World Bank. http://ppp.worldbank.org/public-private-partnership/sites/ppp.worldbank.org/files/documents/Prospectus0for1cty0Access0Programme.pdf.

South Africa

Government of South Africa. 2008. *Electricity Pricing Policy.* http://www.info.gov.za/view/DownloadFileAction?id=94204.

Government of South Africa, Department of Energy. 2011. "Fact Sheet for the Media Briefing Session on 31 August 2011 Re the Renewable Energy Independent Power Producer (IPP) Programme." Government of South Africa, August 31. http://www.info.gov.za/speeches/docs/2011/reippp.pdf.

NERSA (National Energy Regulator of South Africa). 2011. NERSA Consultation Paper, "Review of Renewable Energy Feed-In Tariffs," NERSA, Pretoria, South Africa.

Republic of South Africa Department of Energy. 2012. "Preferred Bidders—Window 2." *Energy,* May 22. http://www.ipprenewables.co.za/#/blog/post/view/isAjaxRequest/true/id/182.

Sri Lanka

CEB (Ceylon Electricity Board). 2000. *CEB Guide for Grid Interconnection of Embedded Generators.* Ceylon Electricity Board, Colombo, Sri Lanka, December.

———. 2012a. *How Your Bill Is Calculated.* http://www.ceb.lk/sub/knowledge/billcalculation.html.

———. 2012b. "Opportunities for Renewable Energy Development: Present Status of Non-conventional Renewable Energy Sector." *Ceylon Electricity Board: Do Business with Us,* August 31. http://www.ceb.lk/sub/db/op_presentstatus.html.

———. 2012c. "Opportunities for Renewable Energy Development: Standardized Power Purchase Agreements." *Ceylon Electricity Board: Do Business with Us.* http://www.ceb.lk/sub/db/op_sppa.html.

ESD/RERED Government of Sri Lanka. 1999. *Village Hydro Specifications Sri Lanka: Line Distribution.* http://www.energyservices.lk/pdf/techspecs/vh_w_b/line.pdf.

Government of Sri Lanka, Ministry of Power and Energy. 2006. *National Energy Policy and Strategies of Sri Lanka.* October. http://www.energy.gov.lk/sub_pgs/elibrary_policy .html.

Parliament of the Democratic Socialist Republic of Sri Lanka. 2007. *Sri Lanka Sustainable Energy Authority Act.* http://www.documents.gov.lk/Acts/2007/Sri%20Lanka%20 Sustainable%20Energy%20Authority%20-%20Act%20No.%2035/Act%20No.%20 35-E.pdf.

PUCSL (Public Utilities Commission of Sri Lanka). 2010a. *Cost Reflective Methodology for Tariffs and Charges: Methodology for Charges.* September 30. http://www.pucsl.gov.lk /english/wp-content/themes/pucsl/pdfs/methodology_for_charges.pdf.

———. 2010b. *Non Conventional Renewable Energy Tariff Announcement: Purchase of Electricity to the National Grid under Standardized Power Purchase Agreements (SPPA).* November 25. http://www.pucsl.gov.lk/download/Electricity/PUCSL_NCRE _Advertisement.pdf.

SEA (Sri Lanka Sustainable Energy Authority). 2009. *Purchase of Electricity to the National Grid under Small Power Purchase Agreements (SPPA): Explanatory Notes to the Non Conventional Renewable Energy Tariff Announcement Dated 24th April 2009.* April 24. http://www.energy.gov.lk/pdf/explanatory_note_april_2009.pdf.

———. 2011. *On-Grid Renewable Energy Development: A Guide to the Project Approval Process for On-Grid Renewable Energy Project Development.* Sri Lanka Sustainable Energy Authority, July. http://www.energy.gov.lk/pdf/guideline/Grid_Renewable.pdf.

Tanzania

EWURA (Tanzanian Energy and Water Utilities Regulatory Authority). 2008. *Standardized Tariff Methodology for the Sale of Electricity to the Main Grid in Tanzania under Standardized Small Power Purchase Agreements.* http://www.ewura.go.tz/pdf /public%20notices/SPP%20Tariff%20Methodology.pdf.

———. 2009a. *Guidelines for Grid Interconnection of Small Power Projects in Tanzania: Part A: Mandatory Requirements and Test Procedures.* http://www.ewura.com/pdf/SPPT /PROPOSED%20GUIDELINES/TECHNICAL%20GUIDELINES/Guidelines%20 for%20Grid%20Interconnection%20-%20Part%20A.pdf.

———. 2009b. *Guidelines for Grid Interconnection of Small Power Projects in Tanzania. Part B: Technical Guidelines.* http://www.ewura.com/pdf/SPPT/PROPOSED%20 GUIDELINES/TECHNICAL%20GUIDELINES/Guidelines%20for%20Grid%20 Interconnection%20-%20Part%20B.pdf.

———. 2009c. *Guidelines for Grid Interconnection of Small Power Projects in Tanzania. Part C: Appendix: Studies to Be Conducted, Islanding and Protection.* http://www.ewura .com/pdf/SPPT/PROPOSED%20GUIDELINES/TECHNICAL%20GUIDELINES /Guidelines%20for%20Grid%20Interconnection%20-%20Part%20C.pdf.

———. 2009d. *Standardized Tariff Methodology for the Sale of Electricity to the Mini-Grids in Tanzania under Standardized Small Power Purchase Agreements.* http://www.ewura .com/pdf/public%20notices/Tanzania%20STM%20for%20Mini-grids%20under%20 SPPA-2009.pdf.

———. 2010. *The Electricity Act (CAP 131): The Electricity (Development of Small Power Projects) Rules, 2010.* http://www.ewura.com/pdf/SPPT/PROPOSED%20RULES /The%20Electricity%20(Development%20of%20Small%20Power%20Project)%20 Rules-2010.pdf.

———. 2011a. *Detailed Tariff Calculations for Year 2011 for the Sale of Electricity to the Main Grid in Tanzania under Standardized Small Power Purchase Agreements in Tanzania.* http://www.ewura.go.tz/pdf/SPPT/PROPOSED%20GUIDELINES /PROCESS%20GUIDELINES/2011%20SPPT%20Calculation%20for%20Main%20 Grid.pdf.

———. 2011b. *Detailed Tariff Calculations for Year 2011 for the Sale of Electricity to the Mini-Grids in Tanzania under Standardized Small Power Purchase Agreements in Tanzania.* http://www.ewura.go.tz/pdf/SPPT/2011%20SPPT%20Calculation%20 for%20Mini-Grid.pdf.

———. 2012a. *Detailed Tariff Calculations for Year 2012 for the Sale of Electricity to the Main Grid in Tanzania under Standardized Small Power Purchase Agreements in Tanzania.* http://www.ewura.go.tz/pdf/SPPT/2012/2012%20SPPT%20Calculation%20for%20 Main%20Grid.pdf.

———. 2012b. *Detailed Tariff Calculations for Year 2012 for the Sale of Electricity to the Mini-Grids in Tanzania under Standardized Small Power Purchase Agreements in Tanzania.* http://www.ewura.go.tz/pdf/SPPT/2012/2012%20SPPT%20Calculation%20for%20 Mini-Grid.pdf.

———. 2012c. *The Electricity (Development of Small Power Projects) Rules.* Proposed for Public Consultation. Dar es Salaam, Tanzania.

———. 2013. *The Electricity (Development of Small Power Projects) Rules, 2013.* Proposed for Public Consultation, June. Dar es Salaam, Tanzania.

Mwenga Hydro Limited. 2012. *Application for Tariff Approval by Mwenga Hydro Ltd (MHL) Submitted to EWURA.* Mwenga Hydro Limited, Dar es Salaam, Tanzania.

TANESCO (Tanzania Electric Supply Company). 2010. *TANESCO Tariff Review Application.* http://www.ewura.go.tz/pdf/Notices/Tariff%20Application%202010%20 -%20With%20Covering%20Letter.pdf.

———. 2011. *TANESCO Manual for SPP Application Approval and Interconnection.* Unpublished draft circulated to electricity stakeholders. Dar es Salaam, Tanzania.

———. 2012a. *Electricity Charges.* http://www.tanesco.co.tz/index.php?option=com _content&view=article&id=63&Itemid=205.

———. 2012b. *In Review. Draft—Tanesco Grid Code for Embedded Generation.* Unpublished draft circulated to electricity stakeholders.

United Republic of Tanzania. 2008. *Electricity Act, 2008.* http://polis.parliament .go.tz/PAMS/docs/10-2008.pdf.

United Republic of Tanzania, Ministry of Energy and Minerals. 2009a. *Standardized Power Purchase Agreement for Purchase of Capacity and Associated Electric Energy to the Isolated Mini-Grid Namely, between (The Buyer) and (The Seller).* EWURA. http:// www.ewura.go.tz/pdf/public%20notices/Tanzania%20SPPA%20Isolated%20 Grid%20Connection%20-%202009.pdf.

———. 2009b. *Standardized Power Purchase Agreement for Purchase of Grid-Connected Capacity and Associated Electric Energy between (The Buyer) and (The Seller).* EWURA. http://www.ewura.go.tz/pdf/public%20notices/Tanzania%20SPPA%20MainGrid%20 Connection-2009.pdf.

All EWURA documents can be downloaded from the website of EWURA at http://www .ewura.go.tz/sppselectricity.html.

Thailand

EPPO (Energy Policy and Planning Office), Ministry of Energy. n.d. *Model Power Purchase Agreement for the Purchase of Power from a Very Small Power Producer (for the Generation Using Cogeneration System) between [...] and the Provincial Electricity Authority/Metropolitan Electricity Authority.* http://www.eppo.go.th/power/vspp-eng /PPA%20Model%20-VSPP%20Cogen%20-10%20MW-eng.pdf.

————. n.d. *Model Power Purchase Agreement for the Purchase of Power from a Very Small Power Producer (for the Generation Using Renewable Energy) between [...] and the Provincial Electricity Authority/Metropolitan Electricity Authority.* http://www.eppo .go.th/power/vspp-eng/PPA%20Model%20-VSPP%20Renew%20-10%20MW-eng .pdf.

————. 2006. *Application for Sale of Electricity and System Interconnection (For Generators with Net Output under 10 MW).* http://www.eppo.go.th/power/vspp-eng /Application%20Form%20-VSPP%2010%20MW-eng.pdf.

————. 2009a. *Regulations for the Purchase of Power from Very Small Power Producers (for Generation Using Cogeneration System).* http://www.eppo.go.th/power/vspp-eng /Regulations%20-VSPP%20Cogen-10%20MW-eng.pdf.

————. 2009b. *Regulations for the Purchase of Power from Very Small Power Producers (for Generation Using Renewable Energy).* http://www.eppo.go.th/power/vspp-eng /Regulations%20-VSPP%20Renew-10%20MW-eng.pdf.

————. 2010. *Distribution Utilities' Regulations for Synchronization of Generators with Net Output under 10 MW to the Distribution Utility System.* http://www.eppo.go.th/power /vspp-eng/VSPP%20Synchronization%2010%20MW-eng.pdf.

————. 2011. *Status of Purchase of Electricity from VSPP September 2011 (PEA).* http:// www.eppo.go.th/power/data/index.html.

Government of Thailand, Provincial Electricity Authority. 2000. *Electricity Rates.* http:// www.pea.co.th/th/eng/downloadable/electricityrates.pdf.

All documents can be downloaded from the EPPO website at http://www.eppo.go.th /power/vspp-eng/index.html.

Uganda

ERA (Electricity Regulatory Authority). 2011. *Uganda Renewable Energy Feed-in Tariff (REFIT) Phase 2 Approved Guidelines for 2011–2012.* Government of Uganda. http://www.era.or.ug/Pdf/Approved_Uganda%20REFIT%20Guidelines%20V4%20 (2).pdf.

Vietnam

Socialist Republic of Vietnam, Ministry of Industry and Trade. 2006. *Technical Regulations for Rural Electrification/Electric Network.* http://ppp.worldbank.org/public-private -partnership/sites/ppp.worldbank.org/files/documents/Vietnam11 Technical1standards10Part0I.pdf.

————. 2008. *Regulation on Avoided Costs Tariff for Small Renewable Energy Power Plants.* Attachment to Decision No 18/2008/QD-BCT, July 18. http://www.erav.vn.

Socialist Republic of Viet Nam, the Prime Minister of Government. 2011. *Decision on the Mechanism Supporting the Development of Wind Power Projects in Vietnam.* Decision 37/2011/QD-TTg, June 29. http://www.erav.vn.

Reports, Studies, and Presentations

Adama, Sissoko, and Alassane Agalassou. 2008. "Mali's Rural Electrification Fund." Presentation at the Sustainable Development Week, Washington, DC, February. http://siteresources.worldbank.org/INTENERGY2/Resources/presentation8.pdf.

Adkins, Edwin, Sandy Eapen, Flora Kaluwile, Guatam Nair, and Vijay Modi. 2010. "Off-Grid Energy Services for the Poor: Introducing LED Lighting in the Millennium Villages Project in Malawi." *Energy Policy* 38: 1087–97.

Africa Electrification Initiative. 2012. *Institutional Approaches to Electrification: The Experience of Rural Energy Agencies/Rural Energy Funds in Sub-Saharan Africa.* Washington, DC: International Bank for Reconstruction and Development/World Bank Group. http://siteresources.worldbank.org/EXTAFRREGTOPENERGY /Resources/717305-1327690230600/8397692-1327690360341/AEI_Dakar _Workshop_Proceedings_As_of_7-30-12.pdf.

ARE (Alliance for Rural Electrification). 2011. "Hybrid Mini-Grids for Rural Electrification: Lessons Learned." Brussels. http://www.ruralelec.org/fileadmin/DATA/Documents /06_Publications/Position_papers/ARE_Mini-grids_-_Full_version.pdf.

ASER (Agence Sénégalaise d'Electrification Rurale), and Columbia Earth Institute Energy Group. 2007. *Costing for National Electricity Interventions to Increase Access to Energy, Health Services, and Education: Senegal Final Report.* Report prepared for the World Bank, August 15. http://modi.mech.columbia.edu/wp-content /uploads/2013/04/Senegal_WorldBank_Report_8-07.pdf.

Assani, Mansour. 2011. "Regulatory and Technical Issues in Operating Hybrid Mini-Grids." AEI Practitioner Workshop, Dakar, Senegal, November 15. http://siteresources .worldbank.org/EXTAFRREGTOPENERGY/Resources/717305-1327690230600 /8397692-1327691245128/Regulatory_TechnicalIssues_Operating_Hybrid _MiniGrids.pdf.

Bacon, Robert, Eduardo Ley, and Masami Kojima. 2010. "Subsidies in the Energy Sector: An Overview." Background Paper, World Bank Group Energy Sector Strategy, Washington, DC. http://siteresources.worldbank.org/EXTESC/Resources/Subsidy _background_paper.pdf.

Bakovic, Tonci, Bernard Tenenbaum, and Fiona Woolf. 2003. "Regulation by Contract: A New Way to Privatize Electricity Distribution?" Working Paper No. 14, World Bank, Washington, DC, September. http://rru.worldbank.org/Documents/PapersLinks/2552 .pdf.

Banerjee, Sudeshna Ghosh, Douglas Barnes, Bipul Singh, and Hussain Samad. 2013. *Power for All: Electricity Access Challenge in India.* India Power Sector Diagnostic Review, World Bank, Washington, DC.

Barnard, Wolsey. 2011. *Renewable Energy and Rural Electrification: South Africa Experience.* http://www.energy.gov.za/files/media/presentations/2011/20111206 _WolseyBarnard_RE_electrificationPresentationv2.pdf.

Barnes, Douglas F. 2007. *The Challenge of Rural Electrification: Strategies for Developing Countries.* Washington, DC: Resources for the Future.

Barnes, Douglas F., and Voravate Tuntivate. 2009. "The Challenge of Grid Rural Electrification: Experience of Successful Programs." Presentation at the AEI Practitioners Workshop, Maputo, Mozambique, June 9. http://siteresources .worldbank.org/EXTAFRREGTOPENERGY/Resources/717305-1264695610003 /6743444-1268073442212/1.1.Overview_bestpractice_institutional_issues.pdf.

Bhatia, Mikul, and Heather Adair Rohani. 2013. "Defining and Measuring Access to Energy." Presentation at Learning Days, SDN Forum, World Bank, March 7.

Bhatia, Mikul, Nicolina Angelou, Elisa Portale, Ruchi Soni, Mary Wilcox, and Drew Corbyn. 2013. *Defining and Measuring Access to Energy for Socio-Economic Development.* Washington, DC: World Bank and ESMAP.

Bhattacharyya, Subhes C. 2013. *To Regulate or Not to Regulate: Off-Grid Electricity Access in Developing Countries.* Pre-publication version. Leicester, England: OASYS South Asia Research Project, Montfort University.

Bhattacharyya, Subhes C., and Leena Srivastava. 2009. "Emerging Regulatory Challenges Facing the Indian Rural Electrification Program." *Energy Policy* 37: 68–79.

Breyer, Stephen G. 1982. *Regulation and Its Reform.* Cambridge, MA: Harvard University Press. http://site.ebrary.com/id/10313884.

Brown, Ashley C., Jon Stern, Bernard Tenenbaum, and Defne Gencer. 2006. *Handbook for Evaluating Infrastructure Regulatory Systems.* Washington, DC: World Bank.

Bryce, Donnella, and Chin Ching Soo. 2004. "Bulelavata Women Speak." *ENERGIA News* 6 (2): 19–20.

Cabraal, Anil. 2011. "Empowering Communities: Lessons from Village Hydro Development in Sri Lanka." Presentation at the World Bank, Washington, DC, May 25.

Chanthan, Ky, and Pascal Augareils. 2013. "Potential for Increasing the Role of Renewables in Mekong Power Supply: Cambodia." Presentation at the CPWF Mekong Basin Development Challenge, Hanoi, Vietnam, February 20.

Chowdhury, Nazmul Hossain. 2009. "Rural Electrification: Bangladesh Experience." Presentation at the African Electrification Practitioner Workshop, Maputo, Mozambique, June.

CORE International. 2008. "Study on Tariff Setting Principles and Issues Surrounding Tariffs and Electricity Pricing in Southern Africa." Submitted to the Southern African Power Pool, Harare, Zimbabwe, April 23.

Couture, Toby, Karlynn Cory, Claire Kreycik, and Emily Williams. 2010. *Policymaker's Guide to Feed-in Tariff Policy Design.* Technical Report, National Renewable Energy Laboratory, Colorado, July. http://www.nrel.gov/docs/fy10osti/44849.pdf.

Daunghom, Kitsanapol. 2010. "Thailand Grid Code for VSPP." Presentation at the Solar Business Bangkok 2010, Bangkok, Thailand, March 23.

DB Climate Change Advisors. 2009. *Global Climate Change Policy Tracker: An Investor's Assessment.* Deutsche Bank Group, New York. http://www.dbcca.com/dbcca/EN /_media/Global_Climate_Change_Policy_Tracker_Exec_Summary.pdf.

Desmukh, Ranjit, Juan Pablo Carvallo, and Ashwin Gambhir. 2013. *Sustainable Development of Renewable Energy Mini-Grids for Energy Access: A Framework for Policy Design.* Berkeley, CA: Lawrence Berkeley National Laboratory, University of California at Berkeley, Prayas Energy Group at Pune, India.

Deutsche Bank. 2010. *GET FiT Program: Global Energy Transfer Feed-in Tariff Program for Developing Countries.* http://www.ipfa.org/news/12564/db-climate-change-advisors.

Dixit, Shantanu. 2012. "Powering 1.2 Billion People: Case of India's Access Efforts." Presentation at the World Bank's Energy Days 2012, Washington, DC, February 23.

Dixit, Shantanu, and N. Sreekumar. 2011. *Rural Electrification Program: Urgent Need for Mid-Course Correction.* Prayas Energy Group, Pune, India.

Dorji, Chhimi. 2012. *Smart Grid Technology: GridShare Project in Rukubji, Bhutan.* http://www.sari-energy.org/PageFiles/What_We_Do/activities/BhutanCross BorderWorkshopAug2012/PResentations/GridShare_SARIE_CD_Final.pdf.

Dou, Charlie, Sicheng Wang, Luying Dong, Winfried Rijssenbeck, Zhizhang Liu, Ian Baring-Gould, Zhongying Wang, and Jingli Shi. 2005. *China Village Power Project Development Guidebook: Getting Power to the People Who Need It Most—A Practical Guidebook for the Development of Renewable Energy Systems for Village Power Projects.* United Nations Development Programme. http://siteresources.worldbank.org /EXTRENENERGYTK/Resources/5138246-1237906527727/5950705 -1239305592740/China0Village01ple0Who0Need0it0most.pdf.

du Preez, Jaap. 2011. "Design of Low-Cost Options for Distribution Networks in South Africa." Presentation at Africa Electrification Initiative (AEI) Practitioner Workshop, Dakar, Senegal, November 14.

Eberhard, Anton. 2013. *Feed-in Tariffs or Auctions: The Renewable Energy IPP Procurement Process in South Africa.* Viewpoint Number 338. April. Washington, DC: World Bank Group, Financial and Private Sector Vice Presidency.

Eberhard, Anton, Vivien Foster, Cecilia Briceño-Garmendia, Fatimata Ouedraogo, Daniel Camos, and Maria Shkaratan. 2008. "Underpowered: The State of the Power Sector in Sub-Saharan Africa." Background Paper, Africa Infrastructure Country Diagnostic, World Bank, Washington, DC. https://openknowledge.worldbank.org /handle/10986/7833.

Eberhard, Anton, Orvika Rosnes, Maria Shkaratan, and Haakon Vennemo. 2011. *Africa's Power Infrastructure: Investment, Integration, Efficiency.* Directions in Development: Infrastructure. Washington, DC: World Bank. http://www.ppiaf.org/sites/ppiaf.org /files/publication/Africas-Power-Infrastructure-2011.pdf.

Ehrhardt, David, and Rebecca Burdon. 1999. *Free Entry in Infrastructure.* Washington, DC: Castalia Strategic Advisors and World Bank. http://www.castalia-advisors.com /files/1857.pdf.

Electricity Authority of Cambodia. 2008. "Report on Power Sector of the Kingdom of Cambodia. 2009 Edition." Compiled by Electricity Authority of Cambodia from Data for the Year 2008 Received from Licensees. Annual Report of the Electricity Authority of Cambodia, Phnom Penh. http://www.eac.gov.kh/pdf/reports/Annual%20report%20 2008.en.pdf.

———. 2009. "Report on Power Sector of the Kingdom of Cambodia. 2010 Edition." Compiled by Electricity Authority of Cambodia from Data for the Year 2009 Received from Licensees. Annual Report of the Electricity Authority of Cambodia, Phnom Penh. http://www.eac.gov.kh/pdf/reports/Annual%20report%202009.en.pdf.

———. 2010. "Report on Power Sector of the Kingdom of Cambodia. 2011 Edition." Compiled by Electricity Authority of Cambodia from Data for the Year 2010 Received from Licensees. Annual Report of the Electricity Authority of Cambodia, Phnom Penh. http://www.eac.gov.kh/pdf/reports/Annual%20Report%20 2010%20En_final.pdf.

———. 2011. "Report on Power Sector of the Kingdom of Cambodia. 2012 Edition." Compiled by Electricity Authority of Cambodia from Data for the Year 2011 Received from Licensees. Annual Report of the Electricity Authority of Cambodia, Phnom Penh. http://www.eac.gov.kh/pdf/reports/Annual%20Report%20 2011En_%20Final2.pdf.

Elizondo Azuela, Gabriela, and Luiz Augusto Barroso. 2011. "Design and Performance of Policy Instruments to Promote the Development of Renewable Energy: Emerging Experience in Selected Developing Countries." Energy and Mining Sector Board Discussion Paper, World Bank, Washington, DC, April. http://siteresources.worldbank .org/EXTENERGY2/Resources/DiscPaper22.pdf.

Energy and Mining Sector Board, World Bank Group. 2007. "Technical and Economic Assessment of Off-Grid, Mini-Grid, and Grid Electrification Technologies." ESMAP Technical Paper, Energy Sector Management Assistance Program, World Bank Group, Washington, DC. http://siteresources.worldbank.org/EXTENERGY/Resources /336805-1157034157861/ElectrificationAssessmentRptSummaryFINAL17May07 .pdf.

European Commission. 2008. "The Support of Electricity from Renewable Energy Sources. Accompanying Document to the Proposal for a Directive of the European Parliament and of the Council on the Promotion of the Use of Energy from Renewable Sources." Commission Staff Working Document, Brussels, January 23. http://ec.europa.eu/energy/climate_actions/doc/2008_res_working_document _en.pdf.

Faulhaber, Gerald R., and Stephen B. Levinson. 1981. "Subsidy-Free Prices and Anonymous Equity." *American Economic Review* 71 (5): 1083–91.

Ferrey, Steven, and Anil Cabraal. 2005. *Renewable Power in Developing Countries: Winning the War on Global Warming.* Tulsa, OK: PennWell Books.

Foster, Vivien, and Cecilia Briceño-Garmendia, ed. 2010. *Africa's Infrastructure: A Time for Transformation.* Washington, DC: World Bank.

Foster, Vivien, and Jevgenijs Steinbuks. 2008. *Paying the Price for Unreliable Power Supplies: In-House Generation of Electricity by Firms in Africa.* Africa Infrastructure Country Diagnostic, International Bank for Reconstruction and Development, Washington, DC, January. http://www.infrastructureafrica.org/system/files/WP2 _Owngeneration_2.pdf.

Gencer, Defne, Peter Meier, Hung Tien Van, and Richard Spencer. 2011. *Vietnam's Rural Electrification Story: State and People, Central and Local, Working Together.* Washington, DC: World Bank.

Gipe, Paul. 2011. "Model Advanced Renewable Tariff Policy." *Wind-Works.org,* February 7. http://www.wind-works.org/FeedLaws/USA/Model/ModelAdvancedRenewable TariffLegislation.html.

Golumbeanu, Raluca, and Douglas Barnes. 2013. *Comparisons of Grid Connection Costs and Electricity Access in Developing Countries.* Africa Electrification Initiative. Washington, DC: World Bank.

Graves, Frank, Philip Hanser, and Greg Basheda. 2006. *Electric Utility Automatic Adjustment Clauses: Benefits and Design Considerations.* Washington, DC: Edison Electric Institute. http://www.eei.org/whatwedo/PublicPolicyAdvocacy/StateRegulation/Documents /adjustment_clauses.pdf.

Greacen, Chris. 2004. "The Marginalization of 'Small Is Beautiful': Micro-hydroelectricity, Common Property, and the Politics of Rural Electricity Provision in Thailand." PhD thesis, University of California, Berkeley. http://palangthai.org/docs/Greacen Dissertation.pdf.

Greacen, Chris, Richard Engel, and Thomas Quetchenbach. 2013. *A Guidebook on Grid Interconnection and Island Operation of Mini-Grid Power Systems Up to 200 kW.* Schatz

Energy Research Center and Palang Thai. Lawrence Berkeley National Laboratories report LBNL-6224E. Berkeley, CA.

Greacen, Chris, Sirikul Prasitpianchai, Tawatchai Suwannakum, and Christoph Menke. 2007. *Renewable Energy Options on Islands in the Andaman Sea: Hybrid Solar/Wind /Diesel Systems*. Study for the Tsunami Aid Watch Programme of the Heinrich Böll Foundation Southeast Asia Regional Office. http://www.palangthai.org/docs /KohPoKohPuEng.pdf (associated HOMER file: http://www.palangthai.org/docs /KohPo.hmr).

Greacen, Chris, and Sopitsuda Tongsopit. 2012. *Thailand's Renewable Energy Policy: FiTs and Opportunities for International Support*. Palang Thai, May 31. http://www .palangthai.org/docs/ThailandFiTtongsopit&greacen.pdf.

Hanley, Christina. 2010. "Feed-in Tariff Readiness." Presentation at the Renewable Energy Policy Workshop, World Resources Institute, Washington, DC, November 22. http:// powerpoints.wri.org/repw_hanley_fit_readiness_panel.pdf.

IEA (International Energy Agency). 2008. *Deploying Renewables: Principles for Effective Policies*. http://www.iea.org/publications/freepublications/publication/name ,34727,en.html.

———. 2012. "Access to Electricity." In *World Energy Outlook* (online). http://www .worldenergyoutlook.org/resources/energydevelopment/accesstoelectricity/.

IEA PVPS (International Energy Agency, Photovoltaic Power Systems Programme). 2013. *Rural Electrification with PV Hybrid Systems: Overview and Recommendations for Further Deployment*. Report IEA-PVPS T9-13. http://www.iea-pvps.org/index .php?id=1&eID=dam_frontend_push&docID=1590.

IFC (International Finance Corporation). 2012. *From Gap to Opportunity: Business Models for Scaling Up Energy Access*. http://www1.ifc.org/wps/wcm/connect/ca9c22004b5d0 f098d82cfbbd578891b/EnergyAccessReport.pdf?MOD=AJPERES.

INENSUS GmbH. 2011. *The Business Model of Micro Power Economy*. http://www .inensus.com/download/MicroPowerEconomy.pdf.

Interstate Renewable Energy Council. *Model Interconnection Procedures 2013 Edition*. Lantham, New York. http://www.irecusa.org/wp-content/uploads/2013-IREC -Interconnection-Model-Procedures.pdf.

IRENA (International Renewable Energy Agency). 2013. *Renewable Power Generation Costs in 2012: An Overview*. IRENA Report, International Renewable Energy Agency Bonn, Germany.

Irwin, Timothy C. 2007. *Government Guarantees: Allocating and Valuing Risk in Privately Financed Infrastructure Projects*. Directions in Development: Infrastructure. Washington, DC: World Bank. http://siteresources.worldbank.org/INTSDNETWORK/Resources /Government_Guarantees.pdf.

Jhirad, David J. 2013. "SPEED: Smart Power for Environmentally-Sound Economic Development." Presentation at the Incubating Innovation for Off-Grid Rural Electrification: London Investors' Conference, London, United Kingdom, March 21.

Johnston, L., K. Takahashi, F. Weston, C. Murray, and Gary Nakarado. 2006. *Rate Structures for Customers with Onsite Generation: Practice and Innovation*. PIER Final Project Report, California Energy Commission, April. http://www.energy.ca .gov/2006publications/CEC-500-2006-038/CEC-500-2006-038.PDF.

Joshi, Balawant. 2012. "Policy and Regulatory Interventions for Community Off-Grid Projects." Presentation at the Workshop on Off-Grid Access Systems in South Asia,

The Energy and Resources Institute, New Delhi, India, January 5. http://www
.oasyssouthasia.info/docs/oasyssouthasia_Jan2012_ppt9.pdf.

Joskow, Paul L. 2010. "Comparing the Costs of Intermittent and Dispatchable Electricity
Generating Technologies." Discussion draft, Department of Economics, Massachusetts
Institute of Technology, September 27. http://economics.mit.edu/files/5989.

Kahn, Alfred E. (1970) 1988. *The Economics of Regulation: Principles and Institutions.*
Vol. 1. Cambridge, MA: MIT Press.

Kapika, Joseph, and Anton Eberhard. 2013. *Power Sector Reform and Regulation in Africa:
Lessons from Ghana, Kenya, Namibia, Tanzania, Uganda and Zambia.* Cape Town,
South Africa: HSRC Press.

Karhammar, Ralph, Arun Sanghvi, Eric Fernstrom, Moncef Aissa, Jabesh Arthur, John
Tulloch, Ian Davies, Sten Bergman, and Subodh Mathur. 2006. "Sub-Saharan Africa:
Introducing Low-Cost Methods in Electricity Distribution Networks." ESMAP
Technical Paper, World Bank, Washington, DC.

Keosela, Loeung. 2013. "Status of Power Sector in Cambodia." Presentation at the
Renewable Energy Workshop, Chiang Mai, Thailand, January 22.

Kirubi, Charles. 2009. "Expanding Access to Off-Grid Rural Electrification in Africa: An
Analysis of Community-Based Micro-Grids in Kenya." PhD dissertation, University of
California, Berkeley.

Klein, Arne, Benjamin Pfluger, Anne Held, Mario Ragwitz, Gustav Resch, and Thomas
Faber. 2008. *Evaluation of Different Feed-in Tariff Design Options: Best Practice Paper for
the International Feed-in Cooperation.* Fraunhofer Institute Systems and Innovation
Research, Munich, Germany.

Komives, Kristin, Vivien Foster, Jonathan Halpern, and Quentin Wodon. 2005. *Water,
Electricity and the Poor: Who Benefits from Utility Subsidies?* Directions in Development
Series. Washington, DC: World Bank. http://siteresources.worldbank.org/INTWSS
/Resources/Figures.pdf.

Kumar, Geeta, and Yogita Mumssen. 2010. "Output-Based Aid and Energy: What Have We
Learned So Far?" *OBApproaches* Note 39, World Bank, Washington, DC, November.

Lemaire, Xavier, and Daniel Kerr. 2010. *SERN Literature Review 2010—An Annotated
Bibliography and Reference Guide on Off-Grid and Rural Electrification.* Renewable
Energy and Energy Efficiency Partnership (REEEP), Vienna, Austria.

Lilienthal, Peter. 2013a. *Hybrid Mini-Grids.* Webinar, Clean Energy Solutions Center,
March 6. http://cleanenergysolutions.org/training/hybridrenewable-mini-grids.

———. 2013b. "The Problem with 100% Renewable Energy." *HOMER Energy* (blog), June
30. http://blog.homerenergy.com/the-problem-with-100-renewable-energy/?utm_
source=Microgrid+News+by+HOMER+Energy&utm_campaign=18f8238c1b
-Microgrid_News_June_20136_13_2013&utm_medium=email&utm
_term=0_0f7f799f46-18f8238c1b-164784662.

Long, Bridget. 2011. "Presentation to SAIPPA: Recent NERSA COFIT and REFIT Public
Hearings." South Africa, May 24. http://0101.nccdn.net/1_5/1d3/1a7/263/Bridget
-Long-Presentation.pdf.

Lovins, Amory, and Rocky Mountain Institute. 2011. *Reinventing Fire: Bold Business
Solutions for the New Energy Era.* White River Junction, VT: Chelsea Green.

Madrigal, Marcelino, and E3 (Energy and Environmental Economics). 2010. *Creating
Renewable Energy Ready Transmission Networks: A Survey of 14 Jurisdictions Highlights
Emerging Lessons.* Washington, DC: World Bank and E3.

Marboeuf, Guy. 2011. "Mini Grids and Regulatory Issues: EDF's Experience in Mali." Presentation at the Practitioner Workshop, Dakar, Senegal, November 14. http://siteresources.worldbank.org/EXTAFRREGTOPENERGY/Resources/717305 -1327690230600/8397692-1327691245128/Mini_grids_And_RegulatoryIssues _Guy_Marboeuf.pdf.

Matly, Michael. 2010. "Best Practice of Rural Electrification Funds in Africa." Review Paper, ICTS-NTUA and SOFRECO, Clichy Cedex, France.

Meier, Peter. 2010. *Economic and Financial Analysis of Grid-Connected Renewable Energy Generation*. World Bank and the Government of Vietnam, Ministry of Industry and Trade, Hanoi, Vietnam.

Meier, Peter, Voravate Tuntivate, Douglas F. Barnes, Susan V. Bogach, and Daniel Farchy. 2010. *Peru: National Survey of Rural Household Energy Use*. Energy and Poverty. Washington, DC: World Bank. http://www.esmap.org/sites/esmap.org/files/ESMAP _PeruNationalSurvey_Web_0.pdf.

Moncef, Aissa. 2011. "Techniques to Reduce Costs of Rural Distribution Networks in Tunisia." Presentation at Africa Electrification Initiative (AEI) Practitioner Workshop, Dakar, Senegal. http://go.worldbank.org/WCEDP90SZ0.

Mostert, Wolfgang. 2008. *Review of Experiences with Rural Electrification Agencies: Lessons for Africa*. Draft Report, European Union Energy Initiative—Partnership Dialogue Facility, August 24. http://www.mostert.dk/pdf/Experiences%20with%20Rural%20 Electrification%20Agencies.pdf.

————. 2010. *Publicly Backed Guarantees as Policy Instruments to Promote Clean Energy*. Report, United Nations Environment Programme Sustainable Energy Finance (SEF) Alliance, New York. http://fs-unep-centre.org/sites/default/files/media /guaranteesweb.pdf.

Mukherjee, Mohua. 2013. *Lessons Learned from Two Decades of Experience with Private Sector Participation in the Indian Power Sector*. Washington, DC: World Bank, India Power Sector Diagnostic Review.

————. 2013. "Private Sector Led Off-Grid Energy Access: The A-B-C Business Model and How Third Parties Can Support the Development of Mini-Grids." Presentation at the Incubating Innovation for Off-Grid Rural Electrification, London Investors' Conference, London, United Kingdom, March 21.

Mumssen, Yogita, Lars Johannes, and Geeta Kumar. 2010. *Output-Based Aid: Lessons Learned and Best Practices*. Directions in Development: Finance. Washington, DC: World Bank. https://openknowledge.worldbank.org/bitstream/handle/10986/2423 /536440PUB0outp101Official0Use0Only1.pdf?sequence=1.

Murthy, Ramachandra, and Ramalinga Raju. 2009. "Electrical Energy Loss in Rural Distribution Feeders—A Case Study." *APRN Journal of Engineering and Applied Sciences* 4 (2): 33–37.

Mutambi, Benon M. 2012. "How to Make Energy Financing (and Opportunities) a Reality in Uganda." Presentation at the Energy Business Dialogue Uganda: Creating Commitment and Momentum for Increased Energy Access, Kampala, Uganda, December 13.

Nagendran, Jayantha. 2001. *Sri Lanka Energy Services Delivery Project Credit Programme: A Case Study*. Sri Lanka Energy Services Delivery Project. Colombo: DFCC Bank. http://www.martinot.info/Cases/Sri_Lanka_ESD_case_Nagendran.pdf.

NRECA International Ltd (National Rural Electric Cooperative Association). 2005. *Bangladesh Rural Electrification at the Crossroads*. Report Submitted to U.S. Agency for International Development, Arlington, VA, January.

————. 2012. *Affordability Analysis and Options for a Program to Make the Cost of Rural Household Grid Connections Affordable.* Unpublished draft report, Arlington, VA, June.

Palit, Debajit, and Akanksha Chaurey. 2011. "Off-Grid Rural Electrification Experiences from South Asia: Status and Best Practices." *Energy for Sustainable Development* 15 (3): 266–76.

Pandey, Bikash, and Ratna Sansar Shresthi. 2007. "Micro-Mini Hydro (<5 MW) Feed-In Tariff Study for Pakistan." Presentation at LFA Workshop, Islamabad, Pakistan, December 18.

Peon, Rodolfo, Ganesh Doluweera, Inna Platonova, Dave Irvine-Halliday, and Gregor Irvine-Halliday. 2005. "Solid State Lighting for the Developing World—The Only Solution." *Optics and Photonics 2005, Proceedings of SPIE* 5941: 109–23.

Pigaht, Maurice, and Robert J. van der Plas. 2009. "Innovative Private Micro-hydro Power Development in Rwanda." *Energy Policy* 37 (11): 4753–60.

Practical Action. 2012. *Poor People's Energy Outlook 2012—Energy for Earning a Living.* UK: Practical Action. http://cdn1.practicalaction.org/p/p/4f1ea5d5-024c-42a1-b88d -026b0ae4f5bb.pdf.

Quetchenbach, T. G., M. J. Harper, J. Robinson IV, K. K. Hervin, N. A. Chase, C. Dorji, and A. E. Jacobson. 2013. "The GridShare Solution: A Smart Grid Approach to Improve Service Provision on a Renewable Energy Mini-Grid in Bhutan." *Environmental Research Letters* 8 (1): 014018.

Radecsky, Kristen. 2009. "Understanding the Economics behind Off-Grid Lighting Products for Small Businesses in Kenya." Thesis, Humboldt State University. http:// humboldt-dspace.calstate.edu/xmlui/bitstream/handle/2148/508/RadecskyThesis .pdf?sequence=1.

Raghunathan, Krishnan, Anjali Garg, Gevorg Sargsyan, and Mikul Bhatia. 2010. *Empowering Rural India: Expanding Electricity Access by Mobilizing Local Resources: Analysis of Models for Improving Rural Electricity Services in India through Distributed Generation and Supply of Renewable Energy.* Washington, DC: South Asia Energy Unit, Sustainable Development Department, World Bank. http://online.wsj.com/public /resources/documents/WorldBankreport0215.pdf.

Raj, Anil. 2012. "The Micropower Opportunity: Paving the Way for Rural Electrification." Presentation, World Bank, Washington, DC, November 15.

Reiche, Kilian, Bernard Tenenbaum, and Clemencia Torres de Mästle. 2006. "Electrification and Regulation: Principles and a Model Law." Energy and Mining Sector Board Discussion Paper, World Bank, Washington, DC. http://siteresources.worldbank.org /EXTENERGY/Resources/336805-1156971270190/EnergyElecRegulationFinal.pdf.

Reiche, Kilian, and Witold Teplitz. 2009. *Energy Subsidies: Why, When and How? A Think Piece.* Eschborn: GTZ. http://www.medemip.eu/Calc/FM/MED-EMIP /OtherDownloads/Other_Energy_Topics/201008_Energy-Subsidieswhy_when _and_how.pdf.

Ren21. 2011. *Renewables 2011 Global Status Report.* July 12. http://bit.ly/REN21 _GSR2011.

————. 2012. *Renewables 2012 Global Status Report.* http://www.map.ren21.net/slider /index.html#fragment-1.

Revolo, Miguel. 2009. "Mechanism of Subsidies Applied in Peru." Presentation at the AEI-Maputo Workshop, World Bank, Peru, June. http://siteresources.worldbank.org /EXTAFRREGTOPENERGY/Resources/717305-1264695610003/6743444 -1268073611861/11.3Mechanism_subsidies_applied_in_Peru.pdf.

Rickerson, Wilson. 2012. "Feed-in Tariffs as a Policy Instrument for Promoting Renewable Energies and Green Economies in Developing Countries." Technical Paper, United Nations Environment Programme, Washington, DC. http://www.unep.org/pdf /UNEP_FIT_Report_2012F.pdf.

Rickerson, Wilson, Christina Hanley, Chad Laurent, and Chris Greacen. 2012. "Implementing a Global Fund for Feed-in Tariffs in Developing Countries: A Case Study of Tanzania." *Renewable Energy* 49, Special Issue: Selected Papers from World Renewable Energy Congress—XI (March 20): 29–32.

Rikos, Evangelos, Stathis Tselepis, and Aristomenis Neris. 2008. *Stability in Mini-Grids with Large PV Penetration under Weather Disturbances: Implementation to the Power System of Kythnos*. Fourth European PV-Hybrid and Mini-Grid Conference, Center for Renewable Energy Sources, Glyfada, Greece, May 30. http://www.cres.gr/kape /publications/photovol/new/Stability%20in%20Mini-Grids%20with%20Large%20 PV%20Penetration%20under%20Weather%20Disturbances-Im_.pdf.

Roach, Mary, and Charlotte Ward. 2011. *Harnessing the Full Potential of Mobile for Off-Grid Energy*. London: GSMA. http://www.gsma.com/mobilefordevelopment/programmes /community-power-from-mobile/.

Rodriguez, Sebastian, and Vanessa Lopes Janik. Forthcoming. *Case Studies on Gender and Electrification from Mali*. Washington, DC: World Bank.

Sargsyan, Gevorg, Mikul Bhatia, Sudeshna Ghosh Banerjee, Krishnan Raghunathan, and Ruchi Soni. 2010. *Unleashing the Potential of Renewable Energy in India*. Washington, DC: ESMAP and the South Asia Energy Unit, Sustainable Development Department, World Bank. http://siteresources.worldbank.org/EXTENERGY2/Resources /Unleashing_potential_of_renewables_in_India.pdf.

Sawe, E. N. 2005. "Rural Energy and Stoves Development in Tanzania." Conference presentation at the Workshop on Rural Energy, Stoves, and Indoor Air Quality in China, Beijing, China, January 14.

Searchinger, Timothy D., Steven P. Hamburg, Jerry Melillo, William Chameides, Petr Havlik, Daniel M. Kammen, Gene E. Likens, Ruben N. Lubowski, Michael Obersteiner, Michael Oppenheimer, G. Philip Robertson, William H. Schlesinger, and G. David Tilman. 2009. "Fixing a Critical Climate Accounting Error." *Science* 326: 527–28.

Siyambalapitiya, Tilak. 2001. *Study on Grid Connected Small Power Tariff in Sri Lanka*. Final Report, Resource Management Associates (Pvt.), Sri Lanka.

———. 2007. *Standardised Small Power Purchase Tariffs for Tanzania*. Tanzania Ministry of Energy and Minerals, Final Report, September. Dar es Salaam, Tanzania.

Smith, Nigel. 1995. "Low Cost Electricity Installation." Overseas Development Administration, June. http://www.dfid.gov.uk/R4D/PDF/Outputs/R5613-TRL211 .pdf.

Sundqvist, Thomas. 2000. "Electricity Externality Studies: Do the Numbers Make Sense?" Licentiate thesis, Lulea Tekniska Universitet, Sweden.

Szabo, Sandor, K. Bodis, T. Huld, and M. Moner-Girona. 2011. "Energy Solutions in Rural Africa: Mapping Electrification Costs of Distributed Solar and Diesel Generation Versus Grid Extension." *Environmental Research Letters* 6. http://iopscience.iop .org/1748-9326/6/3/034002/fulltext/.

TERI (The Energy and Resources Institute). 2007. *Evaluation of Franchise System in Selected Districts of Assam, Karnataka and Madhya Pradesh*. Draft Final Report, the

Energy and Resources Institute, New Delhi, India. http://recindia.nic.in/download /Franchisee_Eval_TERI.pdf.

Terrado, Ernesto, Anil Cabraal, and Ishani Mukherjee. 2008. *Operational Guidance for World Bank Group Staff—Designing Sustainable Off-Grid Rural Electrification Projects: Principles and Practices.* Washington, DC: World Bank and Energy and Mining Sector Board. http://siteresources.worldbank.org/EXTENERGY2 /Resources/OffgridGuidelines.pdf.

UNIDO (United Nations Industrial Development Organization) and REEEP (Renewable Energy and Energy Efficiency Partnership). 2012. *REEEP/UNIDO Training Package: Sustainable Energy Regulation and Policymaking for Africa.* Online Training Package. http://africa-toolkit.reeep.org/.

UPDEA (Union of Producers, Transporters and Distributors of Electric Power in Africa). 2009. *Comparative Study of Electricity Tariffs Used in Africa.* General Secretariat, UPDEA: Abidjan, Côte d'Ivoire, December http://www.updea-africa.org/updea /DocWord/TarifAng2010.pdf.

U.S. DOE (U.S. Department of Energy). 2007. *The Potential Benefits of Distributed Generation and Rate-Related Issues That May Impede Their Expansion.* A Study Pursuant to Section 1817 of the Energy Policy Act of 2005, U.S. Department of Energy, Washington, DC, February. http://www.ferc.gov/legal/fed-sta/exp-study.pdf.

U.S. Foreign Commercial Service. 2010. *Doing Business in Tanzania: 2011 Country Commercial Guide for U.S. Companies.* http://photos.state.gov/libraries/tanzania /231771/PDFs/Country_Commercial_Guide_2011_Tanzania.pdf.

Van Leeuwen, Richenda. 2013. *The Role of Hybrid Renewable Mini-Grids in Providing Energy Access.* Webinar, Clean Energy Solutions Center, March 6. http://cleanenergysolutions .org/training/hybridrenewable-mini-grids.

Vernstrom, Robert. 1995. *Published Small Power Purchase Tariff for Sri Lanka.* Study, World Bank, Washington, DC.

———. 2010. *Long-Run Marginal Cost of Service Tariff Study.* Final Report to Tanzania Electric Supply Company, Menlo Park, CA. http://www.ewura.go.tz/pdf/Notices /Tariffs%20COSS%20Final_Annex%20to%20the%20TA.pdf.

Wijayatunga, Priyantha D. C. 2012. "Regulation for Renewable Energy Development: Lessons from Sri Lanka Experience." *Renewable Energy,* May 17. http://www .sciencedirect.com/science/article/pii/S0960148112002674.

World Bank. 2005. *Electricity for All: Options for Increasing Access in Indonesia.* Washington, DC: World Bank, Energy and Mining Unit, Infrastructure Department, East Asia and Pacific Region. http://siteresources.worldbank.org/INTINDONESIA/Resources/Publi cation/280016-1106130305439/Electricity-for-All-Options-for-Increasing-Access-in -Indonesia.pdf.

———. 2008. *Doing Business 2009: Comparing Regulation in 181 Economies.* Washington, DC: World Bank. http://www.doingbusiness.org/~/media/GIAWB/Doing%20 Business/Documents/Annual-Reports/English/DB09-FullReport.pdf.

———. 2009. *Doing Business 2010: Comparing Regulation in 183 Economies.* Washington, DC: World Bank.

———. 2011. *One Goal, Two Paths: Achieving Universal Access to Modern Energy in East Asia and the Pacific.* Washington, DC: World Bank. https://openknowledge.worldbank .org/handle/10986/2354.

———. 2012. *World Databank*. October 31. http://databank.worldbank.org/ddp
/home.do.

World Bank and IEA (International Energy Agency). 2013. *Sustainable Energy for All:
Global Tracking Framework*. Washington, DC: World Bank and IEA.

Other Works or Communications Referenced

A-E-S Europe GmbH. 2012. "Price Trend PV Modules: Price Trend Photovoltaic
Modules—Updated Weekly." *Clean Energy Investment*, October 12. http://www
.europe-solar.de/catalog/index.php?main_page=page_3.

Agalassou, Alassane. 2011. Personal communication. March.

BERD (Bureau d'Électrification Rurale Décentralisée). 2011. Personal communication.
November 10.

Bert and Rich (blog). 2011. "Buy Prepaid Electricity by MPESA." May 9. http://www
.bertandrich.com/blog/how-to/buy-prepaid-electricity-by-mpesa/.

Chanthan, Ky. 2013. Personal communication. February.

Daily Mirror. 2012. "Signing of Delegated Cooperation Agreement between the
Government of Norway and KFW for the GET FiT Program," December 21.

Dunnison, David. 2011. "US Residential PV Market Driven by More than Price." *D-bits*,
April 5. http://d-bits.com/residential-pv-price-sensitivity/.

Gaddis, Isis. 2012. "Only 14% of Tanzanians Have Electricity: What Can Be Done?"
End Poverty (blog), October 31. http://blogs.worldbank.org/africacan/node/2187.

Gonzalez, Angel, and Keith Johnson. 2009. "Spain's Solar-Power Collapse Dims
Subsidy Model." *Wall Street Journal*, September 8. http://online.wsj.com/article
/SB125193815050081615.html.

Gosling, Melanie. 2011. "Government's U-turn on Wind Energy Rates." *Cape Times*,
June 20.

Gratwicke, Michael. 2012. Personal communication. June.

Lakshmi, Rama, and Simon Denyer. 2012. "Lack of Power Symbolizes India's Inequalities."
Washington Post, Kataiyan, India, August 6, sec. Asia & Pacific. http://www
.washingtonpost.com/world/asia_pacific/lack-of-power-symbolizes-indias-inequalities
/2012/08/06/ecdbef64-df20-11e1-a19c-fcfa365396c8_print.html.

Long Island Power Authority. 2012. *Common Commercial Electric Rates*. http://www
.lipower.org/pdfs/account/rates_comm.pdf.

Mafia Island. 2011. Personal communication. June.

Mahato, Rubeena. 2010. "Power Sharing, Nepali Style." *Nepali Times*, July 23.

Mills, Rob. 2013. Personal communication. March 11.

Mwenga Hydro. 2011. Personal e-mail communication. November 7.

Ostrom, Elinor. 1990. *Governing the Commons: The Evolution of Institutions for Collective
Action*. The Political Economy of Institutions and Decisions. Cambridge, United
Kingdom: Cambridge University Press. http://www.amazon.com/Governing
-Commons-Evolution-Institutions-Collective/dp/0521405998.

Rekhani, Badri. 2011. Personal communication. November.

———. 2012. Personal communication. March.

Reuters. 2012. "Kenya Regulator Cuts Diesel Price, Petrol, Kerosene Up." Nairobi, Kenya, May 14. Africa edition. http://af.reuters.com/article/investingNews/idAFJOE84D08Z20120514.

Revolo Acevedo, Miguel. 2011. Personal communication. September 14.

———. 2013. Personal communication. February.

Samuelson, Robert J. 2011. "Why We Need to Fix Social Security, and Other Year-End Reflections." *Washington Post* (blog), December 28. http://www.washingtonpost.com/blogs/post-partisan/post/why-we-need-to-fix-social-security-and-other-year-end-reflections/2011/12/28/gIQAiJyUMP_blog.html.

Sguazzin, Antony. 2011. "South Africa Cuts Proposed Renewable Energy Prices, Business Day Says." *Bloomberg*, South Africa, March 23. http://www.bloomberg.com/news/2011-03-23/south-africa-cuts-proposed-renewable-energy-prices-business-day-says.html.

Shrestha, Binod. 2012. Personal communication. September.

Siyambalapitiya, Tilak. 2012. Personal e-mail communication. August 22.

Supreme Court of Judicature of Jamaica. 2012. *Meadows vs. Blaine et al.*, JMSC Civ 110 (Civil Division).

Tan, Rauf. 2012. Personal communication. April.

Todeschini, Luca. 2011. Personal communication. February.

Touré, Nava. 2011. Personal communication. October.

TPC. 2012. Personal communication. February.

Van Couvering, Jim. 2011. Personal communication. March.

Van Tien, Hung. 2011. Personal communication. June.

Van Tien, Hung, and Beatriz Arizu. 2011. Personal communication. January.